FREEDOM'S LABORATORY

Audra J. Wolfe

FREEDOM'S LABORATORY

The Cold War Struggle for the Soul of Science

JOHNS HOPKINS UNIVERSITY PRESS Baltimore

Johns Hopkins University Press
2715 North Charles Street
Baltimore, Maryland 21218-4363
www.press.jhu.edu

Library of Congress Cataloging-in-Publication Data

Names: Wolfe, Audra J., author.
Title: Freedom's laboratory : the Cold War struggle for the soul of science / Audra J. Wolfe.
Description: Baltimore : Johns Hopkins University Press, 2018. | Includes bibliographical
    references and index.
Identifiers: LCCN 2018004445 | ISBN 9781421426730 (hardcover : alk. paper) | ISBN
    9781421426747 (electronic) | ISBN 1421426730 (hardcover : alk. paper) | ISBN 1421426749
    (electronic)
Subjects: LCSH: Science and state—United States—History—20th century. | Science and state—
    Europe, Western—History—20th century. | United States—Relations—Europe, Western. |
    Europe, Western—Relations—United States. | Cold War—Social aspects.
Classification: LCC Q127.U6 W654 2018 | DDC 338.973/0609045—dc23
LC record available at https://lccn.loc.gov/2018004445

A catalog record for this book is available from the British Library.

*Special discounts are available for bulk purchases of this book. For more information, please contact
Special Sales at 410-516-6936 or specialsales@press.jhu.edu.*

Johns Hopkins University Press uses environmentally friendly book materials, including recycled
text paper that is composed of at least 30 percent post-consumer waste, whenever possible.

For my parents

Contents

Abbreviations

AAAS	American Academy of Arts and Sciences
ACLU	American Civil Liberties Union
AEC	Atomic Energy Commission
AIBS	American Institute of Biological Sciences
ASHG	American Society of Human Genetics
ASTA	American Scientists Traveling Abroad
ASTP	Apollo-Soyuz Test Project
BSCS	Biological Sciences Curriculum Study
CCF	Congress for Cultural Freedom
CERN	European Organization for Nuclear Research
CIA	Central Intelligence Agency
FBI	Federal Bureau of Investigation
GSA	Genetics Society of America
HUAC	House Un-American Activities Committee
ICSU	International Council of Scientific Unions
IGY	International Geophysical Year
JRDB	Joint Research and Development Board
NAS	National Academy of Sciences
NASA	National Aeronautics and Space Administration
NATO	North Atlantic Treaty Organization
NRC	National Research Council
NSA	National Student Association
NSC	National Security Council
NSF	National Science Foundation
OCB	Operations Coordinating Board
OPC	Office of Policy Coordination
OSI	Office of Scientific Intelligence
OSRD	Office of Scientific Research and Development
PSAC	President's Science Advisory Committee
PSB	Psychological Strategy Board

PSSC	Physical Science Study Committee
SADS	Soviet-American Disarmament Study
TAF	The Asia Foundation
USAID	US Agency for International Development
USIA	US Information Agency
VASKhNIL	Lenin All-Union Academy of Agricultural Sciences
VOA	Voice of America

Introduction

The American herpetologist Arnold Grobman once told me a story about earthworms. I had already encountered this story in various manuscript drafts and correspondence; Grobman told it frequently. But since I had come all the way to Gainesville, Florida, to meet the crotchety nonagenarian, I was happy to hear it firsthand. In the late 1950s, Grobman began, he traveled to Hong Kong to observe high school science teaching. The teachers invited him to watch the students dissect earthworms. As the students began their work, Grobman noticed something peculiar. The anatomy of the students' specimens—dug up from local soil—did not match that in their British textbooks. The students nevertheless dutifully labeled their worms according to the diagrams in their texts. Since they were studying for British exams, they prioritized fealty to the text over observed reality.

A contemporary observer might draw any number of conclusions from this set piece, from the tyranny of standardized tests to the insidious ways that colonialism distorted even the most mundane aspects of daily life. Grobman absorbed a different lesson. As a humid Florida breeze blew across the patio, Grobman explained to me that forcing students to choose between empirical observation and received authority deprived them of one of the central benefits of laboratory instruction. Instead of basing their understanding of the natural world on scientific knowledge derived from their own experience and curiosity, they did what they were told. This was dangerous, he told me, because it left the students vulnerable to the influence of Communism.[1]

With the distance of sixty years, Grobman's claim that mislabeled earthworms exposed students to the risk of Communist indoctrination is strange, even comical. To midcentury US educators like Grobman, however, the links between scientific observation and liberal democracy were transparent and urgent. They believed that political freedom depended on scientific freedom and that scientific freedom emerged from unobstructed encounters with the natural world. Nor was this assumption specific to educators. From the late 1940s through the late 1960s, the US foreign policy establishment saw a particular

1

way of thinking about scientific freedom as essential to winning the global Cold War—and not just because science created weaponry. Throughout this period, the engines of US propaganda amplified, circulated, and, in some cases, produced a vision of science, American style, that highlighted scientists' independence from outside interference and government control. Science, in this view, was apolitical.

Many Cold War ideologies collapsed alongside the Berlin Wall in 1989, but the idea that science is apolitical has had remarkable staying power. Scientists in the United States continue to lean on the language of science and freedom to defend their funding streams and research agendas, even as political leaders display less and less interest in their claims of expertise or even in the existence of facts. The claim sticks, even though a society's decisions about how science should be conducted are inherently and obviously political, involving choices about access, representation, compensation, and expertise. Appeals to apolitical science arguably made even less sense during the Cold War than they do now. Scientific research consumed a larger portion of the US federal budget during the Cold War than during any other peaceful period in US history. The fervor of anti-Communism subjected researchers to loyalty oaths and security checks. In the conformist culture of the Cold War, the specific kind of scientific freedom on offer was primarily available to college-educated, straight, able-bodied white men with impeccable anti-Communist credentials.[2]

Scientific freedom, in other words, had to be constructed and maintained through a series of political choices. Those choices involve power—politics by another name. And yet, for twenty years, leading US scientists and government officials alike attempted to convince audiences both at home and abroad that American science had uniquely transcended politics through its commitment to scientific freedom. This book is an attempt to understand why.

The actors in the story that follows occasionally attempted to work out exactly what they meant when they referred to "scientific freedom" or the "US ideology." More frequently, however, they defined American science through the foil of Communist science. As depicted in US propaganda, Communist science forced its practitioners to adhere to party dogma, favored political loyalties, prioritized practical achievements and technological spectacle over basic science, and advanced nationalistic goals. In response, the politics of freedom fused with anti-Communism to create a vision of science in the United States that highlighted empiricism, objectivity, a commitment to pure research, and internationalism. This development took place against a larger backdrop in which the United States declared itself (somewhat unconvincingly) to be a beacon of freedom to oppressed peoples around the world. The Cold War ended long ago, but the language of science and freedom continues to shape public debates over the relationship between science and politics in the United States.

: :

Historians quibble over the exact beginning and end dates for the Cold War, but I follow the consensus dates of 1947 to 1989. I adopt this periodization because I understand the Cold War to be an ideological phenomenon that pitted capitalism against communism. Of course, the Cold War was not the only major political movement of the second half of the twentieth century. More than forty countries in Africa, the Middle East, and Asia gained independence between 1940 and 1962, representing the hopes and dreams of more than 800 million people. That leaders of newly independent nations had their own goals for nationalism and independence, however, did not deter US and Soviet leaders from attempting to recruit them to their own side. By defining the Cold War as a bipolar struggle, leaders in Moscow and Washington hoped to cajole, coerce, or force peoples throughout the world to choose either communism or capitalism. In the Global South, the Cold War turned hot with depressing regularity.[3]

Soviet and US strategists used other countries' civil wars and independence movements as proxy battles—substitutes for the nuclear apocalypse that they hoped to avoid. Sometimes they intervened in these conflicts directly; in other cases, they attempted to orchestrate their desired outcomes from afar. The United States armed and trained paramilitary groups, attempted to remove leaders it deemed inconvenient, and conducted counterinsurgency campaigns. But particularly in the first two decades of the Cold War, the United States backed up live bullets with psychological warfare campaigns. Policymakers saw psychological campaigns as a kinder, gentler form of domination that might avert physical violence through persuasion. These "hearts and minds" offensives included programs easily recognized as propaganda—for instance, Voice of America broadcasts—as well as activities better characterized as "propaganda of the deed," such as sponsoring academic conferences or student exchanges.[4]

Historians frequently describe the wide variety of activities in this category of psychological warfare as "cultural diplomacy," an umbrella term that can encompass everything from consumer culture to orchestral tours. At heart, this is the idea that a country can use its cultural resources—its cultural products as well as its cultural values—to strengthen its alliances and build bridges to those not yet in its camp. The United States' commitment to cultural diplomacy during the Cold War meant that its international image depended, at least in part, on its treatment of its own citizens. Given the state of civil rights and civil liberties in the United States in the 1950s, the nation's claims to moral authority on matters of freedom during the Cold War have to be taken with a grain of salt, or perhaps an entire salt cellar. But somewhat surprisingly, the imperatives of cultural diplomacy sometimes inspired US politicians to endorse appeals for equality and justice, even in an era of anti-Communism. Most contemporary historians agree that President Dwight D. Eisenhower would never have taken even the limited measures that he did on behalf of civil rights, for example, if segregation hadn't become a bludgeon for Communist propaganda. President Lyndon Johnson's Great Society programs might never have passed Congress had the United States not needed to show the rest of the world that capitalist societies could feed and clothe their most vulnerable citizens and heal their sick.[5]

Something similar happened with science. Scientists in the United States experienced real and troubling limitations on their political and intellectual freedoms during the Cold War. J. Robert Oppenheimer famously lost his US Atomic Energy Commission (AEC) security clearance in 1954 because of allegations that he had been involved with a Communist Party cell in Berkeley, California, in the 1930s. Physicist Melba Phillips (a former Oppenheimer student) got pushed out of the Federation of American Scientists in 1948 because she refused to renounce the Progressive Party. Other scientists lost their passports, their funding, or their jobs. On an institutional level, funding agencies selectively funded the work they wanted to see done and ignored the work that they did not. And yet, scientists who successfully avoided these political land mines enjoyed a remarkable level of control over their own work during the Cold War, in part because US authorities understood the importance of drawing a contrast with the situation of science in the Soviet Union. Funding for so-called basic research flourished, and new institutions like the National Science Foundation gave scientists a fiefdom where they could shield research from partisan interference. The descriptions of scientific freedom that US scientists and propagandists produced fell somewhat short of reality, but they nonetheless described a set of conditions that would have been worth aspiring to, had they applied to all Americans equally.[6]

Early Cold War psychological warfare campaigns consistently contrasted US individualism with Soviet collectivism. Given the near impossibility of conveying this message through government-sponsored programs, most of this work was carried out by private individuals, not all of whom realized they were participating in US psychological warfare campaigns. This brings us to one of the US government's more curious choices in the fight against Communism. From 1950 until 1967, when its covers were blown, the Central Intelligence Agency (CIA) funded and supervised a number of nominally private organizations engaged in the work of cultural diplomacy. The CIA used convoluted funding streams to obscure its support for US and foreign artists, musicians, writers, students, and, yes, scientists who lent their voices and pens to the international fight against Communism.

The best known of the CIA's covert cultural diplomacy operations was the Paris-based Congress for Cultural Freedom (CCF), which conducted cultural festivals, held academic conferences, and published literary magazines (most famously, *Encounter*). The CCF, however, was dwarfed by the Asia Foundation, a poorly understood organization that spent tens of millions of the CIA's dollars developing nongovernmental, pro-democracy organizations in countries surrounding the People's Republic of China. The CIA's rapidly evolving apparatus for carrying out covert operations allowed the individuals and organizations involved in these campaigns to claim independence from the government—an alleged psychological benefit that, as we shall see, would come to haunt all those involved.[7]

Over the past twenty-five years, historians and journalists have produced dozens, perhaps even hundreds, of books and articles documenting the extraordinary range of this cultural offensive, in both its overt and covert forms. Science is oddly absent from these accounts. Over time, I have come to believe that this peculiar silence results from a profound misunderstanding of what US policymakers meant when they referred to "culture." Most accounts of cultural diplomacy, with their focus on artists, musicians, philosophers, and other writers, implicitly replicate midcentury intellectuals' preoccupation with the relationship between cultural production and social change. One line of analysis associated with the influential Frankfurt School of critical theory suggested that the "culture industry" placated the masses, replacing the shared class interests of the proletariat with the easy pleasures of capitalism. In the postwar period, many European intellectuals retained a Marxist suspicion of the power of popular culture, particularly the American forms (Hollywood films, rock music) that threatened to overtake European achievements on stage and screen. In 1951, political theorist Hannah Arendt's *The Origins of Totalitarianism* took this critique of mass culture in a different direction, suggesting that the average citizen's susceptibility to manipulation by mass media contributed directly to the horrors of World War II.[8]

Many of the writers and theorists who participated in the cultural Cold War, particularly via the conferences and publications of the Congress for Cultural Freedom, were deeply steeped in the analytical traditions of critical theory. The CCF sponsored music festivals, art exhibits, and literary magazines designed to show *either* that capitalism could also produce high culture (in response to critiques of capitalist aesthetics) *or* that middlebrow culture didn't inevitably lead to totalitarianism (in defense of rock-and-roll). It is a mistake, however, to read the obsessions of the individuals involved with organizations like the CCF as expressions of US foreign policy objectives. Historians' confusion on this point is a direct consequence of the archival record. While the products of psychological warfare campaigns (magazines and books produced by the CCF, for example) have never been classified, many of the records explaining the strategic thinking behind such programming have only been declassified in the past five to fifteen years. Until recently, then, historians hoping to understand *why* the CIA would fund, say, European art exhibits had little to go on beyond the statements of key participants (who may or may not have been aware of the CIA's role at the time). Those participants tended to emphasize the importance of "culture." Even today, some materials associated with the CIA's covert cultural programming remain classified.

The (partial) opening of the archive has finally allowed historians a glimpse of the logic that drove policymakers to adopt cultural diplomacy as a psychological warfare strategy in the first place, and it shows a rather different story. Declassified documents make abundantly clear that the architects of US psychological warfare policy grounded their vision of culture in anthropology, not in the Frankfurt School. During World War II, the US war effort drew heavily on the expertise of anthropologists. Margaret Mead's new notion of "national character" offered a way to translate anthropologists' traditional knowledge of family, community, and social order into policy-based statements on morale, political leadership, and national goals. Mead herself spent most of the war shuttling between New York and Washington, DC, where she advised colleagues working at the Office of Strategic Services, the State Department, the

Army, and the Office of War Information. Anthropologists and policymakers continued their partnership into the uneasy peace that followed, with entrepreneurial figures like Clyde Kluckhohn converting the government's hunger for cultural insight into a funding bonanza for the social sciences.[9]

The makers of US foreign policy quickly gained fluency in this social scientific language. When NSC 68, a guiding document of US Cold War strategy, described the conflict between the United States and the Soviet Union as a clash of "civilizations," readers would have understood this term not just as shorthand for "geopolitical spheres of influence" but also as a reference to competing systems of culture, in the midcentury, anthropological sense. NSC 68, like many other policy documents of its time, repeatedly circles back to the role of freedom and authority in the "structure of society"—a clue to policymakers' intent. If, as many mid-twentieth-century social scientists believed, a nation's political culture could be explained by its social institutions, then the best way to combat the global spread of Communism was to encourage the development of social structures similar to those in the United States.[10]

Scientific institutions played a larger role in US life in the immediate postwar years than at any other time in US history. And for that reason alone, science had a place in US cultural diplomacy from the very start. Indeed, this broader understanding of what policymakers meant by "culture" explains several aspects of US propaganda campaigns that make no sense whatsoever if culture is understood to refer primarily to the world of arts and letters. The US campaign for global hearts and minds included everything from sporting events to church groups, from 4-H clubs to the YMCA. Focusing on science helps to reveal the logic behind what might otherwise seem an odd mishmash of campaigns aimed at intellectuals, technocrats, and children. This book, in other words, is not only the first to integrate science into the history of cultural diplomacy but also a retelling of the history of cultural diplomacy that takes seriously US policymakers' vision of civilizational change.

As a practical matter, my more comprehensive definition of culture means that I devote as much attention to the Asia Foundation's emphasis on developing nongovernmental institutions as to the CCF's literary magazines. For reasons explained in more detail later in the book, the Asia Foundation managed to avoid most of the recriminations leveled at the CIA's other proprietary organizations in the late 1960s and early 1970s. Even today, few scholars outside the narrow field of intelligence history are aware of the Asia Foundation's complicated history as an instrument for Cold War propaganda. Following the twisted path of science as cultural diplomacy reveals the scale and scope of the Asia Foundation's role within the broader framework of US psychological warfare—a story presented here for the first time.

No single agency directed the United States' cultural offensive. Instead, a mix of acknowledged (overt) and unacknowledged (covert) government agencies and private philanthropic organizations collaborated in sharing the US vision of science with the world. This intensely political effort to change the balance of Cold War power depended on partnerships with US scientists who preferred to think of themselves as apolitical actors. Hindsight presents us with the opportunity as well as the responsibility to understand what was at stake in such appeals to scientific objectivity.

::

Ideologies have consequences. When US actors suggested that science transcended politics, they were also establishing practices that benefited the United States in concrete ways. The American notion of scientific freedom suggested that knowledge should operate in a frictionless environment, without regard to party dogma, political affiliation, or national borders. In recent years, historians of science have adopted somewhat similar language to describe the global circulation of knowledge. But, as historian Fa-ti Fan writes in an essay mildly critical of the term, "the image of smooth circulation probably doesn't encourage a critical analysis of, say, power relations in science." Knowledge doesn't circulate in the absence of social, political, and legal structures—and in the unbalanced Cold War power system, the United States was in a unique position to build and maintain structures of scientific exchange that benefited participants unequally.[11]

I'm referring, of course, to intelligence collection. It is no coincidence that the US notion of scientific freedom crystallized in the late 1940s, precisely the moment that the US government was attempting to set up a permanent peacetime structure for gathering scientific intelligence. While scientific knowledge and technological know-how have always been intertwined with state power, the logic of the Cold War elevated the strategic importance of both scientific knowledge and scientific intelligence. This knowledge might be as concrete as the technical specifications necessary to build a lighter, more powerful atomic bomb or as abstract as social scientific theories predicting the outcomes of a negotiation. Both the United States and the Soviet Union desperately wanted insights into the scientific and technological abilities of their opponents. This might be information with direct military application, such as an estimate of when the Soviet Union would detonate an atomic weapon, or information with political importance, such as the structure of science education in Japan. Scientific knowledge might be scientific intelligence, but scientific intelligence extended much further than scientific knowledge.[12]

The concept of "scientific internationalism" bridged the ideology of scientific freedom and broader national security goals. As articulated by US scientists in the postwar period, scientific internationalism meant many different things. It might refer to international scientific cooperation, or the international control of atomic energy, or perhaps a universal language of science. At heart, this was the idea that science knows no borders. A laudable goal, yes— but not a utopian one. The historical record shows, over and over again, that the loudest Western voices in favor of scientific freedom and scientific internationalism were at least as interested in advancing US foreign policy as in promoting civil liberties.[13]

The National Academy of Sciences (NAS), a nongovernmental entity deeply associated with the values of scientific internationalism, facilitated many of these campaigns in science. Founded by congressional charter in 1863 to advise the US government on "any subject of science or art," the NAS elects its own members from the scientific community. Unlike the national academies in many other countries, however, the NAS is not quite a government entity. While the bulk of its work is undertaken at the government's request and at the government's expense, it does not have a standing appropriation. It was a sleepy, elite institution for the first half-century of its existence. This changed in 1916, when a new operational arm, the National Research Council (NRC), allowed the NAS to involve nonmembers in its operations and to fund its work through government contracts. The academy's footprint grew tremendously during World War II, as the US government turned to the NRC's various committees for advice on any number of defense questions.[14]

By the mid-1950s, as we shall see, the NAS had also become the US government's de facto contractor for international science. The NAS selected US delegates for international conferences, vetted scientists for travel, managed US contributions to worldwide scientific databases, and sponsored conferences that brought together scientists from around the world. Starting in 1959, the NAS administered the United States' formal scientific exchanges with the Soviet Union. The associated overhead charges secured the academy's financial footing, but this was not the only reason the NAS's leaders embraced the arrangement. The community of scientific administrators who had risen to power during World War II associated scientific greatness with scientific independence. Men like Vannevar Bush and Detlev Bronk credited the United States' wartime technological achievements, including the atomic bomb and the proximity fuse, to the government's partnerships with university scientists, whom they viewed as objective, apolitical, and forward-thinking. They contrasted these virtues both with the thinking of military brass, which they characterized as fixated on interservice rivalries and secrecy, and with the structures of research in Nazi Germany, where racist ideology warped research agendas. The gap between the NAS's contractual relationship to the US government and its institutional identity as a nongovernmental body controlled by scientists created an ambiguity that proved increasingly useful as Cold War propaganda turned to science.

∷

That many of the most outspoken advocates for scientific freedom were also deeply committed to US policies that exacerbated global inequality has made this book, at times, troubling to write. In looking for heroes who stood up to McCarthyism, historians of science have overlooked the ways that these same scientists were upholding a system of privilege from which they stood to benefit. In the 1990s and 2000s, many historians of science in the United States—myself included—assumed that American scientists severely restricted their political activities from the late 1940s through the mid-1960s. According to this school of thought, the climate of anti-Communism, combined with a growing dependence on federal funds, meant that scientists who valued steady employment either retreated to their laboratories or tried to effect limited change through federal advisory channels.[15]

Put another way, mainstream US historians wrote accounts suggesting that American institutions valued scientists who eschewed politics. And in a Cold War context in which American institutions seemed to be more interested in enforcing political consensus than in advancing justice, those few scientists who bucked the trend stood out as heroes. Here and there among the Cold Warriors and organization men, historians identified a handful of scientists who dared talk back. These oppositional figures defended their colleagues against McCarthyism, spoke out about the dangers of excessive secrecy and the military-industrial complex, and, on rare occasions, even voiced support for women and minorities. Historians writing about such persons typically presented them as champions for scientific freedom.

I made the same mistake myself. The project on which this book is based started off as an investigation into the life of Johns Hopkins geneticist H. Bentley Glass, a man who did all of the things just described. I wanted to know how he got away with it. His first brush with public controversy came when he agreed to serve on the Genetics Society of America's Committee to Counteract Anti-Genetics Propaganda, a group created in 1948 to combat the growing influence of Soviet agricultural geneticist Trofim Lysenko. Glass opposed Lysenko, but he also opposed censorship. He earned his former advisor H. J. Muller's ongoing distrust when he defended an Oregon college instructor who promoted Lysenko's views in his classroom. This episode would prove to be only the first of many incidents in which Glass supported the freedom of expression, both as a spokesman for professional societies and as a private citizen. In 1955, he was elected president of the Maryland chapter of the American Civil Liberties Union (ACLU), a position he held for ten years.[16]

In 1954, Glass won a seat on the Board of School Commissioners of the City of Baltimore. Newspaper and magazine reports suggest that he used this position, along with his authority as a geneticist, to argue for an accelerated pace of desegregation in the city's schools. In 1960, he advised the Democratic National Committee on the scientific planks of the party platform. The collection of his papers at the American Philosophical Society in Philadelphia includes a hand-signed thank-you letter from John F. Kennedy, as well as the pen used by the mayor of Baltimore to sign that city's Civil Rights Ordinance. Nor were his political activities limited to the domestic arena: by 1958, he had become an outspoken voice on the dangers of radioactive fallout and one of the leading American figures in the Pugwash Movement for international disarmament.[17]

Perhaps not surprisingly, given his political activities, Glass was the subject of several investigations by the Federal Bureau of Investigation (FBI) during the early Cold War. Glass's 228-page dossier is typical for a postwar scientist, consisting mostly of background checks for visa applications and AEC activities. Page after page of testimony from colleagues and neighbors describe both Glass and his wife as "staunch and loyal American citizens." More than one informant remarked on Glass's deep religiosity. Even so, Glass found himself labeled a "possible security risk" on the basis of a five-dollar donation to the Baltimore chapter of the American Youth for Democracy, an organization with alleged ties to the Communist Party. Glass had attended a student luncheon where he spoke against segregation in the military; in 1944, he signed a petition sponsored by the group on the same topic. Astonishingly, Glass had come to the FBI's attention not for his outspoken support of academic freedom but for his rather moderate views on racial equality. FBI agents filed updated reports on Glass's loyalty every six months from 1951 to 1955. Although his record was finally cleared in 1955, the FBI continued to collect information on his views on genetic mutation, nuclear proliferation, and the Soviet threat throughout the 1960s.[18]

And yet—and this is what caught my eye—Glass's professional career flourished. During Glass's time at Johns Hopkins, his research received generous funding from the AEC, even as he roundly criticized the agency's position on fallout. In 1959, he was inducted into the National Academy of Sciences. And, beginning with the American Institute of Biological Sciences (AIBS) in 1954, he assumed the presidency of one professional group after another, culminating in the presidency of the American Association for the Advancement of Science in 1969. His appointment as academic vice president of SUNY–Stony Brook in 1965 recognized both his administrative prowess and his breadth of mind.

Glass's experience simply did not line up with the existing narrative that US scientists who wanted to protect their funding and institutional arrangements had to engage in self-censorship. How, I wondered, had Glass managed to navigate the minefield of anti-Communism so successfully? How did those around him come to understand, or at least tolerate, his insistence on civil liberties during an age of suspicion? Were his experiences unique, or did they indicate a need for historians to rethink what constituted acceptable behavior for a scientist during the Cold War?

About six months into the project, I realized I had fundamentally misunderstood two aspects of Glass's public persona. First, he himself did not consider his scientific activities "political." Even after holding public office, advising the Democratic National Committee, and refusing to sign a loyalty oath, Glass saw himself as an objective scientist who brought that perspective into the public arena. When he criticized the AEC's position on fallout, he claimed to do so from a position of objectivity. Here was a man who researched blood-type differences among racial groups and, in his capacity as president of the Maryland ACLU, defended the rights of African American men, yet refused to see any connection or conflict between these activities. Glass ensured his political freedom by distancing himself from politics.

Still, it was one thing for Glass to make this claim and another for authority figures to believe it. That's when it began to dawn on me that Glass had a less confrontational relationship with the government than his copious FBI file might suggest. The clues came slowly at first. In 1957, Hiden Cox, the director of the AIBS, asked a potential translator to do some work for the organization "at the request of the Central Intelligence Agency." Glass recommended the consultant by name, but, more to the point, Glass was a member of the AIBS's governing board at the time that the organization was apparently entering into a relationship with the CIA. Once I saw that link, I noticed that Glass seemed to spend a lot of time with people with close ties to the intelligence community. These included his brother, David, who had worked in Army intelligence in China during World War II, and Harold Coolidge, an Office of Strategic Services veteran who ran the NAS's Pacific Science Board.[19]

Bentley Glass, it turns out, had a fairly comfortable relationship with various US government agencies, particularly those involved in science diplomacy. In 1950, he wrote to his department chair, asking for two months' leave to take up a post with the State Department in Occupied Germany. Glass asked his chair to keep his whereabouts vague: "The State Department does not care to have the business widely publicized and permits me to say only that I am going to Germany for two months as a scientific consultant." A decade later, as part of his association with Pugwash, Glass received briefings in Washington. During the Cuban Missile Crisis, his name appeared on a cable sent to Moscow urging restraint, apparently sent with the State Department's blessing.[20]

The real moment of clarity came once I finally understood the nature of the Asia Foundation's relationship to the CIA. *Most* of the organizations Glass was involved with, including the NAS, received support in one form or another from the Asia Foundation. Without insight into the nature of the Asia Foundation, it appeared that these organizations were doing politically courageous work on behalf of scientific freedom and scientific internationalism. But now that I had insight into the real purpose of the Asia Foundation, I began to wonder what else I had missed.[21]

The interesting question wasn't how Glass had flourished while maintaining an oppositional stance toward the government. The question was why so many historians of science, myself included, had failed to see that, in 1955, support of scientific freedom was not oppositional at all. Notwithstanding the FBI's suspicion of anyone and everything associated with civil rights, other branches of the US government embraced advocates of scientific freedom as spokespersons for American values. The State Department found Glass's commitment to scientific objectivity politically useful, as did the CIA and the US Information Agency. Historians of science have greatly misread the uses of apolitical science in the Cold War.

Once seen, the links between Cold War propaganda and scientific freedom cannot be unseen. Philosopher of science Michael Polanyi, whose 1962 article, "The Republic of Science," famously compared the "free cooperation of independent science" to a "highly simplified model of a free society," published that piece in the first issue of *Minerva*, a journal sponsored by the CIA's Congress for Cultural Freedom. Detlev Bronk, who as president of the NAS and chair of a presidential advisory committee on the role of science in foreign affairs declared that "science is apolitical," made this claim in the context of protecting the NAS's role in facilitating scientific intelligence collection. Physicist Eugene Rabinowitch, who used his long-time post as editor of the *Bulletin of the Atomic Scientists* to promote the idea of scientific internationalism and is generally regarded as a champion of nongovernmental scientific activism, sought partners at the State Department and the CIA to advance his cause of disarmament.[22]

Bentley Glass, in other words, was no anomaly. The personnel who implemented cultural diplomacy involving science circulated freely between meetings and appointments at overt, covert, and private organizations. For most of the 1960s, for example, the Asia Foundation, the US Agency for International Development, the Rockefeller Foundation, and the Ford Foundation all supported separate efforts to adapt Biological Sciences Curriculum Study (BSCS) textbooks—the textbooks that Arnold Grobman hoped would counter Communism—for different parts of the globe. Neither the BSCS nor its funders seem to have made much of a distinction between the sources of this support.[23]

When we spoke at his home in Florida, Grobman denied any knowledge of the CIA's involvement in funding biology textbook translations. He continued to do so even after I showed him photocopies of letters, bearing his signature, that suggested as much. As the BSCS's executive director, Grobman had been eager to identify funding partners willing to underwrite the program's expansion abroad. When a colleague suggested in 1961 that the BSCS contact the CIA, which had "funds in abundance" for the kind of work the group wanted to do, Grobman readily assented. The resulting partnership between the BSCS and the Asia Foundation continued well into the 1970s, several years after the Asia Foundation's relationship with the CIA made minor headlines in the *New York Times*.[24]

I showed Grobman a series of letters he and his colleagues had written. I drew diagrams showing how the CIA's pass-through funding system worked. I assured him that the Asia Foundation's ties to the CIA were documented on the CIA's own Freedom of Information Act website and that he wouldn't be violating any security measures if he told me all about it. I sent him a long letter outlining my various theories about the BSCS and its relationship to the CIA and invited him to contradict me, but he only replied, "I can't tell you that." Whether he was referring to his failing memory or a government oath, I'll never know. He died several months after our interview.

At the point in the project when I interviewed Grobman, I still thought it important to understand whether scientists realized they were working with covert government agencies. I was, in other words, still looking for heroes and goats. It was only once I got deeper into the archival and declassified records that I understood how far a distinction between overt, covert, and private propaganda activities missed the mark. Grobman, like Glass, Polanyi, Rabinowitch, and Bronk, saw no conflict in accepting *any* kind of US government funds to promote his version of scientific freedom, so long as he retained direct control over the message being disseminated. Put another way, these scientists' belief in their autonomy *as scientists*—their commitment to scientific freedom—limited their ability to recognize that their ideas were being put to use in campaigns over which they had no control. If I had that interview to do again, I would press him on why he thought scientific thinking, rather than political sovereignty or economic independence, offered the best hope for freedom for Hong Kong's ambitious youth.

∷

All these stories, and others like them, are told in the pages that follow. Many more remain to be told. For now, I want to plant in readers' minds a healthy skepticism about postwar US scientists' claims of scientific autonomy, scientific freedom, and scientific internationalism. Existing narratives on science and the Cold War have been preoccupied with questions of guilt and innocence, collaboration and resistance. The process of researching this book has convinced me that this black-and-white approach merely reproduces the same ideological divisions that drove the Cold War in the first place. In 1954, a scientist who spoke out against loyalty oaths was *both* taking a brave stand against political witch hunts *and* demonstrating to an international audience that (some) scientists in the United States had the freedom to criticize their own government. In the 1960s, US scientists who participated in international exchange programs with the Soviet Union were *both* building genuine friendships with their Soviet counterparts *and* collecting scientific intelligence.

Contrary to what the preceding paragraphs might imply, I do not want this book to be read as a broadside against scientific freedom and scientific internationalism. To say that the relationship between science and freedom had to be constructed and maintained is not to say that scientific freedom is not desirable. Of course collaboration is better than competition; of course humans, including scientists, fare better under conditions of freedom than under totalitarianism. In this strange political moment in which an advisor to a US president can describe obvious falsehoods as "alternative facts," real facts are worth defending. My point is instead that these ideas—that science is apolitical, that science knows no borders—were popularized (and indeed institutionalized) with the help of state power during a highly polarized and unique historical moment, and they have not always been used in the cause of justice.[25]

It is my hope that unearthing the tangled roots of this story will help all of us—scientists, historians, politicians, journalists, members of the general public—develop more meaningful and inclusive visions of science's role in democracy. Scientific thinking alone cannot save us from the myriad challenges we face. Putting science-based perspectives in the service of justice just might.

1 : Western Science vs. Marxist Science

In 1939, the British crystallographer J. D. Bernal published a remarkable book, *The Social Function of Science*. The book's thorough accounting of scientific institutions, research salaries, career trajectories, educational systems, and national priorities makes it a landmark publication in the sociology of science, but Bernal's careful research wasn't what sparked a public controversy and political backlash. Bernal, a Marxist and a member of the Communist Party of Great Britain, premised his book on the idea that the scientific process was intimately tied to social and economic conditions. Bernal acknowledged the concept of "science as a pursuit of pure knowledge for its own sake," but he described the attitude as only one end of a spectrum bounded at the other by "science as power." He relentlessly pointed out how capitalism shaped the production of knowledge in the United States and repeatedly referred to scientists as "scientific workers." The book was, in many ways, a brief on behalf of Soviet-style planning as the surest and swiftest path to transform society and improve the human condition.[1]

Bernal's book became a flashpoint for public debates in the English-speaking world about the relationship between science and the state. But within the narrower community of scientists, radicals had already been debating the virtues of "Western science" versus those of "Marxist science" for nearly two decades. In the 1920s and 1930s, a number of young American socialists and communists dreamed of translating their radical political ideas into a program for scientific progress. One of the most important of these was the geneticist H. J. Muller, who spent much of the 1930s searching for fertile Soviet soil in which to grow a socialist, eugenic utopia. But unlike Bernal, who remained a loyal Communist Party member for the rest of his life, Muller's firsthand experiences of life in Stalin's Russia convinced him to abandon both socialism and his admiration for scientific planning. By the late 1940s, Muller had become a rabid anti-Stalinist who willingly lent his name and time to cultural campaigns against communism.[2]

For Muller, as for many Western scientists, the problem with Marxism (and Communism) boiled down to a problem with Soviet biology. Starting in the late 1920s, thanks in large part to the ambitions of a Ukrainian agronomist named Trofim Lysenko, something went deeply wrong with genetics in the Soviet Union. At first, Lysenko's denunciations of the bourgeois tendencies of classical genetics proved only a minor irritant to Soviet geneticists. But starting in the mid-1930s, as Lysenko consolidated his power in the Soviet scientific bureaucracy, Soviet geneticists started disappearing from their labs. The scientists were exiled, imprisoned, or shot. In August 1948, Lysenko announced that Soviet Communist Party officials had endorsed his unorthodox views on inheritance as the official interpretation of genetics in the Soviet Union. For observers in the West, the phenomenon that came to be known as "Lysenkoism" symbolized the failures of the system Bernal so lavishly praised.[3]

Lysenko's triumph came at a transitional moment in the history of US cultural diplomacy, when US officials deeply feared the psychological appeal of Communism to European elites yet possessed few formal channels from which to launch a counteroffensive. For Muller and his like-minded colleagues, the tragedy of Soviet genetics offered an unrivaled opportunity to denounce Communism and the dangers it posed to scientific freedom. Their approach, however, was not universally admired. Muller's outbursts against Soviet science struck many moderate and liberal US scientists as unabashedly political, subjective, and dogmatic—the mirror image of Communist science rather than a true alternative. Although largely conducted without government support, US scientists' campaigns against Lysenkoism, including their internal debates over the best way to conduct those campaigns, set the terms of US propaganda involving science for the next twenty years.

∷

Mid-twentieth-century geneticists operated within unusually tight international networks bound together by the humble fruit fly. As young researchers fledged the nest of Thomas Hunt Morgan's laboratory at Columbia University in the 1910s and 1920s, they took their stocks of *Drosophila melanogaster*—the experimental basis for studies of Mendelian inheritance—with them. Throughout the 1920s and into the 1930s, researchers seeking special fruit fly cultures had to write directly to one of a handful of laboratories in New York, Berlin, and Texas. When geneticists visited one another's labs, they brought gifts of flies, sometimes relying on subterfuge to slip their stocks past customs agents. Besides exchanging personal letters and test tubes of precious stock, geneticists built connections by spending time in each other's labs—often on fellowships sponsored by the Rockefeller Foundation—and by attending international meetings.[4]

Muller was a central, if prickly, node in this network. Born in New York City in 1890, Hermann Joseph Muller (rhymes with "duller") obtained both his undergraduate and graduate degrees from Columbia University. During his undergraduate years, the frugal Muller earned money working as a hotel and bank clerk; as a graduate student, he took the subway downtown to teach English to immigrants enrolled in night school. Muller secured a desk in Morgan's famous fly room in 1912, at which point he immediately started building his reputation as a brilliant theorist and difficult colleague. In 1920, he struck out for Austin, Texas, where he would build the University of Texas into a major site for mutation research.[5]

For Muller, the move to an independent laboratory represented political and scientific freedom. He took his first trip to Europe in 1922, visiting labs in England, France, Germany, and Russia. The Russians made a particularly good impression on Muller, who had been a socialist since his college days. He spent most of August touring nearly a dozen of the state institutes run by the new Bolshevik government, including the Institute of Experimental Biology in Moscow and the Institute of Applied Botany in Petrograd (St. Petersburg). While some of these institutes had been established with private patronage during the reign of the tsar, scientific research in Russia had experienced rapid growth under the Bolsheviks. Within a decade of the October Revolution, the Soviet Union had more than 1,000 scientific institutions, all funded directly by the state.[6]

The leaders of the new Communist government saw science and scientists as essential resources in winning the civil war and stabilizing the economy, and they delighted in showing off their newly socialized system to foreign visitors. Muller was one of approximately 100,000 foreigners to visit the Soviet Union in the 1920s and 1930s, usually drawn by some combination of idealism and curiosity. In an account of his trip, published in *Scientific Monthly*, Muller recorded his admiration for Russian scientists in the face of very real material challenges. Instead of the "starving creature in rags, hiding in some attic," that Americans might have imagined, Muller found the typical Russian scientist "busily engaged in his laboratory, actively interested in his scientific work and getting results, and desirous above all else of getting in touch with the work that the west is producing." After years of civil war, uncertainty, and food shortages, Russian scientists now drew food rations and wages from the state "just for being a scientist." Scientists had initially rejected the Communists, but "on the whole," Muller reported, "the scientists whom I met do not actively concern themselves with politics at all; they are too busy with science." Everywhere he found geneticists, Muller left flies.[7]

Back in Austin, Muller continued his research. In 1926, he shocked the small community of *Drosophila* researchers by using X-rays to create new, heritable mutations. He began assembling a talented team of young scientists and opened his doors to international researchers. Two of these, Israel Agol and Solomon Levit, were students whom Muller had met in Moscow. But to Muller, who suffered from depression, these accomplishments offered little comfort. He left Texas permanently in 1932, after increasingly erratic behavior and a failed suicide attempt damaged his relationships with colleagues. The Federal Bureau of Investigation, meanwhile, notified university officials of Muller's involvement with *The Spark*, a leftist student newspaper, leaving his future employment in doubt. In September, Muller left for Berlin.[8]

Muller's plans for Berlin centered on the laboratory of Nikolay Timofeev-Ressovsky, a Russian *Drosophila* researcher who had come to the Kaiser Wilhelm Institute for Brain Research for the unlikely purpose of studying Lenin's brain. Once there, Timofeev-Ressovsky assembled an international team of researchers that soon gained a reputation as the most talented group of mutation researchers in continental Europe. Their flies included those descended from stocks that Muller had left on his 1922 tour. Muller initially found the company in Timofeev-Ressovsky's lab congenial, but when Nazi storm troopers attacked the institute one evening in March 1933, he decided the time had come to leave. He accepted an invitation from Nikolai Vavilov, one of the Soviet Union's leading scientists, to direct a genetics laboratory at the Institute of Applied Botany. Timofeev-Ressovsky, having closer knowledge of the state of Soviet genetics at the time—his teacher, Sergei Chetverikov, had been arrested and exiled from Moscow in 1929—remained in Berlin.[9]

Muller threw himself into Soviet scientific life. As Muller's biographer (and former student) Elof Carlson put it, Muller embraced the Soviet system "with the conviction of a convert." With Vavilov's sponsorship, the Soviet Academy of Sciences recognized Muller as a foreign (corresponding) member in the spring of 1933. When Vavilov's newly renamed Institute for Genetics moved to Moscow in late 1934, Muller moved with it.

As late as 1935, Muller, the most prominent US scientist to relocate to the Soviet Union, remained passionate about the socialist experiment. Even such a booster as he, however, recognized that not all was well in Soviet science. Scientists working in the Soviet Union were increasingly isolated. Censors monitored scientists' correspondence and reviewed their publications. In 1933, the Politburo squashed researchers' access to Rockefeller-funded fellowships.[10] Vavilov, especially, began to have trouble with the authorities. His frequent trips abroad to collect specimens for his All-Union Institute of Plant Industry attracted unwanted attention from the press as well as from the secret police. In 1932, he had been the sole Soviet representative to attend the Sixth International Congress of Genetics in Ithaca, New York. In 1936, not a single Soviet scientist attended an international conference.[11]

∷

The social and economic experiment underway in the Soviet Union included a commitment to dialectical materialism, a philosophy of science committed to explaining the natural world through matter and energy. One possible reading of dialectical materialism is that scientists should base their interpretations on objective reality. But as dialectical materialism developed under the Bolsheviks, the concept grew to include the possibility of different natural laws for different "dialectical levels." Explanations of the physical world, for example, must range beyond physical reality to include the biological and social realities that shape the observer's perception.[12]

For the first decade or so of the Soviet Union's existence, dialectical materialism developed alongside, but largely separate from, scientific and technical work being conducted at the new, state-run research institutes. All this changed in 1929, when Joseph Stalin, having consolidated his power over the Politburo, committed to industrializing the country at a breakneck pace. Party ideologues began dismissively referring to something called "bourgeois science," which generally meant science concerned with knowledge for its own sake rather than science conducted on behalf of the needs of the people. In 1931, the Communist theoretician Nikolai Bukharin shared this scientific worldview with scholars gathered at the Second International Congress for the History of Science in London. Science, he argued, is a social product that develops differently under socialism than under capitalism. Another delegate from the Soviet Union, Boris Hessen, delivered a paper that interpreted the work of Isaac Newton as a byproduct of seventeenth-century mercantilism.[13]

In the decade between Muller's first visit to the Soviet Union and his decision to move there in 1933, ideologues had begun to target Mendelian genetics as bourgeois, theoretical, and incompatible with Marxism. One line of criticism came from followers of Trofim Lysenko, who promised to transform Soviet agricultural production by shortening the growing period for winter wheat by exposing seeds to cold conditions. Lysenko denied the physical existence of genes and claimed that a real materialist understanding of inheritance would focus on the environmental conditions surrounding organisms. While *Drosophila* researchers tinkered with fruit flies and theorized about chromosomes, Lysenko and his allies promised to revolutionize Soviet grain production in just a few short years. A second line of criticism emerged in the mid-1930s in response to the rise of Nazi "racial hygiene." Soviet antifascists correctly pointed out that Nazis based their eugenic theories on a Mendelian understanding of human genetics. Both groups accused Mendelian genetics—frequently referred to in the Soviet Union as "Morganism"—of being anti-Darwinist and, thus, favoring an unchanging view of the human condition, in violation of the precepts of dialectical materialism.[14]

Upon arriving in Leningrad (Petrograd had acquired yet another new name after Lenin's death in 1924), Muller attempted to reconcile Mendelian genetics with party doctrine. His 1934 essay, "Lenin's Doctrines in Relation to Genetics," argued that chromosomes represented a true materialist basis for inheritance, and it claimed a "direct Marxian influence" for the "younger coworkers occupying the official position of 'students'" in Morgan's *Drosophila* lab. He made the case for a socialist program of human genetics, uncorrupted by "bourgeois prejudices," that would eventually allow "every individual to reap the benefits of the biological fruits of socialism." Muller would later distance himself from this doctrinaire essay, omitting it from his list of publications; in truth, it had little effect on the official status of Mendelian genetics (in either direction) within the Soviet Union. In the mid-1930s, Lysenko continued to expand his power base in Soviet agriculture, but geneticists continued breeding their flies in their own, separate institutes. A "public discussion" on the merits of Mendelian genetics in 1936 ended in a draw.[15]

Muller doubled down on the socialist possibilities for eugenics in a short English-language tract, *Out of the Night*, published shortly after his Lenin essay. He was so convinced of his approach's appeal that, in 1936, with Levit's encouragement, he sent a copy of the book to Stalin. In an accompanying letter, Muller outlined a "really sound policy with regard to human genetics," namely, using the sperm of "the most transcendently superior individuals" to impregnate the patriotic women of the socialist republic. In Muller's unforgettable phrasing, "Many a mother of tomorrow, freed of the fetters of religious superstitions, will be proud to mingle her germ plasm with that of a Lenin or a Darwin, and to contribute to society a child partaking of his biological attributes." Apparently this phrasing was no less shocking to Stalin than it is to modern ears; word came back to Muller that Stalin was "displeased." With the purges under way, it was once again time to move on.[16]

The next few years were full of uncertainty for Muller, who had managed the impressive feat of alienating authorities in Texas, Berlin, and Moscow. A brief stint volunteering for the International Brigade in the Spanish Civil War in 1937 rehabilitated him to the point that he could retrieve his belongings from Moscow before leaving the country for good. Though their fates probably had little to do with Muller, both Levit and Agol were arrested during his absence; another colleague, W. P. Efroimson, was sent to Siberia. Muller, disillusioned and ostracized, spent the next few years in a series of temporary appointments at the Institute of Animal Genetics in Edinburgh and at Amherst College in Massachusetts. He eventually landed at Indiana University in 1945, but only after Rockefeller Foundation officials promised the university $95,000 in support of its genetics group.[17]

With Muller's departure in 1937, Soviet geneticists had lost their only face-to-face contact with Western scientists. Many international communists, like J. D. Bernal, dismissed rumors of purges—in genetics or otherwise—as propaganda intended to undermine the Soviet experiment. Left to read between the lines of censored letters and garbled newspaper reports, more critical Western geneticists speculated that science had been politicized in the Soviet Union in much the same way that it had been in Nazi Germany. These skeptics were closer to the truth. In 1938, Lysenko was appointed president of the Lenin All-Union Academy of Agricultural Sciences (VASKhNIL), one of the most powerful organizations in Soviet science. Those classical geneticists who remained in their positions faced increasing pressure to justify their work. When not a single one of the fifty geneticists scheduled to present at the Seventh International Congress of Genetics Meeting in Edinburgh in August 1939 appeared, their US colleagues feared the worst.[18]

∷

Historians have now spent more than fifty years trying to understand what exactly happened in Soviet genetics, and why. We know for sure that at least eighty Soviet researchers involved in the genetics controversy, including Vavilov, were arrested and subsequently imprisoned, exiled, or executed between

1936 and 1942. The number who lost their positions is unknown but certainly ranks in the hundreds. The situation improved somewhat during and shortly after World War II, when the leaders of Soviet science hoped to convert their wartime alliance with the United States to a permanent international partnership. Those hopes were dashed, utterly and finally, on August 7, 1948, when Lysenko announced at a public meeting of the VASKhNIL that the Communist Party had endorsed his scientific theories. With that announcement, the remaining strongholds of Mendelian genetics were dissolved, the stocks of *Drosophila* ordered destroyed.[19]

Stalin himself intervened in this decision to issue a Communist Party fiat in the name of a scientific theory. Genetics fell victim to a broader anti-Western cultural campaign in the Soviet Union in 1947 and 1948, with Lysenko's theories now declared fundamentally Soviet and therefore superior to anything the West had to offer. What's so curious about the Lysenko affair, however, is that the murders of geneticists happened *before* this fiat, not after. A handful of geneticists were arrested and imprisoned after, but all were rehabilitated within six years. And, after a surprisingly short interlude, classical genetics began to thrive in the Soviet Union, camouflaged under such names as "radiobiology," "experimental biology," and "biophysics," even while Lysenko maintained a powerful force in Soviet science. As Soviet dissident Zhores Medvedev put it in his *samizdat* account of the controversy, "the more cruel type of repression was still limited."[20]

With the clarity of hindsight and the opening of Soviet archives, most historians now believe that the murder of geneticists in the 1930s had little to do with their interpretation of a scientific theory. Historian Nikolai Krementsov, who has investigated this issue more thoroughly than anyone, distinguishes between a Communist Party position on Lysenko's theories, which he found no evidence for prior to 1948, and Lysenko's ability to appeal to party operatives. As Muller's essays made clear, the facts of Mendelian genetics fit as well into a Marxist philosophy of science in 1948 as they had in 1928, 1936, and 1939—if anything, better, as biologists had begun to work out an understanding of natural selection that reconciled genetic mutations with evolutionary change.[21]

So, why were the geneticists purged, if not for their scientific beliefs? Their fate cannot be separated from the larger tragedy that befell Soviet citizens in the 1930s. Between 1936 and 1938, more than 8 million people were arrested, and somewhere around 1 million executed, in Stalin's Great Terror. Physicists, astronomers, and engineers suffered, as did writers, artists, bureaucrats, and military leaders. The geneticists were not martyrs for scientific freedom; they just got caught up in the maws of Stalin's rage. We know now that the problem with genetics wasn't dialectical materialism or the relationship of Mendelian inheritance to agriculture; the problem with genetics was Stalin. By the time Lysenko consolidated his power over Soviet genetics in the late 1940s, however, the most brutal phase of Stalinism had ended. Those geneticists who publicly criticized Lysenko in the days immediately before his announcement suffered, but they lived to tell the tale.[22]

This is the great irony of Lysenkoism. A political opportunist *did* manage to wrest control over an entire branch of Soviet science in the late 1940s, but not quite in the way that most Western observers believed. Lacking the ability to foretell the future, disillusioned anti-Stalinists like Muller assumed the worst. Recall that several of Muller's close associates had been arrested during the purges; he only narrowly escaped this fate himself. Muller had every reason to believe that Lysenko's announcement signaled Stalin's intent to liquidate an entire field of knowledge, including all its practitioners. If Stalin's Soviet Union represented Communism, then Lysenko's rise to power meant that Communism, including its scientific practices, was rotten to its core.

Muller's dark premonitions of what might yet come to pass in Soviet science resonated with anti-Communist ideologues. It was not, however, the only way that outside observers understood the crisis facing Soviet genetics. Westerners who remained loyal to the Communist Party, including Bernal, either tried to explain away Lysenko's scientific theories as a misunderstanding of genetics or doubled down on the right of the state to dictate research priorities. More moderate and even conservative geneticists worried less about the situation facing scientists in the Soviet Union and more about Bernal's defense of science-by-fiat as a legitimate approach to science policy. Each of these interpretations of what it meant to politicize science played a part in the Western response to Lysenkoism in the late 1940s and early 1950s.[23]

∎∎

The first semiorganized US response to Lysenko's rise came just after the end of World War II in Europe. A visit from Anton Zhebrak, a Soviet geneticist, to the United States in June 1945 brought rare and reliable news. Zhebrak's presence as a Soviet delegate to the San Francisco conference on the organization of the United Nations was itself evidence that Lysenko lacked a stranglehold on Soviet science: Zhebrak had publicly opposed Lysenko's views on inheritance since the late 1930s. Among the geneticists Zhebrak met during his stay was I. Michael Lerner, a Russian émigré who served as an unofficial translator at informal scientific gatherings. When the Soviet authorities unexpectedly cut Zhebrak's trip short, Lerner spread what news he had learned through the genetics grapevine. More than anything, he reported, Soviet geneticists needed books, journals, and reprints; any "tangible support" from the Americans "would be of great help" in Soviet geneticists' "fight to regain the prestige they once held within the Soviet Union itself." Lerner was encouraged by Zhebrak's visit. "Indications are very favorable that full rehabilitation of genetics in Russia can and will be accomplished in time," he wrote to corn geneticist Barbara McClintock.[24]

Theodosius Dobzhansky drew gloomier conclusions from Zhebrak's visit. Another Russian émigré, Dobzhansky had arrived in Morgan's fly room in 1927 on a Rockefeller Foundation fellowship; when the fellowship ran out, he stayed. By 1945, Dobzhansky, now a US citizen, was a professor at Columbia University. Although based in New York, Dobzhansky made regular summer trips to the San Jacinto Mountains in California, where he collected the *Drosophila pseudoobscura* flies essential to his work in population genetics. That summer, Dobzhansky extended his trip to Berkeley with the hope of catching up with news from home. Echoing Lerner, Dobzhansky reported to his friend and Columbia University colleague Leslie Dunn that "Lysenko's position is less secure than it has been." At the same time, Dobzhansky passed along the news that the geneticists Vavilov, Georgii Karpechenko, and Grigorii Levitskii were most definitely dead.[25]

Dobzhansky wanted to help. He reported to Dunn that the Soviet geneticists "personally and *confidentially* ask for support of those US colleagues who are known to be friendly to Russia, yourself *in particular*." Dunn, a man of the left, could criticize Lysenko in a way that Dobzhansky, a *nevozvrashchenets* ("non-returner"), never could. A self-described advocate of "socialist or anarchist solutions," Dunn had organized aid for displaced German refugee scientists and served as president of the wartime American-Soviet Science Society. Dobzhansky's letter—which Dunn kept despite Dobzhansky's request that it be destroyed—kicked off an unlikely campaign to undermine Lysenko's authority by publicizing his scientific claims. In essence, the two geneticists hoped to criticize Lysenko without criticizing the practice of science in the Soviet Union. Unlike Muller, they had no interest in ideological warfare.[26]

Drawing on ideas of scientific objectivity and disinterestedness, Dunn and Dobzhansky believed that the best way to combat Lysenkoism was to expose Lysenko's ideas to public scrutiny. They therefore translated and published an

English-language version of Lysenko's best-known work, *Heredity and Its Variability*, in 1946. Dobzhansky acted as translator, Dunn as agent. When Simon and Schuster turned them down, Dunn convinced King's Crown Press, a division of Columbia University Press, to publish 4,000 copies of the work in pamphlet form, priced at fifty cents each.[27] Dobzhansky resisted Dunn's offers to write an introduction or to solicit one from noted progressive and former US vice president Henry Wallace. (Wallace, also a former secretary of agriculture and now serving as secretary of commerce, had followed the controversy closely, sending reprints and translations of works in Soviet genetics to Dunn when he had them.) In the end, at Dobzhansky's insistence, the published volume included only a minimalist translator's note from Dobzhansky. As Dobzhansky put it, "Let him stand on his own feet."[28]

Having secured a publisher for the work, Dunn and Dobzhansky set about orchestrating its reception. But ensuring the right kind of coverage, the kind that condemned Lysenko and his science but not the Soviet Union, proved more difficult than either man anticipated. In the growing postwar climate of anti-Communism, the story of a Ukrainian huckster who had hoodwinked corrupt Communist officials proved difficult for journalists to resist. In December, Dunn learned that the Hearst Press newspaper empire planned to use the book as anti-Soviet propaganda. He urged his fellow leftist scientists and science writers to review the book from a purely scientific perspective, to somehow condemn Lysenko's genetics without bringing up politics. To his dismay, most of them declined. The response of Lewis Stadler, a corn geneticist, was typical: "I cannot find anything to say about the Lysenko job that seems to me worth printing . . . It is a useful example of unscientific method, and I think I shall want a few copies to give to graduate students." But as a lesson in objectivity and open-mindedness, he said he could think of many "indigenous examples" and could "see no advantage to Russian-American relations in choosing one from Russia." J. B. S. Haldane, a prominent Communist and evolutionary biologist, similarly declined a request passed along via Muller, claiming that the book was not yet available in Britain. In the end, Dunn himself wrote the review for *Science*, and Dobzhansky reviewed the book for the *Journal of Heredity*.[29]

Having secured a review that damned Lysenko's work in just the right way by writing it himself, Dunn attempted to amplify its impact by reaching out to journalists, including Waldemar Kaempffert, the science writer for the *New York Times*. He forwarded to Kaempffert some materials on the controversy, including his and Dobzhansky's reviews. But, much to Dunn's chagrin, Kaempffert's column on *Heredity and Its Variability* repeated the same familiar blanket characterizations of Soviet science. Genetics in the Soviet Union was "under fire," Kaempffert wrote, "partly for pseudo-scientific reasons, mainly for ideological reasons." He explained that the story mattered to Americans because it provided "an example of what a State-imposed ideology can do to science." And in case the message wasn't entirely clear, Kaempffert concluded that "Lysenko's monograph has to be read in the light of this State control of science in Soviet Russia." This was hardly the "objective," apolitical coverage that Dunn had in mind.[30]

Dunn and Dobzhansky drew the line at actually condemning either Lysenko's scientific theories or his political power, hoping instead that an informed general public would draw its own conclusions. This hope was naive, as Kaempffert's review demonstrated. This was a private campaign conducted by scientists on behalf of their colleagues, not a large-scale, state-driven attempt at ideological warfare. But Lysenko's announcement of his triumph at the August 1948 VASKhNIL meeting shattered any lingering doubts among US geneticists that Lysenko posed anything other than a political problem. Within a month's time, as widely reported in the US press, the Soviet Academy of Sciences had removed the remaining leaders of Soviet genetics from their posts at prominent research institutes. H. J. Muller—former Communist, natural propagandist—thought it was time to step up the fight.[31]

: :

Muller's time in the wilderness ended in 1946. He had been at Indiana University for less than a year when a reporter called his Bloomington home to tell him that he had been awarded the Nobel Prize in Physiology or Medicine for his work on X-ray mutations. With the arrival of the Atomic Age, Muller became an oracle for the genetic dangers of radiation, a favorite of science editors and writers looking for marquee quotes on everything from mutations to eugenics. In a typically pessimistic (and typically offensive) interview with a reporter from *Time* magazine, Muller suggested that it might be "fortunate" "if all those exposed to an atomic explosion (such as the people of Hiroshima) were to be made permanently sterile" because of the strong possibility of genetic damage. A year later, he extended this fear to the entire human race, suggesting that the peaceful uses of atomic energy—medical treatments, atomic power, and the like—might "kill off the human species by loading its germ plasm with too many 'mutated' genes."[32]

Upon hearing of Lysenko's VASKhNIL speech in 1948, Muller promptly resigned his membership in the Soviet Academy of Sciences. In a letter widely quoted by the US press, Muller referred to the actions of the leaders of the Soviet academy as "disgraceful" and charged them with "misusing their positions to destroy science for narrow political ends," as had happened in Nazi Germany. "No self-respecting scientist and more especially, no geneticist, if he still retains his freedom of choice, can consent to have his name appear on your list," he wrote. By now, Muller had no qualms about criticizing the Soviet political system. He tirelessly promoted his cause in popular publications, including a two-part cover story in the *Saturday Review of Literature*.[33]

Even while conducting a one-man campaign in the popular press, Muller sought to spark a broader movement. He forwarded a copy of his Soviet academy resignation letter to the US Department of State, for the agency to use as it saw fit. As president of the American Society of Human Genetics (ASHG), Muller moreover hoped that biologists' professional organizations might be convinced to issue some sort of statement. The State Department encouraged Muller in this, writing that "the existence of political control over scientific thought and activity" in Communist areas "has often been either overlooked or avoided by many US scientists." Muller forwarded a copy of this letter to his Indiana University colleague Ralph Cleland, the chairman of the Governing Board of the American Institute of Biological Sciences, scrawling at the top that "the State Dept would certainly approve of a statement of scientific groups."[34]

As the professional umbrella group for the biological sciences, the AIBS nominally represented the views of 80,000 US university scientists, government researchers, and college and high school teachers. A statement issued on AIBS's behalf would unequivocally demonstrate the US biological community's opposition to Lysenko's power grab. Dobzhansky agreed, having urged his colleagues to issue a statement as soon as he heard the news. Temporarily stationed in Brazil to collect yet more flies, Dobzhansky felt helpless, but he admonished his colleagues with biblical verse: "If we do not speak out then stones shall speak!" With Muller's support, Dobzhansky immediately prepared a statement for the signatures of the executive committees of the Genetics Society of America (GSA) and the ASHG; Muller passed it along to Cleland for the AIBS. But, much to Muller's consternation, his colleagues could not be convinced to sign.[35]

The leadership of the US genetics community marshaled various objections to the statement, ranging from its overtly political character to whether the document went too far in establishing its own kind of scientific dogma. And because it was a "society of societies," the AIBS could not endorse the statement without the approval of the holdouts at the GSA, where it encountered surprisingly fierce opposition from leftists like Dunn. Finally, in the spring of 1949, the GSA's executive committee relented and endorsed a watered-down version of the document. References to "Russian apologists" were changed to "Russian spokesmen." The controversial statement that the "Lysenko position is not science, it is a superstition put forward by politics" was dropped, replaced by the more anodyne, "In our opinion the conclusions of Lysenko . . . have no support in scientific fact." But having waited so long to take a stand, the biologists had a hard time attracting publicity. *Science* at first refused to publish the statement, on the grounds that the journal had already devoted too much space to the topic. Not until July 1949, nearly a year after Lysenko's speech, did the AIBS statement finally appear in print.[36]

∷

With geneticists—except for Muller—largely sidelined, the mainstream press covered the VASKhNIL episode through the lens of ideology and political dysfunction. National magazines, including *Time* and *U.S. News and World Report*, regularly updated their readers on the goings-on at various Soviet academies, confidently asserting that Soviet scientists must "toe the party line" or be subject to "demotion, imprisonment, or worse." Muller's own articles on the topic presented Lysenkoism as a symptom of a larger Soviet cancer that could be cured only through excision. Using the language of Cold War struggle, Muller suggested that only "political overturn" could restore Russian science; in the meantime, Americans must "check the already dangerous spread of infection to countries outside the Soviet sphere." The centralized structure of Soviet science lent itself to opportunists; war hysteria confused patriotism with scientific achievement. Science, Muller suggested, was a "tender plant, requiring a special soil." "Very few people appreciate," he wrote, that the progress of science depends on "complete freedom of inquiry and of criticism."[37]

Muller's fierce rhetoric resonated not only with the new politics of the Cold War but also with an emerging folk philosophy of science and freedom being promoted by a new class of powerful scientific administrators in the United States. In contrast to the leftists at the GSA who had spent the 1930s and early 1940s debating the best way to aid refugees, the physicists, chemists, and engineers running postwar scientific institutions had spent the war either designing weapons or supervising those who did. Even those who returned to their university positions in peacetime continued to make frequent trips to Washington, where they advised military leaders and Congress on the best strategy to maintain the United States' assumed lead in science and technology. In their spare time, they delivered lectures and wrote essays that teased out the relationship between science and democracy in a world of atomic threats. They drew similar conclusions to Muller's about the importance of isolating science from ideology, albeit for different reasons.[38]

For scientific defense advisors in 1945, the most pressing question was why the United States, and only the United States, had managed to produce an atomic bomb. Back in 1942, US scientists and military leaders alike assumed that Germany, with its long tradition of scientific leadership, would almost certainly have the advantage in converting nuclear fission into a weapon. And yet, when a US Army Intelligence mission to investigate German progress on the bomb, code-named Project ALSOS, finally arrived in that country in 1945, its leaders were shocked at how little progress the Germans had made. Asked to explain what went wrong, ALSOS's leader, physicist Samuel Goudsmit, fingered totalitarianism.

The preface of the declassified version of Goudsmit's report, published in 1947, pulls no punches. Goudsmit asserted that "science under fascism was not, and in all probability could never be, the equal of science in a democracy." He blamed German complacency, political interference, and a dogma that rejected all of modern physics as "Jewish science." The Nazis had put politics before science and had appointed party hacks to administrative posts. Last but not least, Goudsmit criticized the cult of personality that had developed around Werner Heisenberg, whose contributions could not be questioned. "Science is not authoritarian," Goudsmit wrote, "nor can scientific thought be dominated by a boss, however gifted he may be." Though written in response to the situation in Nazi Germany and published before Lysenko had fully consolidated power, the parallels would have been unmistakable to readers in 1948.[39]

As head of the Office of Scientific Research and Development, the umbrella agency that oversaw the United States' wartime effort in science and technology, engineer Vannevar Bush heard reports on Project ALSOS even before the mission was complete. He drew on its findings in proposing a structure for what ultimately became the National Science Foundation (NSF). Bush's famous report, *Science, The Endless Frontier* (1945), laid out a case for generous public funding for basic research in science and medicine after the war's end. But even though Bush emphasized the contributions of science and medicine to "the public welfare," this was no brief on behalf of scientific planning. Instead, Bush argued that basic research should be controlled by a new scientific agency, shielded from political oversight, to better preserve "freedom of inquiry." Bush's call for a blank check for government funds, free from government control, unsurprisingly encountered resistance in both Congress and the Truman administration. The version of the NSF that ultimately came into being in 1950 placed the agency solidly under the control of the executive branch, with a director appointed by the president. Nevertheless, Bush's insistence that science could not and should not bend to political control solidified during this period. The NSF, an agency headed by scientists, would select its projects solely on the basis of intellectual merit.[40]

Like Goudsmit and Muller, Bush invoked the concept of a "party line" as a key factor that distinguished Western from totalitarian science. In a best-selling 1949 book, Bush explored the choices facing citizens and scientists in a world "split into two camps." Would they allow the producers of culture—science, music, the visual arts—to thrive under conditions of freedom? Or would they subject them to dictatorial control and management by political fiat? In Bush's telling, Lysenko's rise was the inevitable product of a political system that "can tolerate no real independence of thought." His diatribe is worth reproducing: "A great scientist is torn from his post and sent into cold exile because he dares assert that there is validity in the modern theories of genetics, contrary to the state teaching . . . and he is replaced by a charlatan who will see to it that the state theory is taught to young scientific disciples and that all research is based on a blatant fallacy and an unsupported hypothesis." Such a system could only produce "superstitions and folklore." The more optimistic version of this argument, repackaged for an excerpt in *Life* magazine, assured readers that the West's enduring faith and belief in human dignity, combined with the applications of science, could yet triumph over Marxism. Even as the whereabouts of certain Soviet geneticists remained unknown, their fates had become fodder for tracts on freedom.[41]

Harvard University president James Bryant Conant took a more subtle approach to the problem of science and democracy. As former head of the National Defense Research Committee during the war, Conant had worked hand-in-hand with Bush, and he shared Bush's concerns for protecting US leadership in science and technology. Conant's view diverged from Bush's, however, when it came to scientific authority. While Bush wanted to protect scientific autonomy—to shield science from politics—Conant worried that typical Americans lacked the mindset to evaluate scientific claims. Conant proposed to solve this problem through a case-study approach to undergraduate education that emphasized the messy path of scientific progress. Instead of teaching undergraduates facts about biology, chemistry, or physics, instructors at Harvard (including a young Thomas J. Kuhn) would escort their students through the "fumbling," "stumbling" experience of scientific discovery. In his popular lectures and writings promoting the case-study technique, Conant emphasized that such an approach might make students more comfortable with the uncertainty of the modern age. A current of anxiety about the temptations of following blind authority runs just beneath the surface of his 1947 book, *On Understanding Science*, which modestly suggests that his method would eliminate any risk that students might "have to take on faith" statements that might better be classified as dogma.[42]

Lysenkoism wasn't Conant's primary concern. Like Bush and Goudsmit, Conant was consumed with the dilemma of the atomic bomb: whether scientists had been right to build it, whether the military had been right to use it, whether any one nation should control its use, and whether democracy could withstand the answer to any of those questions. Their books and essays contributed to a much larger public debate about the role of scientists—particularly physicists—in controlling atomic energy. But if the bomb was text, the fate of science under totalitarian regimes was subtext. Why else the frequent references to "dogma," "received authority," and "party lines"? All three scientists agreed that science could not thrive without political freedoms for scientists. By the late 1940s, the absence of scientific freedom had come to be shorthanded as Lysenkoism.[43]

∷

By 1949, H. J. Muller and J. D. Bernal had long parted ways on the virtues of Communism and central planning. The two men could agree on only one thing: the situation of genetics in the Soviet Union had become, in Bernal's words, "a major intellectual weapon in the cold war." Soon enough, Muller would find a way to transform his budding ties with sympathetic government officials like

Bush and Conant into the full-fledged assault on Communism he so desired. Understanding how and why psychological warriors came to adopt Muller's perspectives on Soviet science requires a closer look at the CIA and the State Department, each agency's somewhat fraught relationship with scientists, and the role of each in fighting what the United States increasingly defined as an ideological war.[44]

2 : Ambassadors for Science

In April 1954, the American molecular biologist Jeffries Wyman took a three-week working vacation to Italy. As the US science attaché to Paris, Wyman served as a sort of science ambassador to Western Europe. His job was to get to know scientists and their work, represent the United States at informal gatherings and conferences, and generally grease the wheels of international science. He passed through Milan, Padua, and Florence before making his way south. At each stop, Wyman dined with colleagues, visited their labs, and took in the sights. His twenty-seven-page report on the trip is as much travelogue as administrative briefing, containing about an equal number of references to frescoes as to chromatography. Wyman arrived in Rome just in time for Easter Sunday services, where he expressed wonderment at the "immense demonstration" before the pope at the Vatican.

The trip continued in this vein until the evening of Tuesday, April 20. Wyman spent the morning at the Instituto Superiore di Sanità, where his hosts showed off their centrifuges and complained about the difficulty of securing visas to the United States. After an afternoon admiring Caravaggios at the Galleria Borghese, Wyman meandered back to his *pensione* via the Trevi Fountain and the Pantheon. Upon his return, he found an urgent message from the American Embassy. A security officer informed him that a Mr. Winter from the legal attaché's office would be flying in that evening. Wyman should expect a visit around midnight.

What was all this about? The security officer, a Mr. Brown, couldn't say.

At half-past twelve, Brown arrived at Wyman's hotel, alone. Winter had missed his plane; Wyman should report to the embassy tomorrow. When Wyman arrived for the meeting, he wasn't prepared for Brown and Winter to flash their FBI badges. They wanted to know why Wyman had had contact with both J. Robert Oppenheimer and suspected Soviet informer Haakon Chevalier in the previous month in Paris. Wyman explained the obvious: as the science attaché, he welcomed any US scientists who came to visit and pumped them for their local contacts. Wyman was cleared of any wrongdoing, but the experience rattled him. As he noted in his official diary, "I felt for a moment like a criminal."[1]

Wyman's boss, State Department science advisor Joseph Koepfli, recounted a more dramatic sequence of events in an oral history interview conducted thirty years after the fact. "Poor Wyman . . . [was] awakened at three-thirty in the morning in his hotel and told to put on his clothes, that he's going to the airport, that he's under the care of the FBI. He's hauled back to Paris to be questioned and to make statements on his relationship with Haakon Chevalier." Koepfli's interviewer, Elizabeth Hodes, hadn't asked the former diplomat about Wyman or the FBI. Koepfli brought it up, unsolicited, to make a point about how science attachés were routinely harassed by the FBI for attempting to do their jobs, which sometimes involved communicating with scientists who might have Communist affiliations. "And I said, 'Good God, what's the matter with you people.' Just crazy."[2]

In the early 1950s, the United States experimented with using scientific internationalism as a tool to collect scientific intelligence. By then, the United States had a young intelligence agency, the CIA, with a brand-new Office of Scientific Intelligence. But since the logic of scientific internationalism dictated that science functioned best through open channels, US foreign policymakers hoped to collect this scientific intelligence through persons unaffiliated with intelligence work: namely, State Department attachés and private scientists traveling abroad.

In the United States, the rising wave of anti-Communism both justified expanded intelligence operations and offered a rationale for how to collect that intelligence. This complicated things for the Americans because their particular ideas about how science should operate—freely, without the restrictions that they associated with Communism—conflicted with the realities of defense advising and scientific intelligence gathering. Intelligence collection is, by definition, a government function, yet leading voices in US science policy championed a vision of scientific freedom that renounced direction from the government. If the United States hoped to collect valuable intelligence on the Soviet Union's capabilities in science and technology, its agents would presumably have to talk to Communists—but the free exchange of information might put the United States' own secrets at risk.

The earliest attempts to integrate science into US foreign policy embraced, rather than resolved, these contradictions. Both the CIA and the State Department charged their science advisors with advancing the country's interests while simultaneously acting as if credible scientists would never cooperate with government agencies in peacetime. As Joseph Koepfli understood perhaps better than anyone, this strategy could not be sustained.

::

The concept of scientific intelligence depends on having secrets to protect. World War II and the Cold War brought unprecedented levels of secrecy to the practice of science, and not only in the United States. With the circulation of scientific knowledge increasingly frozen by classification, security clearances, and travel restrictions, the US government needed sophisticated ways to collect and evaluate scientific knowledge. For help, they turned to scientists.[3]

The United States' first scientific intelligence agency grew up hand-in-hand with World War II's scientific projects. While each branch of the armed services maintained scientific and research divisions, their intelligence units did not, for the most part, specifically seek scientific intelligence. The exception to this rule was the Army's Manhattan Engineer District—more commonly known as the Manhattan Project—which maintained its own Foreign Intelligence Branch. Since part of the original logic in building an atomic bomb was to beat the Germans at the same game, it followed that the Americans hoped to learn as much as they could about the Germans' progress. This is why Samuel Goudsmit found himself in Heidelberg in 1945. Project ALSOS partnered the Manhattan Project's Foreign Intelligence Branch with an Office of Scientific Research and Development (OSRD) outpost in Paris and Army and Navy intelligence units. The team would identify the locations of German labs, the size of their uranium stockpiles, and the names of scientists involved.[4]

Project ALSOS largely produced a null finding: the German nuclear program hadn't progressed much past the research phase. As Goudsmit put it, "Sometimes we wondered if our government had not spent more money on our intelligence mission than the Germans had spent on the entire project." Even so, military leaders deemed the operation a tremendous success. Project ALSOS teams captured German documents, equipment, and personnel, including Werner Osenberg, chief scientist of the Reichsforschungsrat, the German equivalent of the OSRD; Otto Hahn, one of the discoverers of nuclear fission, the reaction at the heart of the bomb; and Werner Heisenberg, whom Goudsmit blamed for the Germans' failure. Project ALSOS definitively showed that scientists and military operatives could gather better intelligence together than either could on their own.[5]

During the postwar occupation of Germany, the victorious powers—the United States, the United Kingdom, France, and the Soviet Union—fiercely competed with one another for access to German scientific and technical resources. Osenberg's capture gave the Americans a head start, as he brought with him not only 150 members of his staff but also a list of 15,000 German scientific and technical personnel. Project Paperclip, which relocated hundreds of German rocketeers to Huntsville, Alabama, is justifiably the most famous of these operations, but the more workmanlike Field Information Agency, Technical (also known as FIAT) arguably produced more actionable information. Between 1945 and 1947, FIAT microfilmed millions of pages of intellectual property records from German chemical, pharmaceutical, and manufacturing firms.[6]

The US military also sought scientific intelligence from the Pacific theater. In Occupied Japan, the physicist Karl Compton, former president of the Massachusetts Institute of Technology, supervised a Scientific Intelligence Survey that debriefed 300 Japanese scientists on radar, military balloons, bacteriological and chemical warfare, and broader research practices. The scientists involved worked directly with military agencies in collecting and assessing this intelligence—a practice so familiar from the war that few commented on it at the time.[7]

∷

Project ALSOS, FIAT, and Compton's mission to Japan produced significant amounts of scientific intelligence. But without a central intelligence agency— which the United States lacked from the end of the war through late 1947—this information rarely found its way into the hands of those who wanted or needed it most. President Truman had dissolved the CIA's closest predecessor, the Office of Strategic Services, on October 1, 1945. The State Department took over most research and analysis functions, but intelligence gathering reverted to the military services and the FBI. The new Atomic Energy Commission, which took over from the Manhattan Project in January 1947, also maintained its own intelligence division. In theory, these various agencies and services coordinated their efforts and findings via the makeshift Central Intelligence Group, but the system fell far short of a coherent scientific intelligence program.[8]

The Central Intelligence Group maintained its own, albeit small, office for conducting and assessing scientific and economic research. The scientists and government officials at the Joint Research and Development Board (JRDB), the group effectively in charge of the nation's science policy for defense, avidly consumed what little information the group produced, but they worried at the gaping holes in their knowledge. Vannevar Bush, who now headed the JRDB, especially fretted over the lack of information on Soviet scientific and technical capabilities. By early 1947, Bush and General Hoyt Vandenberg, the director of central intelligence, had agreed to hire a full-time scientific intelligence officer, with the hope of improving the situation.[9]

Hiring a director of scientific intelligence turned out to be more difficult than Bush, Vandenberg, and Vandenberg's successor, Rear Admiral Roscoe Hillenkoetter, anticipated. By the time they located a suitable candidate (and a suitable cover), a more centralized intelligence system had fallen into place. The National Security Act of 1947 completely restructured US national defense, creating the National Security Council (NSC), the position of secretary of defense, the Joint Chiefs of Staff, and the CIA. Just weeks after the CIA's creation on September 18, 1947, Ohio State chemist Wallace Brode arrived to take the reins of the new agency's Scientific Branch.[10]

Brode brought with him personal enthusiasm, a wealth of experience, and a cover appointment that ultimately led to his departure from the CIA. The oldest of triplets, Brode attended Whitman College with his three brothers in Walla Walla, Washington, where their father was a professor of biology. All four sons pursued careers in science, but Wallace remained particularly close to Robert, the middle triplet, whose career in physics included heading up the fuse group for the Manhattan Project. Wallace Brode got his first taste of scientific intelligence collecting during the war, when he headed the OSRD's Paris office. He then moved to California, where he evaluated scientific intelligence at the Navy Ordnance Test Station in Inyokern. In August 1947, still on leave from his job at Ohio State, he accepted an appointment as associate director of the National Bureau of Standards in Washington. This, however, was merely a cover. Brode immediately transferred to the CIA.[11]

In a memo prepared for Hillenkoetter before he even accepted the job, Brode outlined his vision for the CIA's Scientific Branch. He proposed a unit that could integrate the various types of information being collected across the nascent agency: registers of scientists' names and institutions; a reference library of scientific publications; abstracts and translations of foreign scientific documents; and, in some cases, captured physical objects. While Brode acknowledged the Scientific Branch's anticipated roles in coordinating intelligence and predicting future threats, he foresaw its greatest contribution in what he referred to as the "Scientific Order of Battle." By this, Brode meant "the character, location, and quality of scientific manpower, institutes, and organizations." This was a remarkably expansive vision that included appraisals of "all foreign scientific publications, news releases on scientists, membership lists, names and appointments." Brode, in other words, envisioned the CIA's Scientific Branch as a global panopticon for science, absorbing all possible information on all possible topics produced by any scientist or technical professional anywhere in the world.[12]

Probably unworkable under any circumstances, Brode's grand plans faltered in the face of interservice rivalries and lack of support from within the CIA. Brode and the Scientific Branch clashed nearly continuously with the AEC, which refused to share its own scientific intelligence unless promised full reciprocity. He had difficulty obtaining security clearances for his staff and, if one CIA history is to be believed, even for himself. A staff list of the Scientific Branch prepared in the spring of 1948 included fewer than two dozen people, including secretaries and other administrative support. A series of handwritten talking points in Brode's archival papers, available at the Library of Congress, suggests that Brode complained—continuously—about staffing, support, and morale. One suspects that Wally Brode, a man from Walla Walla once described as "6 feet 3 inches of egg-headed masculinity," did not fit easily into the CIA's Eastern establishment culture.[13]

The problems for science at the CIA, however, went beyond Brode's personal management crises. In the spring of 1948, the House Un-American Activities Committee (better known by its acronym, HUAC) opened a security investigation on the director of the Bureau of Standards, physicist Edward U. Condon. Like many veterans of the Manhattan Project, Condon was a fierce advocate for scientific internationalism, particularly the international control of atomic energy. He was also involved with the American-Soviet Science Society, the same group that L. C. Dunn had hoped to use to support Soviet geneticists during and immediately after the war. The American-Soviet Science Society had widespread support among the US scientific community, including a $25,000 grant from the Rockefeller Foundation issued in 1947. But in a growing climate of anti-Communism, even the organization's innocuous mission of exchanging reprints, books, and magazine articles with Soviet scientists attracted suspicion. Perhaps more to the point, the American-Soviet Science Society had not quite severed its ties with the American-Soviet Friendship Council, an alleged Communist front that appeared on the US attorney general's list of subversive organizations. In May, word emerged that congressional investigators had grilled bureau employees over whether Condon had pressured them to join the American-Soviet Science Society.[14]

Condon's case marked a major turning point in US scientists' experience of anti-Communism. Based on virtually no evidence, HUAC's report referred to Condon as "one of the weakest links in our atomic security." Scientists associated with either Condon or the American-Soviet Science Society, like Dunn, found themselves the target of smear campaigns in anti-Communist newspapers. HUAC's treatment of Condon clearly signaled that the emerging security apparatus of the Cold War would not tolerate the free exchange of scientific information. Science, did, in fact, have borders.[15]

Condon's case became a cause célèbre for scientists who thought the climate of suspicion had gone too far, too fast. They pushed back, urging their professional organizations and elected officials to defend scientists' right to scientific freedom. The response from more established organizations such as the National Academy of Sciences was more tepid—a sign of things to come. The NAS's statement, issued at the end of March 1948, deferred to Congress's wisdom on matters of security, recognizing "the necessity not only of loyalty but also of rigorous discretion on the part of any Government employee who possesses information, the disclosure of which would menace the national security." Instead of defending Condon personally, the NAS warned that baseless attacks would discourage scientists from accepting government employment. When Dunn complained to NAS leadership that the statement fell far short of a defense of scientific freedom, the academy's president, Alfred Richards, chastised him for politicizing Condon's case. Richards explained that the statement was "intentionally completely neutral" because "the Academy is not in a position to act either as a judge or as an attorney for defense."[16]

Condon was no Communist, but the firestorm associated with his case created political problems for the Bureau of Standards and, by extension, for Brode, whose cover had him reporting to Condon. When both Condon and the bureau's other associate director planned to be out of the country for two weeks in early fall, Condon asked Brode to temporarily fill in. Brode's appointment as one of the bureau's two associate directors had been widely covered in the scientific press; as the most senior remaining officer, Brode would have been the obvious interim leader for the agency's 3,000 scientists, engineers, technicians, and secretaries. From the CIA's perspective, however, lending Brode back to the bureau risked exposing its leading scientific administrator to the cloud of suspicion hanging over Condon. Hillenkoetter declined Brode's request to spend up to a few hours a week fulfilling the responsibilities of his cover appointment on the theory that the "CIA could only be harmed . . . by association with the National Bureau of Standards." At that point, Brode resigned, moving directly into the job he had supposedly occupied for the past year.[17]

Upon his departure, Brode offered his assessment of what had gone wrong to Karl Compton, now head of the Research and Development Board (the successor to the JRDB). Like Bush, Compton despaired of extracting useful scientific intelligence from the CIA. Brode primarily blamed the organizational structure of the CIA, which divided responsibilities for collecting materials among various divisions and, most importantly, segregated nuclear information and sources from all other types of scientific information. But Brode also warned that neither the CIA nor the various branches of the military trusted scientists. Brode claimed that the CIA regarded scientists as "untrustworthy and requiring excessive controls beyond those applied to other workers," and he warned Compton that this suspicion extended even to members of the Research and Development Board. Brode could, however, claim one small achievement: a NSC Intelligence Directive (NSCID 8) for the CIA to create a registry of biographical data on foreign scientists and technical personnel.[18]

By this point, the problems within the CIA had attracted the attention of two major federal reviews of the intelligence establishment. Both the Eberstadt Committee and the Dulles Committee warned policymakers about the high stakes of the lack of concrete information in the field of scientific intelligence. The Eberstadt Committee's report claimed that "failure" in scientific intelligence might "have more immediate and catastrophic consequences than failure in any other field of intelligence." The Dulles Committee took a more practical tack by recommending a more centralized body for gathering and assessing scientific intelligence. These reports, combined with Brode's departure and continuing complaints from Compton, pressured the CIA to organize a new division, the Office of Scientific Intelligence (OSI).[19]

The creation of the OSI on January 1, 1949, gave scientific intelligence a higher priority within the CIA but otherwise did little to ease the work of Brode's replacement, Willard Machle. A physician and orchid hobbyist, Machle held cover appointments at the AEC and Johns Hopkins' Operations Research Center. Allocated a staff of 100, Machle struggled to recruit scientists willing to work in the relative obscurity and secrecy of an intelligence agency. An internal CIA history of these early years refers to "compromises in the qualifications of many of the individuals hired." Like Brode before him, Machle continued to fight turf battles not only with other departmental intelligence units but also with other divisions within the CIA. The Soviet Union's successful detonation of an atomic bomb in August 1949—several years ahead of US predictions—drive home the possibly catastrophic consequences of intelligence failures. Machle ascribed the "almost total failure of conventional intelligence" regarding the Soviet bomb to the refusal of both defense agencies and the CIA's covert operations staff to cooperate with the OSI. Machle urged Hillenkoetter to grant the OSI a greater say in the CIA's larger operations, including OSI approval for the "technical and scientific objectives of any covert operation" conducted by the CIA's Office of Special Operations.[20]

Machle's battles with both defense agencies and other units within the CIA boiled down to a philosophical question about the nature of science: Where does science stop and R&D begin? Or, to use terms bandied about by officials at the time, where is the line between "basic" and "applied" science? Machle's view of scientific intelligence included "any development in foreign science and technology which could pose a threat, immediate or eventual, to U.S. national security." The Department of Defense, meanwhile, held this definition to be too broad, arguing that scientific intelligence ended when weapons development began—in other words, when basic research became applied. American norms for scientific research, as being defined by outspoken figures like Bush and James Conant, also worked against Machle in his power struggle within the CIA. If one of the things that distinguished basic science from applied science was its open, internationalist character, Machle should have no need for information gathered through the Office of Special Operations' covert tactics.[21]

Machle could not overcome the inherent contradictions of a scientific intelligence program that defined "scientific intelligence" as either weapons development or research published in the open literature—neither of which seemed to require a separate scientific intelligence unit. The open documentary record offers little evidence as to how the climate of anti-Communism affected Machle's work, but one suspects that he must have encountered the same problems as Brode in obtaining security clearances for scientists with just the right amount of interest in scientific internationalism. Machle resigned in March 1950, having lasted, like Brode, barely fifteen months in the job.[22]

:: ::

Machle's OSI had also been undercut by the State Department. In late January 1949, only weeks after Machle took up his post, the NSC issued an intelligence directive (NSCID 10) that assigned the State Department primary responsibility for collecting information on "the basic sciences." Officials at Foggy Bottom did not entirely welcome this task, and they struggled to reconcile NSCID 10's broad intelligence mandate with State's broader diplomatic mission. While NSCID 10 left the task of determining which countries have "informational potential in fields of basic and applied sciences" to the CIA, it made State responsible for collecting actionable scientific intelligence as well as smoothing channels for international scientific exchange. Among other things, NSCID 10 authorized the State Department to appoint "as practicable, specially qualified scientific and technical personnel to selected United States Missions"—aka, science attachés.[23]

State Department officials rightly anticipated competition in this work with both the CIA and the military. More to the point, however, State Department officials worried about the difficulty of reconciling intelligence with diplomacy and the challenge of isolating "outstanding scientists" from the "taint" of intelligence collection. A memo on the problem, classified SECRET, rejected the possibility of housing the program in State's Research and Intelligence division, where it logically belonged, because of "the basic reluctance of scientists in general to associate themselves with intelligence activities." Beyond scientists' supposed "fear" of "restrictions imposed on the flow of information," a plan to use scientists to collect intelligence information confronted a practical obstacle: foreign contacts would not cooperate with scientists suspected of being intelligence operatives.[24]

James Webb, the undersecretary of state (who would later spearhead the National Aeronautics and Space Administration during the Apollo era), accepted temporary responsibility for scientific intelligence collection, with the understanding that he could assign a staff member to "wear two hats"—scientific administration and intelligence coordination—if necessary. For advice on a more permanent solution, Webb hired Lloyd Berkner, a radio engineer known for his administrative acumen, as a consultant on international science. As the JRDB's executive secretary under Bush, Berkner had run the organization's day-to-day operations, recruited scientists to serve on its various committees, and served on the so-called Policy Committee that coordinated the JRDB's relationship with the Army, the Navy, and the Central Intelligence Group. With the possible exceptions of Bush, Compton, and Brode, Berkner knew more than anyone about what sort of scientific intelligence US strategists desired and how little of it they actually possessed.[25]

Berkner's landmark report, "Science and Foreign Relations," released in April 1950, established an ambitious agenda for the role of science in the State Department. Drawing extensively on a survey that Berkner had commissioned from the National Research Council, the study arm of the NAS, the main body of the report not coincidentally focused on the issues of interest to US scientists. It recommended more opportunities for international exchange, fewer restrictions on exchange of technical information, and increased US participation in international conferences. But this was not international cooperation for the sake of internationalism. The so-called Berkner Report forthrightly acknowledged that "certain benefits which are essential to the security and welfare of the United States, stem from international cooperation and exchange with respect to scientific matters." Berkner urged the State Department to create a Science Office to facilitate international scientific cooperation, headed by a science advisor who could make foreign policy recommendations on matters of science and technology. And last but hardly least, the Science Office should station science attachés in embassies in countries and regions of particular interest to the United States.[26]

Berkner and Webb anticipated a major rollout for "Science and Foreign Relations." The State Department planned to print 5,000 copies and prepared an information campaign to ensure the "proper psychological tone" for the new program. Webb emphasized that press materials should highlight the "humanizing" powers of science, avoiding the impression that the United States was "going in for 'atomic diplomacy.'" A public affairs officer suggested that the report be launched with a press conference featuring Dean Acheson, acting secretary of state, accompanied by a leading scientific figure—ideally Bush or Detlev Bronk, president of the NAS—with appropriate coverage on Voice of America and other vehicles for US propaganda.[27]

It's unclear, however, how many—if any—of these plans were carried out. A cover note stapled to the plan recommended caution, noting "strong doubts about extent of implementation." Among other things, senior State Department officials wondered whether the long-delayed creation of the National Science Foundation, finally established in 1950, eliminated the need for science advice at State. A year later, Joseph Koepfli, the newly appointed science advisor, remarked to the head of the US legation to Switzerland that "I rather doubt that you or very many other people other than those directly concerned have read [the report]."[28]

The State Department's ambivalence about promoting its new Science Office had to do with the ambiguous role of the science attachés. Were they diplomatic representatives, or were they intelligence officers? The question has proven surprisingly difficult for historians to resolve, not least because of Koepfli's own misleading statements to earlier historians. The Berkner Report included an appendix, classified "SECRET," that speculated on the possibilities of using science attachés to gather scientific intelligence. Extrapolating from NSCID 10, Berkner distinguished between overt and covert scientific intelligence and asserted that responsibility for all covert intelligence collection rested with the CIA. The State Department, however, remained responsible for collecting *overt* scientific intelligence. As examples of overt scientific intelligence, Berkner listed published materials, unpublished materials, and reporting on information collected in public settings (conferences, meetings, and so on). Science attachés played a critical role in obtaining this information, whether by attending conferences, purchasing materials on the open market, or facilitating US representation at scientific meetings. But because "inquiry for scientific information by an unqualified person leads to suspicion and closing of sources," it was critical that only persons of the highest scientific reputation be appointed as science attachés.[29]

Given all this—a classified appendix on scientific intelligence collection, an NSC directive assigning responsibility for overt intelligence collection to the State Department, and the establishment of a network of science attachés by 1952—it seems fairly obvious, in retrospect, that the first generation of science attachés were part of the US intelligence collection apparatus. But from here, the story gets somewhat murky. In an oral history conducted in 1995 with historian Ronald Doel, Koepfli not only denied knowledge of the classified portion of the Berkner Report but claimed not to have knowledge of Brode's time at the CIA and little-to-no contact with Machle or with his successor at the OSI, Marshall Chadwell. Pressed directly on the topic of the "Secret Supplement to the Berkner Report," Koepfli first responded that "you're talking about something I don't know about," then claimed a "memory lapse," then claimed that the report couldn't exist, because he surely would have seen it if it had:

> I knew Lloyd Berkner extremely well. I mean, he took an interest in everything I was doing. We were there as a result of the Berkner Report, and I saw Lloyd frequently. But no, if there was such a thing, either I've drawn a complete blank on it or I didn't know anything about it...But outside of that, I knew all the top people. I knew George Kennan for God's sake, I knew Paul Nitze very well, and they were as top as you could be other than the Secretary and the Undersecretary themselves... I'm sure that it doesn't exist.[30]

The appendix remained classified at the time that Doel was conducting his interview (most of it was declassified in 1998, but a few key paragraphs are still redacted at the time of this writing). The version of NSCID 10 on the CIA's Freedom of Information Act website wasn't declassified until 2008. This raises the obvious question: Was Koepfli stonewalling Doel to protect classified information, or is it actually possible that the first science advisor to the State Department didn't know that science attachés played an intelligence function and that the State Department was responsible for collecting scientific intelligence?[31]

Koepfli died in 2004, at age 100, so we can't ask him. I did the next best thing: I asked Doel what he made of the exchange. In an email, Doel told me that he "did not have the sense that Joe was fibbing" and that Koepfli's comments suggest "genuine puzzlement." Of course, Koepfli's age can't be discounted, but Doel—an experienced interviewer—believed Koepfli to be "quite sharp" at 90. Doel wondered the same thing that I did: were the Berkner Report and the science attaché program an elaborate ruse for a covert, CIA-based intelligence program, with the head of the Science Office out of the loop? Having interviewed several of the first generation of science attachés, Doel ultimately rejected this theory on the basis that it simply wasn't workable: sending intelligence agents abroad as "science attachés" was simply too obvious to be practicable.[32]

I eventually came to agree with Doel, albeit for somewhat different reasons. The science attachés were collecting scientific intelligence, but they were collecting it for the State Department, not the CIA. The full text of NSCID 10 renders the classified appendix of "Science and Foreign Relations" superfluous. The list of the science attachés' responsibilities as described in the *open* version of the Berkner Report is fairly explicit: "reporting on significant trends and developments in foreign science"; "encouragement and collection and transmittal of foreign scientific and technical information to the United States by private individuals and organizations (public and private)"; "actual collection and transmittal where the situation does not readily permit of direct access"; and "response to the scientific requirements of all United States government agencies." The classified appendix only clarifies that these tasks should be considered "intelligence" and establishes a mechanism for sharing such information with other government agencies. It is entirely possible that Koepfli simultaneously had no idea that the Berkner Report had a classified appendix and that he fully understood that part of the function of the science attachés—indeed, of the Science Office—was to gather overt scientific intelligence.[33]

Two stories that Koepfli told his other oral history interviewer, Elizabeth Hodes, support this interpretation. Before he was appointed science advisor, Koepfli, an alkaloid chemist, had been stationed in London in 1948 as one of the State Department's first science attachés. He stressed that his job was "perfectly in the open" and said that he had been selected, in part, because he was a chemist and not a physicist. "They wanted somebody . . . who had nothing to do with atomic energy, who was just absolutely clear of atomic energy, who would have a sufficient scientific reputation among his colleagues so that he would not be taken as a cover for a covert intelligence operation." Later in the interview, Hodes asked Koepfli whether the science attachés encountered any hostility in their assigned countries. For the most part, Koepfli replied, no, but they did have a problem with "the Russians and their fellow travelers in France and Italy," who constantly accused the science attachés as working undercover for the CIA. The solution? "The only way we could combat that was to get individuals of sufficient academic reputation."[34]

Berkner had used almost exactly the same language to explain why only the "most highly qualified" persons could be considered for science attaché appointments. Both statements betray a somewhat peculiar assumption for men working in the highest echelons of US foreign policy: that overt government service, by definition, damaged scientists' credibility in the eyes of other scientists. Scientists who worked with the government displayed insufficient loyalty to the supposed values of scientific internationalism; they had access to their own government's secrets and reason to desire secrets from others. Koepfli would soon learn at State what Brode had experienced at the CIA: an inherent distrust of scientists willing to work with the government is a weak foundation on which to build a government program for scientific intelligence.

::

As an alternative to putting scientists in government positions, both the open and the classified version of the Berkner Report made a rather remarkable suggestion. The bulk of the US government's scientific activities in foreign relations should and would be carried out by private organizations and individuals. "The greatest benefit will emerge," Berkner wrote, "if the Department of State encourages and facilitates the conduct of privately sponsored programs of exchange of scientific material and persons." As a practical matter, the number of private exchanges dwarfed any possible government effort, even in an age of expanding government. The State Department's role should be to make it easier for scientists to meet, mingle, and share information.[35]

The classified version offered a more calculating assessment of the benefits of this approach. Private scientific meetings "provide frequent opportunity for American science [sic] to make effective contact with the Soviet counterpart." "It must be remembered," Berkner explained, "that the very basis of science is full intelligence, and the normal functioning of contacts during a conference operates to provide the information needed." Private citizens, acting in their roles as private scientists, offered the most reliable source of scientific information. There was, however, a catch. For the scientists' own protection, they should be questioned only *after* their travels. Advance briefings exposed private citizens to danger and made them poor security risks. Instead, scientists should be encouraged to go about their routine business, with the emphasis on "free and open discussion of the content, procedures, and mechanisms of the science involved."[36]

This was a brilliant insight into the nature of science and scientific intelligence, but no one could figure out how to implement it in a democracy. As Walter Rudolph, the assistant to the science advisor, put it in a letter to the chief science officer at the US Embassy in London, "scientists cannot be obligated to give notice of their plans." What was needed was some sort of informal, friendly seeming mechanism to encourage scientists to share their upcoming travel plans with a central clearinghouse for scientific intelligence.[37]

The outline of the solution worked out among a handful of officials at the State Department, the CIA, and the NAS is visible even through the paragraph-long redactions in documents the State Department released to me in 2014. With assistance from the NRC—the most likely private organization to encounter scientists' travel plans—the Office of the Science Advisor at State would each week compile a list of American Scientists Traveling Abroad, or ASTA. The ASTA list, circulated as a series of detachable note cards, included each scientist's affiliation, discipline, home address, purpose of trip, and, when available, travel itinerary. As a carrot to encourage scientists to participate, Koepfli's office placed a small notice in a "widely read scientific journal" offering nonfinancial assistance to travelers who shared their itineraries. Each week's ASTA list would then be circulated to interested parties "outside the State Department," presumably including the CIA.[38]

Other departments at State greeted the initial list, circulated in June 1951, with enthusiasm. The Office of Educational Exchange offered to help prepare the list, as did the Division of International Conferences. Koepfli's offer of free travel assistance yielded at least a few itineraries, including one from geneticist Curt Stern, who straightforwardly informed Koepfli that he would be spending his upcoming sabbatical at the Institute of Genetics at the University of Paris with Rockefeller Funds. By September, however, the Science Office was receiving "next to no ASTA information." The concept seems to have died a quiet death shortly thereafter.[39]

The paper trail for a systematic effort to exploit private citizens' scientific travel ends here, but it is clear that, given the opportunity, the State Department continued to debrief scientists on their travels. Another letter, also recently released at my request, suggests that the CIA attempted to establish a stronger relationship with the NAS in the fall of 1951. What exactly the CIA wanted was deemed sensitive enough to elude declassification in 2014, but the wording of the response from Wallace Atwood, executive secretary of the NRC's Office of International Relations, suggests it involved "travel plans of persons associated with our program." Atwood indicated that the NAS–NRC would be "glad to assist," but noted, "In the past it was believed wise for all information of this nature to be transmitted to you through the Office of the Science Advisor, Department of State . . . For several reasons it would appear to be advisable for us to deal through the Office of the Science Advisor." While the outcome of this exchange is not available in the open archival record, the letter does make clear that, pending the approval of the academy's president, Atwood would make available "information regarding foreign travel of scientists associated with the Academy–Council."[40]

Based on everything else we know about Detlev Bronk, the NAS's new president, it seems highly likely that he would have endorsed such a scheme. He did, after all, give the academy's blessing to the Berkner Report. It's telling, as well, that Atwood's hesitation was procedural, not ideological. Four months earlier, Atwood had offered his "full cooperation" to the ASTA program, which he "heartily endorsed." He just thought it "wise" to provide that information to the State Department, not the CIA.[41]

By 1954, the NAS's Office of International Relations derived its primary support from a contract with the Office of the Science Advisor in exchange for "various" kinds of assistance. At a 1953 meeting of the Office of International Relations's Policy Committee, Koepfli identified the NAS as the "best qualified body" to organize US participation in international meetings. Berkner, also a member of the oversight committee, heartily agreed. When Atwood asked Berkner and Koepfli whether the State Department should formalize this relationship in some way, both men made clear that "this responsibility was already understood." Whatever objections the US scientific community supposedly had about serving the needs of US foreign policy wasn't on display at the highest levels of the NAS in 1953.[42]

፡ ፡

The short-lived ASTA program was meant to supplement, not replace, a science program within the State Department. And yet, as Brode had warned Compton, the US national security establishment of the early 1950s simply did not trust scientists working for the government. Thanks to revelations of atomic espionage—headlined by physicist Klaus Fuchs's confession to British authorities in January 1950—scientists stood out as especially suspicious, even against a background of intense anti-Communism. The CIA responded by isolating its scientific officers from the rest of the US scientific establishment, in Brode's case demanding that he distance himself from the agency supplying his cover. Such a strategy would never work for the State Department, whose attachés were expected to mingle with neutral and possibly even enemy scientists.[43]

Koepfli faced an uphill battle from the start. Following Berkner's recommendations, Koepfli attempted to recruit the most senior people he could locate who worked outside atomic energy, but, fatigued by war service, few scientists were willing to agree to yet another one- or two-year appointment away from their laboratories. Leading scientific administrators, including Conant and Oppenheimer, refused to recommend their colleagues. Koepfli had somewhat better luck recruiting through the network of organic chemists who had trained under Roger Adams at the University of Illinois, who, not coincidentally, now served as the foreign secretary for the NAS. (Brode himself had trained under Adams; upon becoming science advisor to the State Department in 1958, he, too, drew heavily on his chemistry networks in hiring science attachés.) While documentary evidence of attachés' assignments is fragmented, it appears that, by 1953, at least fifteen scientists—many of them chemists—had accepted postings in London, Paris, Bern, Bonn, Stockholm, Tokyo, and New Delhi.[44]

Since the attachés traveled abroad as scientists, not as foreign service officers, they received relatively minimal training. Richard Arnold, another of Adams's students, by then a professor at the University of Minnesota, recalled a two-week crash course before taking up his post in Bonn in 1952. Arnold, his wife, and his two children checked into the Francis Scott Key Hotel just blocks from the State Department, where each morning someone arrived to ferry him to yet another round of meetings. Wherever he went—the CIA, the Research and Development Board, the Commerce Department—Arnold asked the same question: "What am I, as a science attaché, supposed to do?" Just as consistently, the person he was meeting would tell him to "play it by ear." Perhaps, they suggested, after he returned, "You can tell us about what you think a science attaché ought to do and ought not to do."[45]

The first attachés arrived at their respective posts in Bern and Paris even before embassy staff had been told to expect them. Having correctly guessed that the US legation in Bern might neither desire nor know what to do with a science attaché, Koepfli forwarded a long letter with instructions. Koepfli's letter arrived in Bern after the attaché, a "Dr. Lek," but it made up in detail what it lacked in punctuality. After recounting the by-now familiar list of duties—reporting on scientific trends, collecting scientific information, facilitating scientific exchange and international cooperation, representing US interests at meetings, and so on—Koepfli suggested that the legation could "handle the science staff in any way it cares to administratively." When the Swiss Foreign Office later complained that the science attachés were spending too much time poking around private industry, the legation announced that it preferred not to handle the scientific staff at all—which is why Arnold ended up in Bonn instead of Bern.[46]

Once they settled in, most of the science attachés found a warmer welcome. In London, the ambassador installed an electron microscope in the embassy annex for the use of attaché Ralph Wyckoff, a leading Rockefeller University bacteriologist. According to Koepfli, the instrument was a great boon to Wyckoff's networking, with scientists "standing at the door" to get in to see it. Jeffries Wyman seems to have spent most of his time in Paris attending luncheon and cocktail receptions. During one particularly memorable weekend, Wyman noted in his official diary that he spent the morning reading a "nice old edition" of Louis Antoine de Bougainville's *Voyage autour du monde* (1771) before motoring out to the forest surrounding the Abbaye de Royaumont in microbiologist André Lwoff's Citroen. Lwoff's wife was "very attractive," Wyman noted for the record, and he hoped to "see more of them." Somewhere in that afternoon, which also included a luncheon, tea, and a concert in the abbey, Wyman managed to fit in a conversation with Lwoff about the rumors that the Americans had used biological weapons in Korea—the only topic that Koepfli, in his interview with Doel, claimed to have discussed with the CIA.[47]

Richard Arnold recalled a closer working relationship with the intelligence community. CIA officers frequently dropped by his Bonn office for reports, but always warned him of their impending arrival to protect against unwanted surprises. The possibility of chance encounters outside the embassy posed greater risks, as German scientists had directly accused both Arnold and his co-attaché, William Greulich, of being spies shortly after their arrival. Arnold made a point to steer clear of his CIA contacts when he ran into them at scientific meetings—a situation that, in his memory, occurred "very frequently." On the whole, Arnold recalled, "I was rather impressed with the CIA people I ran into." We have no way of knowing how representative Arnold's views were, but his frank acknowledgment of routine contact with CIA officers makes clear that he, for one, saw little conflict between diplomacy and intelligence, provided everyone behaved appropriately. The CIA's willingness to stay out of his way suggests that the feeling was mutual.[48]

The science attachés faced a more skeptical audience in the FBI. The FBI's fear of Communist infiltration especially created problems for Wyman, who frequently found himself quaffing cocktails with former French Communists. Wyman's attempts to distance himself from high-profile Communist scientists, most notably Frédéric Joliot-Curie, the former head of the French atomic energy program, caught the attention of visiting US scientists, who deemed it counterproductive. As physicist I. I. Rabi put it to Koepfli, "there should be no political acts associated with science . . . we should be big enough to attempt relations with people like Joliot-Curie." Koepfli continued to warn Wyman of "hazardous" connections, but as we saw from the encounter described at the beginning of this chapter, the Paris attaché found it impossible to isolate himself from alleged Communists.[49]

Anti-Communists distrusted everything about the science attaché program. Not only did the attachés spend their days (by design) consorting with foreign scientists, but Koepfli had begun to use the Science Office as a platform to criticize US restrictions on Communists. The anti-Communist McCarran Act, officially known as the Internal Security Act of 1950, established the legal basis for the most egregious violations of civil rights associated with McCarthyism. The act authorized the FBI to investigate persons suspected of subversive behavior and required members of the Communist Party to register with the government. The McCarran Act also posed a public relations problem for US science, in that it restricted passports for US citizens and barred entry visas for members of the Communist Party as well as any foreign national who might "engage in activities which would be prejudicial to the public interest" of the United States. The McCarran Act had an immediate and chilling effect on scientific exchange, by some estimates blocking the entry of half of foreign scientists who sought to enter the country. In effect, it created a "party line" for US science.[50]

Within a matter of weeks, Koepfli's office became the front line of the visa battle. Clearing a single scientist required more than a hundred hours of staff time; Koepfli's full-time staff of seven could not possibly keep up. One of these staff members, Rhodes Scholar and former Navy test pilot Neil Carothers, did nothing but work on the visas. In response, Koepfli lobbied Robert Joyce, a member of State's Policy Planning Staff, to make scientists eligible for so-called Proviso 9 visas, named after a provision in the Immigration Act of 1917 that allowed the attorney general to issue temporary visas to otherwise inadmissible aliens. The Cold War interpretation of Proviso 9 held that former Communists could be issued temporary visas—for instance, to attend a scientific conference—if doing so served the broader national interest.[51]

Koepfli's successful use of Proviso 9 exceptions to bring an unspecified number of scientists with Communist affiliations into the United States earned him enemies at both State and the FBI. In December 1953, as part of a multi-week attack on supposed Communist influence over the State Department, *U.S. News and World Report* quoted an unnamed Republican official who described the Office of the Science Advisor as a "stink hole of out-and-out Communists." The newsmagazine printed a retraction under threat of a million-dollar lawsuit, but the office never recovered from the attacks. By this point, Carothers had already left for a position at the NSF; Koepfli resigned shortly thereafter to return to his position at Stanford. With a new Republican president and Republicans in charge of both houses of Congress, the State Department slashed the Science Office's already lean budget. Koepfli's position remained unfilled until 1958, when the crisis of *Sputnik* forced a reassessment of the role of science in US foreign relations.[52]

: :

As a vehicle for gathering scientific intelligence, the State Department's science attaché program failed spectacularly. So, too, did the attempts of the CIA and the NAS to gather actionable scientific intelligence through overt channels. As a way to build better relationships with potentially sympathetic foreign scientists, however, the programs showed more promise. From the vantage point of

1983, Koepfli described the point of the State Department's science program as to "get good European scientists into this country [to] keep them from going over the Iron Curtain and becoming communist." The Berkner Report makes clear that cultural diplomacy was not, in fact, the original point of the Science Office, but Koepfli's retrospective account makes sense, given how the United States eventually decided to use its science diplomats.

In the early 1950s, as the United States stepped up its psychological warfare programs, it increasingly hoped to capture the loyalties of uncommitted elites, including scientists—hence Koepfli's reference to keeping "good European scientists" from "going over to the Soviets." Especially in Europe, where so many scientists had allied with Communists in the fight against fascism, this work simply could not be done without associating with former Communists. Berkner and Koepfli both had worried that the State Department's science programs would fail unless they attracted scientific names so prominent that no one would suspect them of spying. They did not anticipate that the program's greatest challenges would come from domestic anti-Communists rather than suspicious foreigners. The State Department could shield the extent to which its attachés collected intelligence from foreign eyes, but it could not hide its cultural outreach to Communist scientists from Congress and the FBI.

The CIA, however, could, and did.

3 : The Science of Persuasion

The phrase "Cold War" didn't always refer to a time period. In the late 1940s and early 1950s, the very years that the battle lines between the United States and the Soviet Union were being drawn, US foreign policy strategists used the phrase to invoke a specific kind of conflict, one carried out by "means short of war." If, as NSC 68, a key document of US strategy, asserted in 1950, the United States and the Soviet Union were locked in an ideological clash of civilizations, a battle between "slavery" and "freedom," a victory by force would be hollow. If the United States wanted to defeat Communism, it must do so "by the strategy of cold war," combining political, economic, and psychological techniques. "The cold war," NSC 68 warned, "is in fact a real war in which the survival of the free world is at stake."[1]

This was a new kind of conflict requiring new kinds of weapons: psychological weapons. The question of psychological warfare preoccupied a small but influential group of foreign policy officials during the second administration of President Harry S. Truman (1949–1953). By the time that Eisenhower arrived in office in January 1953, the United States had laid the legal and institutional foundations for overt propaganda campaigns as well as covert action. During this period of experimentation, almost anything US strategists could dream up, short of overthrowing foreign governments (that would come later), was up for discussion. Among other things, the Marshall Plan spent $13 billion to rebuild Western Europe, the Voice of America (VOA) transmitted jazz and news to listeners in forty-six languages in more than a hundred countries, and the CIA sent tens of thousands of balloons filled with anti-Communist pamphlets into China.[2]

Even as State Department, CIA, and Army officials spent countless hours working through the administrative challenges of launching a psychological warfare program more or less from scratch, they spent remarkably little time discussing what kinds of messages might best promote the cause of "freedom." Ideas about science rarely, if ever, explicitly appeared on lists of psychological warfare objectives. Science entered US psychological warfare programs as a stowaway, tucked into the pockets of some of the private individuals to whom the State

Department and the CIA turned to wage the United States' battle against Communism. More subtext than text, ideas about science subtly undergirded policymakers' emerging plans for waging and winning this new kind of war.[3]

■ ■
■ ■

Prior to the Cold War, the United States had never formally mounted psychological warfare campaigns during peacetime. The United States, of course, *engaged* in practices that we might consider psychological warfare, using World Fairs, missionaries, economic policies, and educational exchanges to promote US values. But what changed in the years immediately following World War II was a sense that the United States was engaged in a prolonged battle of civilizations that could not be won through force alone. And, as was so typical throughout the Cold War, US policymakers blamed the Soviet Union for forcing their hand.[4]

On March 12, 1947, President Truman appeared before a joint session of Congress to request $400 million in economic and military aid to Turkey and Greece. In what came to be known as the Truman Doctrine, the president pledged to give such assistance as needed to help "free and independent nations to maintain their freedom" in the face of Communist threats. Three months later, Secretary of State George Marshall announced a program that would allow European leaders—East and West—to request much larger sums, possibly as much as $6 billion to $8 billion annually, to recover from the devastation of World War II. Leaders in the United States didn't consider the so-called Marshall Plan an act of psychological warfare, per se, but the Soviet Union's leaders did and barred its satellite countries from participating.[5]

This turned out to be the opening salvo in a high-stakes game of propaganda. In the fall of 1947, Communist Party officials revived the party's prewar international propaganda network under a new name, the Communist Information Bureau, or Cominform. In mid-1948, the Soviet Union launched a campaign against the United States, targeted at audiences both within its own territories and in the world at large. In Moscow, the authorities celebrated writers, musicians, and scientists—including Lysenko—who promoted seemingly "Russian" values; abroad, the Cominform's agents attacked US aggression and promoted the Communist commitment to peace. Soviet authorities meanwhile cracked down on Soviet citizens' ability to communicate with foreigners and foreign institutions. A dispatch from the US ambassador to Moscow in January 1949 warned of the "near-impregnable barrier between Soviet citizens and foreigners in the USSR" and specifically noted that the new restrictions eliminated exceptions for "scientific and educational institutions."[6]

These developments alarmed US officials. Within weeks of arriving at the helm of the newly instituted CIA in October 1947, Director Hillenkoetter warned that the United States' lack of psychological warfare measures had become a "matter of urgency." The National Security Council agreed. On December 17, 1947, the NSC issued two directives, NSC 4 and NSC 4-A, that for the first time committed the United States to a peacetime psychological warfare program. NSC 4 covered the overt aspects of this program, using the euphemism of "foreign information measures" instead of "propaganda" at the request of the secretary of state. The accompanying NSC 4-A, classified TOP SECRET, explicitly called for "covert psychological operations" to supplement the overt measures. Congress's passage of the US Information and Educational Exchange Act—better known as the Smith-Mundt Act—in January 1948 gave the State Department the legal authority to carry out the overseas radio broadcasts, film productions, and publication programs it already had underway. The CIA, for its part, derived its authority for covert action directly from the NSC. Congress had instructed the agency to "perform such other functions and duties" as directed by the NSC: hence NSC 4-A.[7]

Thus began a period of intense experimentation in psychological warfare. In the spring of 1948, many of these efforts focused on blocking a Communist victory in the Italian general elections. With the help of Hollywood studios, the State Department blanketed Italian movie screens with films promoting the glories of postwar consumer capitalism. The State Department turned to the Advertising Council, a private advertising industry group, for assistance in developing print materials, radio broadcasts, and traveling exhibitions that promoted US aid without offending European sensibilities. In Sicily, the publicity arm of the European Recovery Program staged puppet shows to promote the Marshall Plan. Catholic priests in Newark, New Jersey, and Brooklyn, New York, supplied Italian-American parishioners with form letters to mail to relatives back home. A new special operations group within the CIA, meanwhile, channeled approximately $10 million to prop up Italy's non-Communist parties. F. Mark Wyatt, a CIA operative involved in the operation, later claimed to have delivered "bags of money" to Italian politicians. When the Christian Democrats pulled out a victory in the April 18, 1948, elections, the CIA and the State Department interpreted it as a clear victory for propaganda and covert action.[8]

The State Department, the CIA, and the Joint Chiefs of Staff spent most of the spring of 1948 bickering over how best to coordinate overt and covert operations with US foreign policy objectives. To a surprising extent, given the CIA's subsequent reputation for adventurism, at this point all parties agreed that State needed to retain control over covert psychological operations in peacetime. The resulting guidance, issued as NSC 10/2 in June 1948, nevertheless placed the office responsible for planning and conducting covert operations within the CIA. State's control over this new Office of Special Projects would be limited to nominating its director, subject to the approval of the NSC. In times of war, control of all psychological warfare operations would transfer to the Joint Chiefs of Staff.[9]

NSC 10/2, which replaced NSC 4-A, differed from its predecessor document in another key way. Unlike NSC 4-A, NSC 10/2 offered specific guidance on the nature of US psychological warfare programs. Its expansive list of potential covert actions included "propaganda, economic warfare; preventive direct action, including sabotage, anti-sabotage, demolition and evacuation measures; subversion against hostile states, including assistance to underground resistance movements, guerillas, and refugee liberation groups, and support of indigenous anti-communist elements in threatened countries of the free world." The only thing off the table was direct conflict involving recognized military forces.[10]

In August 1948, the State Department nominated and the NSC approved Frank Wisner as the first head of the CIA's new arm for covert action, now operating under the more discreet name of the Office of Policy Coordination (OPC). Even before taking office on September 1, 1948, Wisner began staking claims for the OPC's independence. With the encouragement of George Kennan, the head of State's Policy Planning Staff, Wisner argued that his urgent mandate required administrative flexibility. The CIA would house the OPC and pay its bills, but Wisner would involve the director only on particularly "important" decisions. Instead, Wisner agreed to keep the State Department, the secretary of defense, and the Joint Chiefs of Staff apprised of the OPC's actions through weekly meetings with their approved representatives. The OPC was Wisner's shop.[11]

A former champion sprinter at the University of Virginia later described by one of his colleagues as "a low hurdler constrained by a vest," Wisner could barely wait for the starting gun. Within its first year of operations, the OPC had a staff of 300, operating on a budget of $4.7 million. Wisner soon started forming the partnerships with private organizations and individuals that would shape the OPC's work for the next decade. The most important of these was the ostensibly private National Committee for a Free Europe, whose network of émigrés broadcast Radio Free Europe for years to come. But Wisner, who had spent the final years of World War II running spies in the Balkans, also immediately involved the OPC in disastrous paramilitary operations. In the most notorious of these, the OPC dropped poorly trained Ukrainian dissidents behind Soviet lines and Albanian rebels onto their home soil. Nearly all of them were captured and killed. It was around this same time, and in this same try-anything spirit, that Wisner's OPC embraced the idea of intellectuals as ideological foot soldiers in the Cold War.[12]

::

The sequential crises of Stalinism, fascism, war, and the Holocaust displaced millions of people between 1932 and 1945. In the postwar period, many of these migrants focused their energies on building new lives in new countries—with any luck, free of the savagery and heartbreak of the 1930s and 1940s. But others hoped to return home or, short of that, to create new communities abroad that replicated all they had lost. Wisner embraced these malcontents, particularly disillusioned former Communists, and hoped to yoke their thwarted ambitions to the US cause. For their part, the most ambitious of the malcontents hoped to use the United States' newfound interest in propaganda to advance their own agendas.[13]

A handful of former Communists brought formidable propaganda skills to the battle against the Soviet Union, having engaged in exactly the kinds of Communist agitprop that Wisner and the OPC now hoped to disrupt. In the 1920s and 1930s, German Communist Willi Münzenberg developed an astonishing media empire to promote the Bolshevik cause around the globe. Münzenberg amplified the impact of his dozens (by some estimates, hundreds) of newspapers, publishing houses, film production companies, and charities by pairing them with front organizations usually associated with "worker aid." With the backing of the Communist International, Münzenberg's fronts promoted antiracism, anticolonialism, antifascism, and famine relief—issues guaranteed to appeal to leftists whose commitments to socialism fell short of revolution.[14]

Münzenberg died under mysterious circumstances in October 1940, but at least a few of his army of propagandists found new purpose after the war. One of the most important of these was Arthur Koestler. Born in Hungary in 1905,

Koestler spent the first few years of his life in middle-class, Jewish Budapest. After some unsettled years during World War I, Koestler and his family landed in Vienna, where Koestler began studying for a mechanical engineering degree at Vienna Polytechnic University. He left before graduating, but scientific and technical themes would season his prodigious writing for the rest of his career.

After brief stints in Palestine and Paris, Koestler, now a journalist, became science editor for the Berlin-based *Vossiche Zeitung* and science advisor for the larger Ullstein Verlag publishing empire. Koestler thrived on the "variety and excitement" of the nascent field of science writing, later recalling with evident zeal how the position allowed him to write on nearly any topic that interested him, "from space travel to a new type of vacuum cleaner, from genetics to diesel engines." In this position, Koestler met leading members of the scientific community in Berlin, which now included his old family friend from Budapest, chemist Michael Polanyi. He also accompanied a team of scientists on a polar expedition aboard the hydrogen-filled *Graf Zeppelin* in 1931—a "majestic" event that Koestler regarded as "the climax of my career as a journalist."[15]

On December 31, 1931, Koestler joined the German Communist Party. His hopelessly bourgeois red Fiat convertible came in handy for transporting members of his Berlin cell on party errands. Like H. J. Muller and so many other converts before him, Koestler toured the Soviet Union to witness the achievements of Communism firsthand. Unlike Muller, however, Koestler stayed only a year; he would contribute to the revolution from afar. Back in Paris in 1933, Koestler met Münzenberg, for whom he would work, off and on, for the next five years. Koestler particularly relished his time at the Institut pour l'Etude du Fascisme, a supposedly scholarly enterprise that produced pamphlets, books, and other documentary materials that would reveal "what Fascism really was." A classic front, the institute benefited from its loose association with leading French leftist intellectuals, including Frédéric Joliot-Curie, but was actually run through the Communist Party.[16]

At the outbreak of the Spanish Civil War, Münzenberg sent Koestler to the front as a "reporter." He was lucky to escape with his life. But even while Koestler's reports from Franco's Spain promoted the Popular Front, he struggled to reconcile his faith in Bolshevism with the realities of Stalin's rule. Koestler was particularly disturbed by the Communist Party's ability to extract confessions from those it accused of crimes against the party. The arrests and purges had reached those close to him, including Eva Striker, his childhood friend and sometime lover (and Michael Polanyi's niece), and her husband, physicist Alex

Weissberg. Koestler had ample time to ponder these questions while sitting in solitary confinement in a Spanish prison in 1937.

In 1938, Koestler resigned his membership in the Communist Party, an experience he dramatized in his best-selling 1941 novel *Darkness at Noon*. An experienced polemicist, Koestler combined Striker's prison experiences with the Moscow show trials to create a gripping account of the ideological horrors of Communism. By the time World War II ended, Koestler, now a British citizen, was lending his considerable skills as a propagandist to the war *against* Communism.[17]

Koestler's impassioned denunciations of Communism became a touchstone for British and US members of what came to be known as the "non-Communist Left." At Koestler's suggestion, the British leftist Richard Crossman, a former psychological warfare specialist, assembled a volume that brought together the voices of prominent socialist intellectuals who had renounced Communism. Not coincidentally, three of the contributors to *The God That Failed* (1949)— Koestler, US journalist Louis Fischer, and Italian writer and politician Ignazio Silone—had worked for Münzenberg. The US occupation government in Germany published each of the essays in German in its literary magazine, *Der Monat*, and made sure the book found its way into libraries and bookstores across Europe.[18]

Koestler's obvious talents as a print propagandist exceeded his skills as a political organizer. Working with like-minded writer George Orwell, in 1946 Koestler drafted a manifesto for a "League for the Freedom and Dignity of Man." In the spirit of Münzenberg's various fronts, the league was intended to inspire intellectuals to join forces in support of unobjectionable causes— human freedom and dignity—while also undermining Communism. But despite Koestler's best efforts to secure the involvement of leading British public intellectuals, the movement did not garner much support. Similar efforts in the United States, variously led by New York intellectuals Mary McCarthy, Dwight Macdonald, and Sidney Hook, also stalled. The intellectuals of the non-Communist Left had yet to find the right formula, or the right sponsors, for their anti-Stalinist crusade.[19]

∷

Sympathetic observers on both sides of the Atlantic watched these moves with curiosity. Pulitzer Prize–winning historian and Office of Strategic Services veteran Arthur Schlesinger, Jr., for instance, urged government officials to embrace the skills and passion of these committed anti-Communists. In July 1946, Schlesinger published an exposé in *Life* magazine of how the Communist Party had infiltrated US political institutions. The article earned Schlesinger the enmity of the *Daily Worker* and the *New Masses* and forced a rupture on the US left, with the anti-Communist Americans for Democratic Action splitting from the Progressive Citizens of America in 1947 over the latter's willingness to work with Communists. During this time, Schlesinger, like Koestler, was learning more about covert operations: he spent the summer of 1948 distributing untraceable funds to European trade unionists on behalf of Averell Harriman, the new head of the Marshall Plan.[20]

Given their collective experience with propaganda and front organizations, members of the non-Communist Left were better situated than most to recognize the hand of the Cominform in various "peace initiatives" that sprung up around the world in 1948 and 1949. In the spring of 1949, an innocuously named professional group, the National Council of the Arts, Sciences, and Professions, announced plans to hold a Cultural and Scientific Conference for World Peace at the Waldorf Astoria Hotel in New York. The peace conference's roster of sponsors and honorary chairs included a long list of prominent leftist Americans, chaired by Harvard astronomer Harlow Shapley. New York University (NYU) philosopher of science Sidney Hook smelled a rat.

Hook's first response was to try to get himself added to the program. He proposed an anti-Lysenkoist paper on the lack of "national," "class," or "party" truths in science. When the organizers confirmed his suspicions by declining to allow him, or anyone who criticized Communism, to speak, Hook attempted to convince (or blackmail) several of the individual sponsors to resign. He sent several sponsors, including writer Thomas Mann, threatening letters claiming that he had identified "almost eighty notorious fellow travelers of the Communist Party" among the conference's supporters. When that tactic also failed to derail the conference, Hook assembled an instant protest group, the Americans for Intellectual Freedom, from the dregs of an earlier, dormant non-Communist Left group, the Committee for Cultural Freedom.[21]

Here, finally, was a successful strategy for the non-Communist Left. Following Hook's lead, anti-Communists disrupted the Waldorf conference, picketing the sidewalk and asking inconvenient questions from the conference floor. From the group's command center in a luxury suite in the Waldorf Astoria, Hook, accompanied by McCarthy, Macdonald, Schlesinger, and several other ex-Communist writers and reporters, engaged in a "war of mimeograph machines and public relations releases." Shapley at one point had to call security when Hook, accompanied by a reporter from the *New York Herald-Tribune*, ambushed the astronomer in his hotel room.

On the last night of the conference, Hook and company staged a counter-conference at Freedom House, where H. J. Muller stole the show. Crowds had filled the room by the time the Nobel laureate stepped up to the microphone. An additional 500 people spilled out onto Fortieth Street, where they listened via loudspeakers set up for the occasion. The crowds hadn't come just to hear Muller, but Hook saw the geneticist's presence as "absolutely essential" to the event's success. He had, after all, repeatedly cited Muller's absence from the Waldorf conference's speakers' list in his campaign to dissuade liberals from participating in what Hook had rightly identified as a Communist propaganda exercise.[22]

Muller's speech, entitled "The Destruction of Science in the U.S.S.R.," surpassed Hook's expectations. As Muller told it, the Communist Party's official endorsement of Lysenko's theories of heredity in 1948 had "repudiated the entire science of genetics," capping a decades-long campaign that had resulted in the deaths of at least two leading researchers and the arrests and disappearances of countless others. The "very foundations of human knowledge" had been "brutally attacked" by "political and official fiat." The result was an insult to science and a testament to the perils of Soviet ideology. And why had the Soviets pushed Lysenko's theories? "The Communist Party regards as a menace any concept that does not fit patly into its scheme for mankind."[23]

Muller's denunciations of Lysenko were joined by equally passionate speeches by Schlesinger and other outspoken figures from the US left, including Max Eastman, former editor of *The Masses* and *The Liberator*; black internationalist Max Yergan; ACLU cofounder Morris Ernst; writer and former Communist organizer Bertram Wolfe; and pacifist A. J. Muste. Koestler and Bertrand Russell sent telegrams of support, which Hook read to an audience transfixed. The counterdemonstrations received glowing coverage in all of the New York papers except the *Times*, whose reporter was—in Hook's telling—a Communist.[24]

Instead of a publicity triumph for the Cominform—a peace conference in New York, the heart of global capitalism!—the event revealed a growing sophistication in the United States' ability to fight propaganda with propaganda. Although the idea for a counter-conference came from Hook, the money to support it most likely came from Wisner's OPC, via David Dubinsky of the Ladies' Garment Workers Union. One of Hook's colleagues in NYU's Department of Philosophy, James Burnham, had by this point already started consulting for the CIA. In the summer of 1949, Burnham would move to Georgetown to take up a full-time appointment as a consultant for the OPC while on sabbatical from NYU.[25]

The State Department, too, was well aware of the nature of both the Cominform's conference and Hook's counter-conference. In the three weeks leading up to the meetings, George Allen, assistant secretary of public affairs, attempted to dissuade Shapley from participating in the Waldorf event. The State Department blocked visas for several Europeans who planned to attend and solicited ongoing reports of the event from yet another official at NYU. Michael Josselson, a polyglot immigrant from Estonia who would soon be working for Wisner, soaked it all in from the back row of Freedom House. He declared it "splendid."[26]

: :

Over the next year, the United States intensified its commitment to psychological warfare and, increasingly, did so publicly. On April 20, 1950, President Truman kicked off a national "Campaign of Truth" with an address before the American Society of Newspaper Editors. In a lunchtime address at the Statler Hotel in Washington, DC, Truman implored the country's leading editors to join the government in meeting "false propaganda with truth all around the globe." "Everywhere that the propaganda of Communist totalitarianism is spread," the president warned, "we must meet it and overcome it with honest information about freedom and democracy." The president briefed the journalists on State's existing information campaigns and promised that the government would do more, but the real point of his address was to—without irony—instruct the nation's private media about their role in the coming conflict. "We shall need to use fully all the private and governmental means that have proved successful so far," Truman said, "and to discover and employ a great many new ones."[27]

Truman's public speech coincided with a new statement of US strategy issued behind closed doors. NSC 68, a TOP SECRET document drafted by a committee chaired by Paul Nitze, the new head of the State Department's Policy Planning Staff, confirmed the US view of the conflict with the Soviet Union as total and ideological. It is not hyperbole to refer to the sixty-six-page document as "apocalyptic," as historians so frequently do, because the document responded directly to a potentially world-ending threat: the Soviet Union's explosion of an atomic weapon in August 1949. The end of the US atomic monopoly, along with Truman's subsequent decision to endorse a hydrogen bomb program in January 1950, dramatically raised the stakes of a potential hot war. The backdrop of nuclear conflict helps explain NSC 68's otherwise perverse conclusion that the United States must engage in a "rapid build-up of political, economic, and military strength" against the Soviet Union "in order that it may not have to be used." Over and over again, NSC 68 called for overt and covert psychological strategies to both strengthen the resolve of allies and foment unrest in the Soviet Union's vulnerable satellites.[28]

This new, explicit focus on psychological warfare, combined with the outbreak of the Korean War in June, had an immediate effect on both overt and covert propaganda programs. Truman requested nearly $90 million from Congress to step up the State Department's information campaigns; Congress agreed to two-thirds of this, $63.9 million, in September 1950. On the covert side, Wisner immediately submitted budget estimates to dramatically expand the OPC's operations through 1957. His request included funds for staff, Washington facilities and overseas supply bases, organizational resources, paramilitary training, and a worldwide communications network.[29]

Wisner also asked for something more difficult to supply than money: expertise. As matters currently stood, the OPC lacked "a significant body of knowledge, personnel reserves, techniques, and philosophy of operations" regarding psychological warfare. For this, the architects of US psychological warfare strategy turned to the scientific community. Undersecretary of State James Webb, who had previously asked Lloyd Berkner for advice on integrating scientists into the foreign policy apparatus, now asked Berkner's help in assembling a crack team of scientists to tackle the problem of psychological warfare. The resulting Project Troy brought together a group of social scientists and physical scientists from MIT and Harvard that either already had or would soon play leading roles in the Cold War. In addition to Berkner himself, the group included electrical engineer (and future Kennedy advisor) Jerome Wiesner, physicist and future Nobel laureate Edward Purcell, and economist Max Millikan, all at MIT; anthropologist Clyde Kluckhohn and psychologist Jerome Bruner, both Office of War Information veterans now at Harvard; and a select few others from outside the universities, including RAND's Hans Speier and Bell Lab's John Pierce.[30]

Webb had specifically asked Project Troy's members to investigate technical obstacles to US information campaigns, especially ways to circumvent the Soviet Union's jamming of Voice of America broadcasts. This ambitious group, however, interpreted its mandate much more broadly, producing an eighty-one-page report (plus appendixes) on all imaginable aspects of political warfare. In addition to the expected chapters on radio transmissions and the use of long-distance balloons, the study group's February 1951 report covered such wide-ranging topics as preparing for Stalin's death and strategies for debriefing Soviet defectors. Nitze, at State, was unimpressed, pointing out that the group "went vastly beyond its original terms of reference and explored a field for which it had no special competence and about which it had little information." Project Troy's biggest impact ultimately turned out to be long-lasting relationships between government officials at the State Department and the CIA and social scientists at MIT and Harvard. In the more immediate future, however, Project Troy's endorsement of some sort of central agency to coordinate the various overt and covert psychological warfare programs already in place sent ripples through the foreign policy establishment.[31]

Despite their Top Secret clearances, the Project Troy members lacked access to information on, or even confirmation of the existence of, some of the OPC's clandestine programs. But even lacking those details, they gleaned the obvious point that having so many government agencies involved in propaganda raised the risk of duplication, crossed purposes, and blown covers. The State Department had its overt information programs, of course, but so did the Economic Cooperation Administration (the agency in charge of implementing the Marshall Plan), the Army, and NATO. The CIA, the Economic Cooperation Administration, and the Army also maintained covert information programs. In Korea, the theater commander controlled psychological warfare operations. None of these programs were being coordinated with the others.[32]

Project Troy recommended a sort of "superboard" that would "plan general strategy for virtually all unconventional warfare measures," including overt propaganda campaigns, covert actions, and economic warfare. In response, on April 4, 1951, Truman created a Psychological Strategy Board (PSB) responsible for "the coordination and evaluation of the national psychological effort." Like Truman's speech announcing a "Campaign of Truth," the creation of the PSB was a public act: in late June, the White House and the State Department issued simultaneous press releases describing the PSB's purpose, membership, and powers. The press releases of course omitted any reference to covert activities, but the US government's broader embrace of psychological strategies was not remotely secret at this point in the Cold War.[33]

As a coordinating body, the PSB proved a disappointment. The wording of the board's mandate suggested that it would oversee psychological programs but not actively participate in them, leaving operational control in the hands of the originating agency. Once again, the State Department, the CIA, and the Joint Chiefs of Staff began sniping over turf. Mired in scheduling conflicts, the board rarely met during its first six months and spent most of the time that it did meet on procedural details. The first meeting confirmed its astonishingly broad mission, covering "every kind of activity in support of U.S. policies except overt shooting and overt economic warfare." In an impressive bit of understatement, Gordon Gray, the PSB's first director, later recalled, "I don't consider this one of the conspicuous successes of my life."[34]

The PSB proved more useful to the new director of central intelligence, former ambassador to the Soviet Union Walter Bedell Smith. Since his arrival at the CIA in October 1950, Bedell Smith had been attempting to bring Wisner's OPC more clearly under the CIA's control. He was alarmed at the scale of the OPC's operations, doubting not only the wisdom of individual missions but also the agency's administrative capacity to supervise them. The existence of the PSB helped; even if the OPC was not yet entirely under the CIA's control, it offered a mechanism to extract covert operations from State. NSC 10/5, issued in October 1951, clarified that the director of central intelligence did, in fact, have authority over covert operations, meaning that Wisner reported to the CIA. The PSB's eventual compromise on how it would evaluate projects—a negotiation that lasted until February 1952, nearly a year after the board's creation—further strengthened Bedell Smith's hand by creating a screening board that kept all but the most controversial projects off the PSB's agenda.[35]

The State Department, for its part, was by now increasingly ready to relinquish control over covert operations. A State Department circular issued in December 1951 carefully distinguished between white, gray, and black propaganda, reminding foreign service officers that neither State nor the Economic Cooperation Administration was authorized to participate in black propaganda. As examples of permitted activities, the circular suggested contracts with publishers and other media producers, with or without attribution to the US government, provided that attribution of material to the United States could be done "without serious embarrassment." Inappropriate activities included direct assistance to foreign newspapers, financial assistance to labor or youth groups, and propaganda campaigns designed to influence foreign elections—all activities notable for being pursued at that very moment by the OPC.

This did not mean, however, that the State Department was out of the covert propaganda business entirely. Either Undersecretary of State Jim Webb or his representative, Robert Joyce, continued to attend the weekly 10/2 Committee meetings, chaired by Wisner, that oversaw the OPC's day-to-day operations. In a memo briefing Webb in November 1951, Edward Barrett, who had replaced George Allen as assistant secretary for public affairs in 1950, stressed the benefits of "subtlety" over "direct, sledge-hammer techniques." He advised Webb that, "to an increasing degree, a major share of the total US funds in this field should go into the gray and covert operations that we have found to pay off so importantly." Barrett noted the benefits of letting "local authorities take the major share of credit for some US aid." Like the CIA, the State Department increasingly came to rely on the passion and initiative of private individuals in carrying out its Cold War propaganda.[36]

: :

So where, exactly, did scientific freedom fit into the emerging US propaganda program? Recall that all of these events were taking place at exactly the same moment that American anti-Communists began to shorthand the experience of science under Communism as "Lysenkoism." The problem of Soviet genetics never reached the level of the National Security Council, but it did attract the attention of senior State Department officials, including John Hickerson, director of the Office of European Affairs. When Muller reached out to the State Department offering his services as a spokesman for scientific freedom in September 1948, Hickerson himself signed the response, and his phrasing is telling. He praised Muller for recognizing that "free science in a democracy like ours is ever increasingly separated from the so-called science dominated by political ideology." He expounded on this theme, at length: "Under an authoritarian system of government the work of scientists, as well as artists, musicians, architects and so forth, is invariably subverted to the political purposes of the ruling group. Thus the work of the scientists in their search for knowledge becomes an instrument of foreign policy." Noting that "the Department" felt that Muller's statement of resignation from the Soviet Academy "should be disseminated as widely as possible," Hickerson indicated that someone from the State Department's Office of International Information—the division for media propaganda, including VOA—would soon be in touch.[37]

As the US government's official, acknowledged outlet for broadcasting overseas, VOA distributed US propaganda throughout the Cold War. Although the Voice would later gain a deserved reputation for journalistic integrity, in the late 1940s it was housed in the State Department, and officials explicitly envisioned it as Foggy Bottom's primary tool for carrying out its psychological warfare obligations (that is, overt "information measures"). When, for example, Bedell Smith (at that time still US ambassador to the Soviet Union) learned in February 1948 that the Soviet Central Committee had banned opera as "depraved and anti-artistic," he shot off a telegram to the secretary of state urging "wide reporting of facts" by VOA, with "comment stressing absence of artistic and thought freedom in USSR and ridiculousness of party CO telling some of the world's greatest living composers how to write music."[38]

Muller's response to Hickerson has been lost to time, but the exchange sparked a long partnership between Muller and Voice of America. The network adapted Muller's two-part article in the *Saturday Review of Literature* into a radio script. At some point in the spring of 1949, Muller recorded a broadcast on Lysenkoism in Russian. In the summer of 1949, Muller agreed to organize and chair a series of six half-hour programs on the situation in Soviet science. This time, the request—marked RESTRICTED—came directly from Karl Compton. Muller signed up several of his most outspoken colleagues, including Ralph Cleland, Robert Cook, and Tracy Sonneborn, for a recording session in February 1950. He continued to answer VOA's call throughout the 1950s, almost always returning to the topic of the impossibility of scientific freedom in the Soviet Union.[39]

Muller's relationship with the Voice of America culminated in the network's presence at a remarkable event held in Columbus, Ohio, in September 1950. The fiftieth anniversary of the rediscovery of Gregor Mendel's laws of inheritance proved an irresistible opportunity for even the most controversy-averse members of the Genetics Society of America to pledge their support for Western genetics. The commemorative efforts of the Golden Jubilee of Genetics, crowned by a four-day event attended by almost 2,500 biologists, included research presentations and public lectures, a scholarly book, a general interest pamphlet with a print run of 100,000 copies, and an attempt to establish a permanent Mendel Museum. With the help of prominent public relations firm Pendray and Leibert, the GSA devised promotional plans ranging from the predictable, such as the edited volume, to the creative, such as an attempt to commemorate Mendel with a postage stamp.[40]

The US government's growing relationship with private groups during this time has complicated the work of historians trying to make sense of events like the Golden Jubilee. Certain aspects of the festivities seem unlikely choices for a group of independent geneticists who, just two years earlier, had balked at issuing a statement condemning Lysenko. Perhaps the most peculiar moment occurred on the jubilee's Thursday night, at the "New World Honors Mendel" ceremony. That evening, Dr. Manuel Elgueta of the Instituto Interamericano de Ciencias Agricolas (Costa Rica) presented Curt Stern, the GSA's president, with a commemorative scroll acknowledging "the debt of science and society to Gregor Johann Mendel." But no one involved in this ceremony could possibly have confused it with an actual expression of "New World attitudes." The "truly impressive" scroll—the engraving alone cost $200—was entirely designed and produced by the GSA. The citation was written by Paul Mangelsdorf, a Harvard corn geneticist whose conference presentation earlier in the afternoon claimed that hybrid corn, by staving off hunger in Europe, deserved credit for "stemming the tide of communism." The event was designed for the press, including the microphones of Voice of America, which recorded the entire Golden Jubilee. And just as its organizers hoped, the coverage highlighted the accomplishments of Mendelian geneticists without once mentioning the event's intended targets: Trofim Lysenko and those who found his arguments on behalf of Soviet agriculture compelling.[41]

The Golden Jubilee's combination of slick publicity, creative accounting, and well-connected individuals bears obvious similarities to other contemporaneous events involving the non-Communist Left, especially Sidney Hook's counter-conference at Freedom House. To me, everything about the Golden Jubilee, from the "New World Honors Mendel" ceremony to its timing in the fall of 1950, when the US government was stepping up its covert cultural operations, felt a little off. What kind of event was this? Had Muller pushed the GSA into a partnership with the CIA?[42]

Researching the history of covert operations can make a person see links that don't necessarily exist. After years of intensive archival searching, Freedom of Information Act requests, and increasingly convoluted database queries, I have yet to find any concrete evidence—beyond the involvement of an unlikely number of people with ties to intelligence work—for Wisner's hidden hand in programming the Golden Jubilee. It's also abundantly clear from the GSA's meticulously organized institutional records that the geneticists came up with their pro-Western, anti-Lysenkoist plans all on their own. Though conspiracy is a tempting explanation, the Golden Jubilee does not appear to have been one, and the State Department's willingness to promote it was entirely in keeping with the United States' commitment to publicizing private anti-Communist propaganda in the early Cold War. Personnel and ideas circulated freely between the private and public sectors during this period, to the point that assigning credit for any individual campaign distracts from the larger pattern. In 1950, the State Department, the CIA, and American geneticists all had ample reason to promote the cause of scientific freedom.

∷

The word "science" is strikingly absent from the key documents that established the parameters for early US psychological operations. Even NSC 68, a notably thorough document, primarily discusses science in terms of weapons technology. But if science itself is missing, rationality is everywhere. The United States would provide "facts" that would allow free-thinking people to decide for themselves between "freedom" and "slavery." NSC 68's sole explicit mention of science is clarifying. Issued only six months after Mao Zedong's Chinese Communist Party claimed victory in the Chinese Civil War, NSC 68 hypothesized that "Asiatics have been impressed by what has been plausibly portrayed to them as the rapid advance of the U.S.S.R. from a backward society to a position of great world power." The Soviet Union offered itself as a "model 'scientific' society," thereby appealing to the "genuine aspirations of large numbers of people." In this light, unmasking the Soviet Union's "scientific society" as a sham based on authoritarianism fit into the broader US objective of "orienting the free world toward the United States."[43]

Despite the dearth of explicit references to science in declassified reports on psychological operations during the Truman era, there is substantial evidence to suggest that US policymakers wanted science to play a larger role, even at this early date. Recall, for instance, the role of Robert Joyce, State's representative on the 10/2 committee, in obtaining Proviso 9 visas for visiting scientists as a favor for Joseph Koepfli. We also know that Marshall Plan administrators directed funds toward rebuilding European scientific research activities, most notably in the form of CERN, the European Organization for Nuclear Research. Approximately 15 percent of the articles published in the State Department's glossy Russian-language publication *Amerika* between 1945 and 1952 covered advances in science, medicine, or technology. In 1950, 20 percent of Fulbright grants to university faculty and teachers went to natural scientists, with an additional 25 percent to social scientists. All of this suggests that scientific programming had a place, if not necessarily a prominent one, in both overt and covert psychological warfare programs in the early 1950s.[44]

Over time, the CIA and the State Department would find ways to incorporate messages about scientific progress more directly into their work, particularly in programming aimed at the sort of elite technocrats in developing nations that NSC 68 proposed to win over. First, though, the US government would have to exercise more direct control over its own psychological warfare campaigns. Under pressure from a series of government advisors to expand psychological operations, but faced with limited staff and organizational capacity, Bedell Smith appointed a "Murder Board" to review every single OPC project in existence as of February 1952. Almost a third of the OPC's projects were deemed "doubtful" or "of marginal value," but, based on what happened next, we know that the work of former Communist intellectuals made the cut. Without access to internal CIA documents, lay historians can't know the specific calculus that protected these sorts of projects or what alternative schemes came to an end with a slash of Bedell Smith's red pen. What we can do, instead, is take a closer look at the origins of scientific programming in what became the most notorious of these fronts: the Congress for Cultural Freedom.

4 : Science and Freedom

In October 1956, Paul Koenig, chair of the physics department at the Université Laval, in Quebec, received an unsolicited copy of a magazine. Koenig didn't mind; he invited the publisher, the Committee on Science and Freedom, to continue to send it, particularly since a letter enclosed with the magazine offered to do so free of charge. But to set his mind at ease, Koenig wanted clarification on an important point. "We are somewhat wondering," Koenig wrote, "as to where the funds come from?" If the publisher responded to this, its letter was not retained.[1]

Koenig had received an issue of *Science and Freedom*, one of the more obscure publications in the Congress for Cultural Freedom's global network of intellectual magazines. Compared with the CCF's other publications, *Science and Freedom* was amateurish and its budget small. While exact figures are hard to come by, the annual budget for the entire operation seems to have hovered around $10,000. Its publication followed no particular schedule, and, more often than not, it merely reprinted versions of papers delivered at various CCF conferences and study groups. At no point did the bulletin's distribution surpass 7,000 copies, despite the editors' near-constant efforts to build the mailing list.[2]

The fate of *Science and Freedom* as a publishing venture, however, is probably the least interesting thing about it. Officially produced in the United Kingdom by the Manchester-based Committee on Science and Freedom, the journal represented a sustained attempt to incorporate science into the CCF's broader cultural offensive against Communism. Science rarely appears in histories of the CCF, but both the organization's intellectual leadership and its supervisors at the CIA originally envisioned scientific freedom as a crucial plank in its anti-Communist platform. The CCF's early science programming portrayed scientific freedom as a specific flavor of academic freedom, with its most prominent scientific spokesmen, including H. J. Muller and Michael Polanyi, arguing that state planning and state oversight of science would inevitably lead to totalitarianism. In Muller and Polanyi, the CCF had located articulate, passionate mouthpieces with successful histories of countering Communist ideology. And yet, by any reasonable yardstick, the CCF's scientific programming, which ranged from a major conference to small study groups to the bulletin itself, failed to spark a larger movement.[3]

From the perspective of the CCF's (CIA-run) headquarters in Paris, the problem with the Committee on Science and Freedom was not its failure to convince European and Asian intellectuals to abandon Communism. Instead, it was that neither Michael Polanyi nor his son George, who carried out the day-to-day work of the Committee and edited its eponymous bulletin, seemed to understand that the project was not a "house organ for the Polanyis," as the CCF's deputy secretary Warren Manshel memorably put it. The power struggle between the Committee on Scientific Freedom and its overseers at the CIA reveals the messy reality of the US government's attempt to outsource its cultural diplomacy to private citizens. This isn't merely a question of "who used whom"; the arm's-length arrangements had practical consequences that ultimately limited the operation's effectiveness as a propaganda exercise.[4]

But I am getting ahead of myself. Before there could be a Committee on Science and Freedom, there had to be a Congress for Cultural Freedom.

∷

The CIA's first attempt to capitalize on Sidney Hook's success in New York bore little relation to the spectacular performances that would follow in years to come. In late April 1949, barely a month after the showdown at the Waldorf Astoria, Frank Wisner's Office of Policy Coordination sent Hook and assorted other anti-Communists to Paris to disrupt yet another Cominform-sponsored peace conference. There they encountered an unpleasant surprise. While the French intellectuals attending the OPC-funded counter-conference reliably opposed Stalin, many of them opposed the United States with equal vehemence. Wisner deemed the event a disappointment, expressing skepticism about the entire concept of a "U.S.-backed 'DEMINFORM'" run by a "nuts folly of miscellaneous goats and monkeys whose antics would completely discredit the work and statements of the serious and responsible liberals." If the anti-Communists were to be useful, they would have to be supervised—with or without their knowledge.[5]

The OPC anticipated a more welcoming reception in Berlin, the divided city that symbolized the Cold War. Divided into French, British, US, and Soviet sectors since World War II, West Berlin stood as an island within the Soviet-controlled eastern portion of Germany. In the summer of 1948, the Soviet Union blockaded the city, cutting off residents' (and occupiers') access to food and fuel. Faced with two unpalatable choices—abandoning Berlin or engaging the Soviet Union in a military conflict—the Western powers identified a third alternative that electrified public sentiment against the Soviets. From May 1948 to September 1949, the Berlin Airlift, officially known as Operation VITTLES, flew more than 2 million tons of milk, flour, coal, gasoline, medicine, and other essential supplies onto runways at Tempelhof, Gatow, and Tegel airfields. The airlift not only signaled the Western powers' commitment to the idea of a "free" Germany but also fanned the flames of anti-Communism among the residents of West Berlin.[6]

In August 1949, while the airlift was still winding down, several figures associated with the US occupation began exploring the possibility of organizing an enormous international gathering of the non-Communist Left in West Berlin, a "Congress for Cultural Freedom." Twenty-nine-year-old Melvin Lasky, the editor of *Der Monat*, took the lead. By the fall, Lasky's enthusiasm had spread to Michael Josselson, who in turn brought Wisner on board. Wisner earmarked $50,000 for a four-day conference, on the condition that US employees, especially Lasky, keep a low profile. That proved impossible. Lasky, whom official accounts portray as "unwitting" of the OPC's involvement, had already started recruiting speakers under *Der Monat*'s sponsorship, listing himself as the "General Secretary" of the meeting. Back on the other side of the Atlantic, Hook and James Burnham recruited a who's-who roster of speakers from the non-Communist Left. The State Department's assistant secretary of public affairs, Jesse MacKnight, supplied visas, travel expenses, and publicity.[7]

On Monday, June 26, 1950, the 200 delegates of the Congress for Cultural Freedom gathered in the art deco splendor of the Titania Palast, a movie complex in southwestern Berlin that had served as the flagship theater for Joseph Goebbels's Ministry of Public Enlightenment and Propaganda under the Nazis. The already tense atmosphere was heightened by news of the North Korean invasion of the South. Each day, the assembled delegates and nearly 4,000 curious audience members watched as anti-Stalinists from across Europe, the United Kingdom, and the United States railed against totalitarianism; each night, Lasky, Hook, Burnham, Arthur Koestler, and Irving Brown met to compare notes and strategize for the fight ahead.[8]

From the beginning, the speakers fell into two camps: those, like Ignazio Silone, who appealed to artists' and writers' cultural sensibilities as a bridge across political differences, and those, like Koestler, who condemned neutralism and urged intellectuals to lend their creative energies to a full-throated assault on Communism. Koestler's view prevailed in the CCF's official "Manifesto for Freedom," which he read to the cheering audience of 15,000 who attended the closing ceremonies in the gardens of the Funkturm Berlin on Thursday afternoon. Ever the propagandist, Koestler proclaimed, "Friends, freedom has seized the offensive!" In the months afterward, however, the CCF adopted Silone's position, with the most vocal anti-Communists, especially Hook and Koestler, sidelined in favor of what both Wisner and his colleagues at the State Department must have regarded as a more "subtle" approach.[9]

Nearly everyone involved deemed the event a tremendous success. Arthur Schlesinger, who was one of a handful of lay participants aware of the CIA's role, wrote to Averell Harriman (now Truman's special assistant for national security affairs) that the organization "could become an invaluable instrumentality in combating the Communist 'peace' drive in Europe and in fighting neutralism in general." General John Magruder, the Defense Department's representative on the 10/2 Committee, declared it "unconventional warfare at its best." Truman reportedly declared himself "very well pleased."[10]

By winter, Josselson and Wisner had settled on a workable institutional arrangement to sustain the Congress for Cultural Freedom. The project received an OPC codename, Project QKOPERA. At Wisner's insistence, Josselson ousted Lasky, whose affiliation with the Occupation government was well known. Josselson's title would change over the years, from administrative secretary to secretary of the executive committee to executive director, but his leadership remained constant. The public face of the CCF, in the role of secretary-general, would be Nicolas Nabokov, a Russian composer (and novelist Vladimir Nabokov's cousin) who had become a US citizen in 1939 and had, like Schlesinger, Lasky, and Josselson, worked in cultural affairs for the US government in Occupied Germany. The main offices of the CCF would move to Paris, with guiding policies set by an executive committee that would liaise with a series of national committees. Josselson and his deputies would keep Wisner's OPC in the loop, and none of the participants need be the wiser.[11]

Back in Washington, Wisner, Walter Bedell Smith, and Allen Dulles, who had joined the CIA as deputy director for plans in January 1951, experimented with an organizational structure to house programs like the CCF. They eventually decided on a new International Organizations Division headed by Thomas Braden to manage the subset of covert operations designed to undermine Communist fronts. Under Braden, a dashing Iowan veteran of the Office of Strategic

Services, the division expanded into a sophisticated apparatus that funded countless cultural campaigns, usually without their participants' knowledge. Braden relied on both real and fake foundations, usually headed by sympathetic US businessmen, to launder the CIA's funds. In the case of the CCF, that willing partner was the Farfield Foundation, a shell foundation headed by yeast magnate and noted philanthropist Julius Fleischmann. The CCF thrived under these conditions. Over the course of its seventeen-year history, the congress published a series of influential magazines, including *Encounter, Preuves, Cuadernos*, and *Quest*; sponsored arts and music festivals around the world; and, at its height, employed nearly 300 people in thirty-five countries.[12]

In concept and execution, the CCF owed its existence to Arthur Koestler. As author of *Darkness at Noon* and prime mover for *The God That Failed*, Koestler had forced the left to reckon with the high price of Stalin's version of Communism. Wisner, Braden, and others at the CIA surely found his essays on the Münzenberg Trust's methods instructive. But Koestler, disillusioned and frustrated with what he saw as the nascent organization's refusal to engage in direct confrontation, had begun asking too many questions. As late as the fall of 1950, Burnham had apparently hoped to share the CCF's secret with Koestler, twice urging him to come to Washington to meet with "some serious persons who have specific interests that intersect our own." In April 1951, Koestler attended a dinner party at Frank Wisner's house, where it became clear that both Wisner and the State Department's Robert Joyce, also in attendance, were well informed about the CCF's operations. But it seems that Koestler, who by now had purchased a farm on an island in the Delaware River, just north of New Hope, Pennsylvania, failed both this audition and a subsequent night of cocktails at Joyce's home. One month later, Burnham sent the CIA a scathing report on Koestler's trustworthiness, calling him "politically unreliable" and "neurotic," with an unhealthy fixation on " 'conspiracy.' " The Congress for Cultural Freedom, the gem of the CIA's cultural campaign against the Communists, would shine without Koestler's light.[13]

::

Science was present at the CCF's creation. The very first full session in Berlin, "Science and Totalitarianism," started off slowly with a lackluster paper by Oxford philosopher A. J. Ayer on the concept of liberty. But the second paper, delivered in H. J. Muller's inimitable style, invigorated the audience. Muller warned of the threats to science of "complete despotism, masquerading under the name of 'national socialism' or 'international communism.'" For his audience in Berlin, Muller customized his usual condemnations of Lysenko and Stalin with a reminder of scientists' role in Nazi atrocities, reflecting that his "own branch of science, genetics, was the most perverted and outraged of all, since in its place a tissue of lies was fabricated in support of the dictator's racist psychosis." After cataloguing Stalin's assaults on genetics, psychology, astronomy, and quantum physics, Muller stepped back to connect the practice of science to the broader project of freedom. In Muller's words, "this right to think differently, to question, and to *express* our disagreements is the primary moral basis for the development of science and, indeed, for all that is valuable in the intellectual life of man."[14]

The combination of Muller's stirring rhetoric and news of the Korean invasion prompted the next speaker, Austrian nuclear physicist Hans Thirring, to withdraw his previously circulated neutralist paper. The original version of Thirring's paper argued that the Soviet Union's propaganda would never translate into military action; now, in the light of events in Korea, he said, "I can no longer be certain that such is the case." Cheers, applause, and some heckling broke out. A paper by biologist Hans Nachtsheim returned to Muller's theme, acknowledging that he and his fellow German scientists "let ourselves be gagged, and bound hand and foot." Nachtsheim, who had experimented on children with epilepsy as a wartime employee of the Kaiser Wilhelm Institute for Anthropology, did not linger on his complicity in Nazi war crimes. Instead, he spent the bulk of his lecture time on what he considered to be a more pressing threat to scientific freedom: a telegram sent from the president of the (East) Berlin Academy of Science to Stalin, on the occasion of the leader's seventieth birthday, without the consent of the academy's members. In response, the Heidelberg economist Alfred Weber stood up and announced his resignation from the academy, again drawing cheers and applause.[15]

Science, according to these speakers, was a sort of canary-in-the-coalmine for the more general condition of freedom. It was both a part of culture and, because of its relationship to state power, remarkably sensitive to changes in the political environment. The CCF's organizers continually returned to these themes as they attempted to convert the momentum of Berlin into a permanent anti-Communist movement. The recruitment letter inviting a "small but representative number of men and women in the arts, professions, and sciences" to join the American Committee for Cultural Freedom, for example, went out over the signatures of Schlesinger, Hook, and Muller. As of November 1950, the list of US signatories to the Manifesto of Freedom included several geneticists, all of whom had participated in the Golden Jubilee of Genetics in September: Robert Cook, Richard Goldschmidt, Milislav Demerec, and Laurence Snyder. In fact, all of the signatories explicitly identified as scientists were geneticists. This was almost certainly Muller's doing, but it nonetheless underlines the central place of genetics in the Cold War struggle as envisioned by anti-Communists in the United States.[16]

Hook repeatedly tried to incorporate scientists more fully into the core of the CCF's US operations. He listed "freedom of science and culture" as the second-most important angle of the American Committee for Cultural Freedom's program, following only "peace and freedom." He had originally hoped to schedule a daylong meeting in Washington, DC, on the topic of scientific freedom, followed by a "nationwide conference of scientists on the question of defense of freedom of science." Neither of these came to pass. Hook did, however, convince Muller to serve as vice chairman of the American Committee. Although Muller attended few of the group's meetings in New York, he traveled to India and Japan on behalf of the CCF, helped recruit sympathetic scientists to the committee's larger work, and arranged for *Science* magazine to reprint his and Nachtsheim's Berlin speeches. With Muller's help, the US group created a Committee on Intellectual Freedom, whose members included Eugene Rabinowitch, editor of the *Bulletin of the Atomic Scientists*. Muller kept up a lively correspondence with Hook's office on various topics relating to scientific freedom for the next decade, feeding him, among other things, confidential committee reports from the Genetics Society of America's wonderfully named Committee on Anti-Genetics Propaganda.[17]

At a meeting in Paris in 1952, the CCF leadership committed to a large-scale meeting on "Science and Freedom" to be held in Hamburg in the summer of 1953. With Muller's help, Hook, Nabokov, and Josselson assembled an impressive roster of anti-Communist scientists willing to lend their names and, in some cases, their time to the CCF's cause, with Polanyi, Theodosius Dobzhansky, and Nachtsheim listed as organizers. This time, the geneticists and social scientists were joined by physicists, including Henry Margenau and Koestler's friend Alex Weissberg, who had ended up in Germany as part of a prisoner exchange between Stalin and Hitler. Margenau, in turn, helped the CCF reach out to Manhattan Project veterans, with J. Robert Oppenheimer, Arthur Compton, and James Franck agreeing to lend their names as "honorary sponsors." Given the growing climate of anti-Communism at home, the organizers had to reassure the physicists, who had closer relationships with the government than did the biologists, about the presence of so many ex-Communists on the guest list. Margenau addressed this concern head-on in his recruitment letters, reassuring Franck, for example, that the meeting in Hamburg would be "entirely non-political" and "hostile to Communism." (He did not explain how a meeting could be both of these things at once.)[18]

The CCF encountered similar skepticism from the Rockefeller Foundation's Warren Weaver, who at first expressed "confusion" over the true "overtones and undertones" of the proceedings. State's Joseph Koepfli cleared up the confusion with a phone call, telling Weaver that "our government is definitely and sympathetically interested in this meeting, that they view it as a constructive move, and that they are going to assure that all delegates going from this country are sound and sensible defenders of domestic liberties." The State Department, Koepfli reported, was "in favor" of the meeting and would be glad to see the Rockefeller Foundation's support. With Koepfli's encouragement, the Rockefeller Foundation contributed $15,000 to the Hamburg conference's $50,000 budget, with the Farfield Foundation (that is, the CIA) and the City of Hamburg making up the balance.[19]

The CCF put this generous support (almost half a million dollars in today's currency) to good use, bringing together 110 scientists from nineteen countries from July 23 to July 26, 1953. The conference program's heavy cream-colored stock announced the CCF's commitment to science and freedom in German, French, and English in bold orange print. An opening address by Michael Polanyi set the tone, announcing an explicit goal to rouse "academic opinion more widely against the treatment of scholars and of scholarship under totalitarianism." For three days, the assembled scientists debated the appropriate balance between pure and applied research, academic freedom in an age of state-supported science, the place of science in dialectical materialism, and scientists' moral responsibility, with simultaneous translators making sure that all present could understand the message. Once again, those who witnessed the event deemed it a tremendous success.[20]

:: ::

Michael Polanyi served as the official chair of both the Committee on Science and Freedom and the Hamburg conference. A Hungarian physical chemist, Polanyi had arrived at the University of Manchester in 1933 by way of Berlin and by 1953 had established himself as one of the leading thinkers on the relationship between science and freedom under liberalism. His interest in this topic had become so consuming, in fact, that in 1948 he switched his affiliation at the University of Manchester from the chemistry department to the social studies department. During and after the war, Polanyi had promoted his views on science and freedom through the British Society for Freedom in Science, a small but influential group of scholars that he and zoologist John Baker had assembled to halt the influence of J. D. Bernal. Its members shared an assumption that government planning, in any form, posed a threat to scientists' ability to control their own research agendas.[21]

Polanyi developed this opposition to planning—scientific or otherwise—into a full-throated defense of scientific freedom that paralleled, but was distinct from, an American tradition of scientific individualism. In Polanyi's telling of the tragedy of Soviet genetics, the problem was less about the influence of a particular individual (Trofim Lysenko) and more about how the Soviet state's insistence on practical agricultural applications had distorted an entire field. Polanyi had explored these ideas in an essay originally published in 1942, but they did not attract much attention outside the United Kingdom until 1953, when he presented a version of the paper at the Hamburg conference. By then, Polanyi had a well-earned reputation as an enemy of Communism—in 1945, Koestler dedicated *The Yogi and the Commissar* to him. Polanyi had also proved useful to the CCF, with his membership on the executive committee a critical element in securing the Rockefeller Foundation's partial funding for the Hamburg conference. Combined with his connections to a preexisting, organic group of scientists opposed to Communism, Polanyi's attempts to articulate a vision of how science might operate in a free society made him an obvious figurehead for the CCF's efforts in this field.[22]

By the summer of 1954, Nabokov had convinced Polanyi that his leadership was essential to the nascent Committee on Science and Freedom's success. On paper, Polanyi would be joined by several non-Communist scientists and philosophers of science, including Hook, Dobzhansky, Rabinowitch, and Edward Shils. In practice, this was no more than a paper committee, with publications and conference planning issuing directly from the cottage of George Polanyi—Michael's son—who served as the organization's secretary and editor.[23]

George Polanyi's acceptance of the position of secretary of the Committee on Science and Freedom in June 1954 marked a major milestone in the committee's operation. For £350 a year, George Polanyi would identify ways to "take up the contacts which were established at the Congress in July 1953 . . . and to foster and extend these contacts so that the initiative taken at that Congress shall not be extinguished." George Polanyi envisioned a monthly bulletin, filled with "short, factual pieces" on topics ranging from government funding of scientific research and interdisciplinarity to obtaining visas to visit the United States. Following in his father's footsteps, Polanyi viewed the group's primary concern as not so much the struggle against Communism, per se, but rather "a struggle against excessive government power and against the 'ivory tower' outlook." He volunteered the services of his wife, Priscilla, as typist and secretary.[24]

In July, a core group of CCF staff and supporters met in Paris to iron out the committee's mandate. The Committee on Science and Freedom, it was agreed, would primarily focus on the themes discussed at the Hamburg conference, especially the different scientific research practices in totalitarian and Western countries, the changing nature of scientific freedom, the "international character" of science, and scientists' moral responsibilities in the atomic age (hence the involvement of Rabinowitch). Instead of a monthly publication schedule, the bulletin would be published irregularly, as needed, and would occasionally be supplemented by pamphlets.[25]

From that point on, Josselson delegated much of the work of supervising the Committee on Science and Freedom to his deputy secretaries, also supplied by the CIA—first Warren Manshel (who would later found both *Public Interest* and *Foreign Policy*), later novelist John Hunt. Michael Polanyi, too, mostly stepped aside, effectively leaving George Polanyi in charge of the committee and its bulletin. From the beginning, Manshel and George Polanyi engaged in a passive-aggressive power struggle over *Science and Freedom*'s management. The correspondence between the two men contains a seemingly endless series of letters in which someone from Paris reminds the younger Polanyi to update him on the committee's progress, with varying degrees of pique. Polanyi's work on the second issue of the bulletin was typical. In December 1954, Manshel inquired, "What progress are you making with your plans for the second issue of the bulletin?" Three months later, Manshel reminded Polanyi that he had not yet answered: "Don't make a mystery of the second issue of the Bulletin . . . Can't you let us have *some* advance information?" By the time Polanyi finally sent Manshel galleys in April, it was too late to make changes. "Frankly," Manshel wrote, "I should very much have liked to have seen the contents of this issue a little sooner."[26]

Whether the Paris office distrusted the Polanyis or simply questioned their editorial skills, the CCF secretariat consistently attempted to steer the Committee on Science and Freedom with a strong hand. Josselson nixed George and Priscilla's original plan for the first issue, which would have consisted solely of a response to a critique of the Hamburg conference that had appeared in the French Communist newspaper *Nouvelle Critique*. After several letters and a phone call from Irving Kristol, Josselson reached "perfect agreement" with George Polanyi as to the eventual content, which included an introductory letter from Michael Polanyi, a reprint of Shils's report on the Hamburg congress that had originally appeared in the *Bulletin of the Atomic Scientists*, and a response to the *Nouvelle Critique* piece. Thereafter, the secretariat sent George Polanyi a steady stream of reprints, newspaper clippings, conference reports, and potential books for review, all for consideration in the bulletin. Polanyi, for the most part, ignored these suggestions, only sometimes bothering to explain why.[27]

The result was that the range of issues addressed in *Science and Freedom* hewed more closely to the Polanyis' political interests than to the secretariat's. Given George Polanyi's resistance to following directions and the (chimeric) committee's overall lack of activity, Josselson and the CCF's executive committee closely examined the documents that Polanyi *did* produce for evidence of his intent. The more they saw, the less convinced they were that the Polanyis' vision meshed with the CCF's larger cultural and political agenda. Instead of either searching for a sustainable left or battling against global Communism, the Polanyis seemed dead set on criticizing restrictions on academic freedom *in the West*.

From the beginning, *père et fils* centered their project on the question of academic freedom. The crisis of Lysenkoism, with its contrast between scientific belief and political authority, crystallized postwar anxiety about whether, and to what extent, scientists controlled their own work. But as Michael Polanyi saw it, the crisis in science was different only in degree, not in kind, from the broader crisis of political authority afflicting intellectuals. The Committee on Science and Freedom would take "the problems of the academic community in maintaining its independent status in society" as its central issue, with the specific problems of science standing in for those of the academy as a whole.[28]

This focus soon became apparent in *Science and Freedom*. The first two issues, published in November 1954 and April 1955, defined the problem of "science and freedom" broadly enough to encompass the role of the scientist in society in both Communist and non-Communist countries. The contents of the third issue, however, on faculty protests at the University of Göttingen following the appointment of an accused neo-Nazi as rector, were more representative of those that followed. Subsequent issues dealt with matters of academic freedom, narrowly defined as either a faculty's right to self-governance or the right of individual faculty to diverge from administrative policy at institutions as far-flung as the University of Alabama, the University of Tasmania, and the University of Cape Town.[29]

By the summer of 1958, the Polanyis had published twelve issues of *Science and Freedom*, each of which typically included four or five articles. Of these fifty or so articles, only a handful dealt with the conditions of scholars working under Communist regimes, usually based on reports of Soviet scholars' presentations at academic conferences. Much more typical were articles that took up questions of scientific freedom and responsibility in a postwar system that assumed the centrality of science in building and maintaining state power. How much self-governance could university scholars expect, for example, at publicly funded universities in democratic systems? Given security considerations in the United States and the United Kingdom, should scientists be subject to loyalty oaths? The two major exceptions were bulletin issues devoted to universities under Franco and the committee's role in organizing a telegram campaign protesting the treatment of scholars during the failed Hungarian Revolution.[30]

The divergences between the Polanyis' goals and those of the CCF became particularly evident in two incidents, one involving a follow-up meeting to the Hamburg congress, and the other involving an article on science in China. In early 1956, George Polanyi proposed a follow-up study group to the Hamburg meeting to more fully "develop a strategy of freedom." His proposed agenda, however, surprised both Josselson and Hook. Of the twenty-four potential topics George Polanyi suggested for discussion, only one dealt with academic freedom in the Soviet Union. The rest involved limitations on academic freedom in the West, whether in the form of security clearances, academic self-governance, or the role of government financing. Josselson balked, noting that the premise went "far beyond the plan which was discussed here in Paris." Hook was more direct, declaring himself "dumbfounded" by the agenda and warning Polanyi against "taking the state of academic freedom of the entire world" for the committee's agenda. George Polanyi nevertheless proceeded to invite high-profile speakers; faced with a *fait accompli*, Josselson found the funds to make the meeting happen. The agenda of the study group that met outside Paris in August 1956, while slightly smaller than originally planned, otherwise hewed fairly closely to Polanyi's original proposal, with papers on scientific freedom and responsibility, the limits of academic freedom in publicly funded institutions, and the importance of faculty self-governance. Critiques of academic freedom under Communism looked no further east than Yugoslavia.[31]

A similar situation unfurled in 1960 when, after much prodding, George Polanyi finally ran an article on the situation of scholars working in Chinese universities. The article, by anthropologist William Newell, had been recommended by Irving Kristol. As was his habit, Polanyi apparently ran it without checking first with anyone in the Paris office. But what Polanyi regarded as an even-handed, perhaps even ironic, take on the nature of academic freedom, Hunt—who had taken over Manshel's role as Josselson's deputy in 1956—perceived as a massive failure of judgment. Nearly beside himself ("I can't believe my eyes"), Hunt took Polanyi to task for running an article implying that "the Chinese academic world is better off under Mao than some of the University teachers in other countries, even non-Communist countries." In the future, Hunt warned, it was "imperative" that Polanyi check with the CCF's Sino-Soviet experts, including Walter Laqueur, Roderick MacFarquhar, and Leopold Labedz, before publishing anything on this topic.[32]

::

Hunt might have taken comfort in the fact that so few people were reading Polanyi's renegade text. The Paris secretariat originally had high hopes for the bulletin's circulation. In January 1955, shortly after publication of the first issue, Josselson, Manshel, Hook, and Michael Polanyi agreed that the ultimate goal should be a circulation of 10,000, distributed free of charge. The first issue was sent to approximately 900 people, including all attendees at the Hamburg conference, people personally recommended by those in attendance, members of the British Society for Freedom in Science, and additional persons suggested by the secretariat. Sol Stein, of the American Committee for Cultural Freedom, requested a standing order of 400 copies; anthropologist Herbert Passin, who represented the CCF in Asia, distributed an additional 600 copies in Japan. But despite George and Priscilla Polanyi's systematic efforts over the next seven years to identify potential subscribers, the mailing list never climbed much above 4,000.[33]

George and Priscilla Polanyi saw their marketing efforts as inextricable from the bulletin's editorial content. Most issues of *Science and Freedom* focused on incidents in a specific geographic area. Priscilla Polanyi used *World of Learning*, a directory of staff and faculty at educational facilities around the world, to identify potential subscribers in the countries or regions mentioned in the bulletin's text. Issue 7, for instance, with its focus on faculty self-governance in British Commonwealth universities, was sent to university deans and department heads in all Commonwealth nations. Similarly, she sent a special mailing of the Göttingen issue to all university and department heads in all universities throughout both East and West Germany—more than 600 in all. Each recipient additionally received a letter querying whether further contact was desired. Those who responded positively were added to the mailing list.[34]

The letters received by potential subscribers (including Paul Koenig) were direct but not particularly informative, simply indicating that future issues would be available gratis. That being said, a surprising number of recipients wrote back to share their impressions of the materials, offering a rare glimpse of how the CCF's targets interpreted its propaganda. As one might expect, most of the readers who bothered to share an opinion agreed with the editors' perspective on science and freedom. The Canadian M. E. McKinnon's comment, "I suppose exposure of weakness and evil is the only means by which one might hope to stem the tide," was unusual only for its bluntness. A stubborn few, however, registered complaints about *Science and Freedom*'s narrow ideological focus. Alexander Szalai, a Hungarian sociologist who had known Michael Polanyi's mother, cautioned that "it is surely a pity that the 'East' is in no way represented in this group." Of all those who took the time to write, Marcel Minnaert, a Belgian-born Dutch astronomer who had been imprisoned by the Nazis during World War II, articulated these objections the most clearly: "It seems to be basically wrong to present 'lack of freedom' as the essential characteristic of the Soviet Union. I have not the impression that your contributors are seriously trying to understand the Eastern point of view and I am afraid that their papers will not contribute toward mutual understanding between the West and the East." Minnaert, in other words, wanted more Silone, less Koestler.[35]

For the most part, the Committee on Science and Freedom lagged behind the CCF's broader attempts to engage the Global South in the mid-1950s, a phenomenon noted by *Science and Freedom*'s readers. A number of subscribers in newly independent countries, for example, wrote in to request clear-headed, empirical studies of the respective virtues of communist and capitalist systems in promoting scientific development. P. Gisbert, a Jesuit priest in the sociology department at St. Xavier's College, Bombay, pointed out that the Soviet Union's "very rapid strides in scientific and technological knowledge and production" were quite attractive to leaders in developing nations. Would it not be wise, he asked, to explore how this was possible "given the lack of freedom for scientific enquiry in those nations?"[36]

Eventually—perhaps inevitably—the tensions between the Polanyis and the Paris secretariat began to interfere with the operations of the Committee on Science and Freedom. Over the years, George Polanyi's "administrative slip-ups" had ranged from inviting the wrong people to participate in meetings, to accusing certain useful German scientists of being neo-Nazis, ignoring the secretariat's request for coverage of certain topics, and refusing to submit draft issues of the committee's publications. By the summer of 1958, George Polanyi was sending Josselson a steady stream of letters complaining about what he and Priscilla saw as the secretariat's lack of support for and interest in the committee's operations. As he pointed out, the committee "as such" had never met. Neither Michael, nor George, nor Priscilla had ever been invited to the CCF's editorial meetings; Josselson's repeated rebuffs to George's requests made clear that he had no intention of changing that policy in the future. Hunt's response to these demands was to demand *more* accountability, not less: quarterly publication of *Science and Freedom*, a higher circulation, and monthly reports.[37]

After yet more misunderstandings and administrative miscommunications, Hunt finally shut down the Committee on Science and Freedom's Manchester operations in May 1961. Hunt and Josselson managed the delicate task of maintaining Michael Polanyi's cooperation by reminding him of the value of science to the CCF's larger operations; since *Science and Freedom* had proved disappointing, a new approach with a new editorial staff was necessary. Hunt installed Edward Shils as the new chair of the committee, and *Science and Freedom* ceased publication. In its stead, Shils developed the journal *Minerva*, which still exists today. To a surprising extent, *Minerva*'s content furthered the editorial agenda of *Science and Freedom*, particularly in its attention to the moral responsibility of the scientific community and the sociology of "Big Science." In terms of the new journal's professionalism, predictability, and sophistication, however, the two publications could not have been more different.[38]

∷

Whatever the secretariat's relationship to the CCF's other publications, it is abundantly clear that Josselson, and therefore the CIA, had hoped to keep *Sci-*

ence and Freedom on a relatively short editorial leash. It is equally clear that the Polanyis just as consistently attempted to evade that direction. One wonders about Josselson's intent in isolating George and Priscilla Polanyi from the Paris office's operations. Did he see them as so hopelessly incompetent that they couldn't be trusted with even cursory information on the CCF's larger operations? Did he hope that they would eventually grow so frustrated that they would quit of their own accord? Or did he simply find the Polanyis' contribution to the fight against Communism worth less than the cost of a couple of plane tickets to Paris?

Given the relative lack of accounts considering the role of science in cultural diplomacy, the last explanation is the most tempting. It is also wrong. Josselson's push to retain Michael Polanyi as a member of the CCF's executive committee, before, during, and after the debacle of the Manchester-based Committee on Science and Freedom, demonstrates the value that the secretariat placed on having a "scientific member," as Nabokov put it, privy to its larger goals. The subsequent establishment of *Minerva* moreover suggests that Josselson and his colleagues saw a role for science in the covert cultural Cold War well into the 1960s. Science never took on the central role in the CCF's operations that Michael Polanyi or Michael Josselson might originally have envisioned for it, but it did play *a* role, and the evidence suggests that policymakers wanted it to play a larger one.[39]

The CCF's insistence on treating the Polanyis as the representatives of a nonexistent organization (the Committee on Science and Freedom) rather than as employees limited the secretariat's ability to discipline their actions. *Science and Freedom* was not a congress publication in the same way that its more successful publications, including *Encounter*, *Preuves*, and *Quest*, were. As of March 1956, George Polanyi was being paid only £350 a year, and Priscilla Polanyi received no compensation whatsoever for her services; by May 1959, George Polanyi was claiming that he, too, was not being compensated for his work. Can you fire a volunteer? While the secretariat understood *Science and Freedom* as a CCF journal, the Polanyis regarded it as a publication by the Committee of Science and Freedom, printed and circulated with the assistance of the congress. The "editors" could not be replaced without ending the publication itself.[40]

Science and Freedom simultaneously was and was not a "house organ of the Polanyis." The one-sided nature of the OPC's relationships with its private partners provided the US government with plausible deniability but limited its options for control. The CIA tolerated this situation until the entire apparatus self-destructed in 1967, but other arms of the US government were less patient. By the mid-1950s, policymakers from President Eisenhower on down increasingly desired a more visibly successful psychological warfare program, especially in science. And with the launch of the Soviet *Sputnik* in 1957, they finally got it.

5 : Science for Peace

The Soviet Union's successful launch of *Sputnik*, the world's first artificial satellite, on October 4, 1957, was announced in the United States by a most unlikely figure: Lloyd Berkner. That evening, Berkner and other space and rocket specialists had gathered for a cocktail reception at the Soviet Embassy to celebrate international scientific cooperation. But before Berkner could take his first sip, a reporter from the *New York Times* pulled him aside and told him that Moscow radio had just reported the launch. With permission from his host, Soviet General Anatoli Blagonravov, Berkner tapped his glass and congratulated his Soviet colleagues on their achievement.[1]

Promoting Soviet space spectaculars was presumably not what Berkner had in mind when he suggested, nearly a decade earlier, that the United States embrace international scientific cooperation to advance its foreign policy interests. Instead, Berkner and his like-minded colleagues at the State Department and the CIA had always intended this stated commitment to scientific openness to foster intelligence collection and to promote the United States' image as a beacon of freedom. The national humiliation of the *Sputnik* launch tempted the US government to change course, to officially devote itself to a race for first place in science and technology. The federal government did, in fact, ramp up funding levels for basic research, science education, rocket and missile development, and nuclear weapons in the late 1950s and early 1960s. But at the same time, the US foreign policy establishment doubled down on promoting scientific internationalism, scientific cooperation, and scientific freedom as a cultural front in the Cold War.[2]

President Eisenhower ended his 1958 State of the Union address, delivered just three months after the launch, with a rousing call for a US-led but international "Science for Peace" program that would demonstrate the US commitment to "waging total peace" throughout the world. "There is almost no limit," Eisenhower proclaimed, "to the human betterment that could result from such cooperation." The US Information Agency (USIA) announced that 1958 would be a "Year of Science." A host of overt government agencies, from the Atomic

Energy Commission to the National Science Foundation, brainstormed ways to signal the United States' commitment to global peace through scientific programming, and nongovernmental groups such as the National Academy of Sciences, the Ford Foundation, and the Rockefeller Foundation asked how they could help. Behind closed doors, the decision-makers at a relatively new government entity, the Operations Coordinating Board (OCB), discussed the best ways to coordinate these overt programs with the CIA's ongoing covert work on behalf of science. With the launch of *Sputnik*, the time had come to fully incorporate science into the United States' global psychological strategy.[3]

Collectively, in the late 1950s and early 1960s, the engines of US cultural diplomacy dedicated themselves to, as one strategic document memorably put it, "Strengthening the Free World Position in Science and Technology." Even as individual agencies argued over the content of these projects, there was general agreement that US-based science programming should continue to appear to be the work of nongovernmental entities. This insistence on arm's-length diplomacy would have dramatic consequences for US diplomacy involving science for the rest of the Cold War. The story of how and why the US government decided to promote the country's own values by hiding its hand, even in propaganda conducted by overt agencies, begins in 1953, when newly elected President Eisenhower attempted to make sense of the array of psychological warfare programs he inherited from the Truman administration.[4]

∷

Dwight D. Eisenhower arrived in office an enthusiastic supporter of psychological warfare. As supreme commander of the Allied Expeditionary Force in Europe during World War II, General Eisenhower had witnessed the power of psychological strategies to shape the outcome of military campaigns, hold alliances together, and undermine enemy morale. He testified in Congress in favor of the Smith-Mundt Act and, as commander of NATO, urged the CIA's Walter Bedell Smith to undertake the kinds of covert political activity that, unbeknownst to Eisenhower, the organization had already begun on its own. Nor was this a quiet passion. Eisenhower made an expanded US program for psychological warfare and propaganda a central theme of his campaign for the presidency, telling audiences that the country "must develop fully . . . every psychological weapon that is available to us."[5]

The new Republican administration inherited a massive but uncoordinated factory for producing psychological weapons that ranged from the CIA's ranks of private partners to the Marshall Plan's vast economic resources and Voice of America's powerful radio towers. Eisenhower installed his wizard of psychological warfare from World War II, *Time-Life* executive C. D. Jackson, as "Special Assistant for Cold War Planning" (informally known as the psychological warfare advisor). He also immediately appointed Nelson Rockefeller as chair of a special study group, the President's Advisory Committee on Government Organization, which recommended that the government's overt propaganda campaigns be moved out of the State Department. Eisenhower agreed, creating the USIA in the summer of 1953.

A second advisory committee—known ambiguously as the Jackson Committee, either for its chair, former intelligence officer William Jackson, or for its most vocal member, C. D. Jackson—recommended in June that the dysfunctional Psychological Strategy Board be replaced with a new entity, the Operations Coordinating Board. Eisenhower accepted this recommendation as well. The OCB would report directly to the National Security Council to "provide for the coordinated execution of approved national security policies," particularly as related to psychological strategy. A smaller 10/2 Committee, now under the authority of the OCB, would continue to oversee the most sensitive covert operations.[6]

The Jackson Committee's report moved psychological operations to the heart of US Cold War strategy. "We find that the 'psychological' aspect of policy is not separable from policy," the report explained, "but is inherent in every diplomatic, economic, or military action." The authors focused on the role of propaganda in establishing a "climate of opinion" that would allow US foreign policy objectives to succeed. Effective propaganda must therefore be "dependable, convincing, and truthful." In its attempt to undermine support for Communism in Eastern Europe and the USSR, the United States had neglected its current and potential allies in the "free world"; the revised program should focus on convincing free people the world over that they stood to benefit from US power. In practice, as outlined in the committee's list of more than forty recommendations, this would mean more unattributed ("gray") propaganda— ideally, produced in partnership with either private organizations or "indigenous" groups—and more propaganda of the deed to back up US rhetoric.[7]

The US government's embrace of an expanded psychological warfare program, as well as the existence of a new propaganda agency, the USIA, forced policymakers to confront a question they had long avoided: what, exactly, did it mean to refer to an "American ideology"? In May 1954, the OCB created a working group on the "US Doctrinal Program" to figure it out. On the advice of the State Department, the working group explicitly stated that the United States did not intend to develop a particular doctrine of freedom. Instead, the program would cultivate a better understanding of "the traditions and viewpoints of America and the Free World" (always capitalized) among target audiences. Accordingly, the program should spend as much time in developing "positive or pro-Western" materials as in distributing anti-Communist literature. A bibliography attached to an earlier USIA proposal gives some specificity to this vague notion of "freedom": the list of thirty-four titles, all written by white men, leaned heavily on US political biography and the history of liberal philosophy. James Conant's *Science and Common Sense*, Frederick Stern's *Capitalism in America*, and Carl Becker's *Heavenly City of the Eighteenth-Century Philosophers* made the list; the *Narrative of the Life of Frederick Douglass* did not.[8]

As a program explicitly focused on the intellectual foundations of the Western world, the US Ideological Program (so renamed in 1955) targeted "present and potential leaders of intellectual and official life," a category that included historians, philosophers, theologians, psychologists, anthropologists, physicists, biologists, chemists, labor leaders, politicians, and journalists. The historians, philosophers, and social scientists might be convinced through books, but persuading scientists and technical professionals—particularly those in newly independent nations—would require compelling actions and personal contact to convince them of the benefits of US-style freedom. To reach this audience, the group considered increasing US participation in international conferences, convincing international scholars to attend conferences in the United States, or briefing influential US citizens on the virtues of freedom before they traveled abroad. Once again, the OCB recommended that as many of these programs as possible be carried out by private agencies, both to reduce costs and to underscore the point that, in the United States, individuals themselves—not the government—controlled their destiny.[9]

The concepts outlined in the Ideological Program drove US propaganda efforts for the rest of the Eisenhower administration. Even as some overseas embassy officials questioned such a broad definition of "freedom," the State Department reported that ideological planning had become part of its "normal operations." The USIA, meanwhile, had distributed almost a million books and pamphlets either critiquing Communism or explaining principles of US democracy. But the objectives in science, which depended on US participation in international scientific events, hit a roadblock: no one wanted to pay for it.[10]

: :

Since 1950, when the Berkner Report recommended that the United States use international scientific contacts to gather scientific intelligence, both the NAS and the NSF had been trying to locate stable sources of funding to send US scientists abroad, with limited success. Officials at the NSF believed that government agencies should pay for activities designed to benefit the government. And by "the government," the NSF's director, Alan Waterman, meant the NSF. But in the summer of 1955, members of Congress made clear to Waterman that they considered international travel a misuse of foundation funds. The NSF should fund scientific research, not international travel.[11]

The NAS, meanwhile, saw its own opportunities. At one point early in Detlev Bronk's tenure as president, the academy suggested it should organize US attendance at every international science conference. The NAS hoped to fund this ambitious program with continuing grants from the State Department and the NSF and a new grant from the Ford Foundation. But by the time the Ford Foundation produced a four-year, $400,000 grant in the summer of 1956, the academy needed the money to pick up the State Department's slack. The position of science advisor had been vacant since early 1954; the last science attaché appointments expired in early 1956. State's entire science operation had withered to a single career service officer, trained as an economist, supported by two secretaries. The NAS advised the government on matters of international science policy, paid US dues to international scientific organizations, answered day-to-day queries from scientists on visas and passports, briefed US scientific delegations prior to their travel, and debriefed them upon their return—all activities that, under any normal circumstances, should have been the responsibility of the State Department. Unless and until Congress was willing to fund these efforts through either the State Department or the NSF, the NAS—a quasi-governmental organization—was the US government's de facto agency for international science.[12]

At least until the launch of *Sputnik*, those in charge of the government's purse strings refused to see the point of a major international effort driven solely by the values of science. The CIA's relatively small-scale cultural operations in science, such as the Congress for Cultural Freedom's Committee on Science and Freedom, bypassed congressional budget scrutiny. There were, however, two major exceptions during this period that suggested that science-based propaganda campaigns could be justified—but only if they simultaneously advanced more concrete national security goals.

On December 8, 1953, President Eisenhower delivered a major speech before the General Assembly of the United Nations proposing a new, cooperative international program devoted to the uses of atomic energy and radioisotopes in medicine, agriculture, and power generation. "Atoms for Peace" became a centerpiece of the Eisenhower administration's psychological warfare operations, the USIA's "top priority" for 1954. In 1955 alone, the USIA claimed that more than 5 million people visited Atoms for Peace–related exhibitions in Western Europe, Brazil, India, and Pakistan; the following year, the agency distributed more than 6 million pamphlets on the benefits of atomic energy in thirty-four languages. The OCB created a working group to coordinate the massive interagency propaganda effort, discussing issues as large as bilateral research agreements and details as small as the best design for an Atoms for Peace postage stamp. One historian has described the effort as "quite possibly the largest single propaganda campaign ever conducted by the American government."[13]

Why would the US government invest millions of dollars in an international propaganda campaign on behalf of nuclear power and medicine if it couldn't even be bothered to appoint science attachés? Quite simply, it did so because Atoms for Peace also advanced goals more directly related to national security. Eisenhower's plans called for nuclear nations to deposit some quantity of fissionable materials—the exact amount subject to negotiation—in a "bank" controlled by a new International Atomic Energy Agency. Working on the assumption that the Soviet Union had limited access to uranium, US intelligence officials saw no potential downsides for Atoms for Peace. If the Soviets cooperated, the program would divert a substantial quantity of fissionable materials from nuclear weapons production; if they refused, the United States would have scored a major propaganda victory. The publicity campaign associated with the practical uses of atomic energy in medicine and agriculture, meanwhile, would distract global critics from the United States' ramped-up nuclear testing program in the Pacific. And all of these benefits would manifest at the same time that international conferences associated with Atoms for Peace would create a new channel for gathering valuable scientific intelligence.[14]

Atoms for Peace combined economic, military, political, and psychological objectives, just as the Jackson Report recommended. And yet, the campaign's success threatened to undermine US credibility in multiple ways. An observer reported that the 1955 Geneva Conference on Peaceful Uses of the Atom "uncorked the security bottle so that it can never be closed tight again," a result that "delighted" US scientists but worried administrators at the AEC, who feared losing control over research on the dangers of fallout. This tension reached a dramatic head when the AEC attempted to prohibit H. J. Muller, of all people, from delivering a paper on the genetic dangers of radiation. Muller therefore delivered his remarks from the floor, to general applause from the assembled scientists and the delight of the international press. The incident left the AEC, the USIA, and geneticists all complaining about different kinds of bad publicity—non-nuclear fallout—that resulted from the incident. Fortunately for the AEC, Muller—always the savvy ideologue—declined George Polanyi's misguided invitation to write up his experience of being muzzled by the US government for *Science and Freedom*.[15]

An even greater challenge to Atoms for Peace came from the program's inability to deliver on its promises. "Popular enthusiasm for power reactors poses a problem," one OCB report noted, because of the "time lag" in visible accomplishments. The USIA's phenomenally successful information campaign on the wonders of cheap electrical power created the risk of "popular disillusionment," particularly in developing countries like India that were "hot" for nuclear power. Authorities in the United States were so concerned about this issue that they considered launching a "World Solar Energy Project" as a more viable alternative. In promising too much, Atoms for Peace had violated an important tenet of the Jackson Report: propaganda should be "dependable, convincing, and truthful."[16]

The Eisenhower administration's second large-scale international science program failed more dramatically. In the spring of 1950, the same time that he was drafting "Science and Foreign Relations," Berkner proposed an international collaborative research project modeled on a prior program known as the International Polar Year. With the support of his contacts at the State Department and the CIA, Berkner began assembling international support at the International Council of Scientific Unions (ICSU) for what became known as the International Geophysical Year (IGY). In 1954, under Berkner's leadership, the ICSU officially endorsed a program that committed sixty-seven nations to a global study of meteorology, geology, geography, oceanography, the polar regions, solar activity, and the earth's atmosphere, which would last from July 1, 1957, to December 31, 1958. Not coincidentally, many of the scientific studies promised to produce the sorts of geophysical measurements on which contemporary warfare depended. Berkner's connections and interests ensured that the IGY's scientific goals were deeply intertwined with national security concerns.[17]

In 1955, both the United States and the Soviet Union announced plans to launch artificial satellites as part of the IGY. Each country assumed it would be first. For the United States, launching a peaceful satellite would serve multiple purposes, from establishing the freedom of the skies (a necessary step toward ensuring the legality of the military's planned reconnaissance satellites) to fine-tuning data needed for launching missiles and demonstrating US scientific prowess. In a discussion designated TOP SECRET, the OCB working group for the satellite program reiterated that the "civilian and scientific" parts of the program should be "maximized . . . in every practicable way"; the military aspects "minimized." "The stake of prestige that is involved," Nelson Rockefeller noted in registering his support for the $20 million effort, "makes this a race that we cannot afford to lose." On that, he turned out to be right.[18]

::

News of the successful Soviet launch of a 184-pound satellite in October 1957 wasn't exactly a surprise at the highest echelons of US power. Intelligence channels had suggested a launch was imminent; over the summer, the OCB had reviewed multiple drafts of a press release to be issued by the NAS in the event that the Soviets beat the Americans to the punch. But Rockefeller's comments aside, Eisenhower's psychological warfare team hadn't anticipated the damage a Soviet satellite launch would do to the United States' international reputation. In marked contrast to the United States' commitment to using civilian equipment for its IGY programs, the Soviets lofted their satellite into orbit atop an intercontinental ballistic missile. Newspapers and television reports both in and outside the United States repeatedly pointed this out in headlines stoking fears that the West was falling behind, in both space and national defense.[19]

Merely two weeks after the launch, the USIA reported disturbing signs of lasting damage to the United States' reputation. News editors in Mexico had already indicated a new preference for Soviet-issued rather than USIA-drafted feature articles on science and technology, and Iranian officials were avoiding their US counterparts to minimize potentially embarrassing encounters. In a more detailed survey of world opinion, the USIA reported that Soviet claims of scientific and technical superiority, and Soviet propaganda in general, had gained credibility and that allies openly wondered whether the balance of military power had shifted to the Soviet Union. "American prestige is viewed as having sustained a severe blow," the USIA reported. The general panic among the US public exacerbated "the disquiet of friendly countries," giving the Soviets an unexpected bonus round of free propaganda.[20]

The situation forced US authorities to take prestige seriously, with lasting consequences for both domestic and international science policy. Both the Truman and Eisenhower administrations had resisted previous attempts to coordinate science and technology, fearing the whiff of centralized planning—that boogeyman of anti-Communism. As of October 1957, the United States lacked a domestic science advisor, a scientific advisory board, or even a formal interagency committee to manage the government's efforts in science and technology. It gained the first two of these within a month of the Soviet launch; the third would take more time. On November 7, Eisenhower named James Killian, president of MIT, as his special assistant for science and technology. Killian's duties included chairing the new President's Science Advisory Committee (PSAC). Eisenhower's new group of scientific advisors spent much of the following year debating the merits of establishing a cabinet-level department of science to manage the scientific work of a spectrum of government agencies. In the end, though, the fear of political control over science doomed plans for a department of science. Instead, in March 1959, nearly eighteen months after *Sputnik*, the president created the less powerful Federal Council on Science and Technology to coordinate interagency science policy.[21]

At the State Department, administrators' new appreciation for international scientific prestige meant appointments in the long-neglected Office of the Science Advisor. On January 13, 1958, Secretary of State John Foster Dulles swore in Wallace Brode, former science advisor to the CIA, as his new science advisor. Brode occupied a more senior role than his predecessor, Joseph Koepfli, who officially advised the undersecretary; Brode understood his role to be "parallel to the position of Dr. Killian as Science Advisor to the President." The appointment was greeted with fanfare, with Brode's deep-set eyes and bushy brows gracing the cover of the *Saturday Review* and newspapers across the country. The most honest of the dozens of letters congratulating Brode came from Koepfli himself. "I know you are going to have your work cut out for you," he wrote. Brode should feel free to call him "any time."[22]

Brode's plans for the reorganized office followed the lines recommended by the 1950 Berkner Report, with one major exception: science at State would now focus on providing scientific advice for foreign policy rather than collecting open intelligence. Whether in response to the (justified) rumors about the prior role of science attachés in intelligence collection or to an unclassified 1955 recommendation that the attachés officially report to the CIA, the State Department encouraged reports that minimized the intelligence aspects of the job. As an editorialist for the chemical weekly *Chemical & Engineering News* put it, "Our science attachés were not and will not be super-duper spies." Still, the appointments lagged. As of September 1958, Brode had not filled a single one of the nine science attaché positions approved in his budget. In a frustrating repeat of his experiences at the CIA, Brode found his hiring held up by unexplained security delays and balky university deans. Finally, in January 1959, the first batch of US science attachés since 1956 began taking up their positions in London, Paris, Rome, Bonn, Stockholm, and Tokyo, with more appointments planned for Moscow, New Delhi, and Rio de Janeiro.[23]

Even the USIA appointed a science advisor, in the person of Harold (Hal) Goodwin, a former Marine and public affairs officer for the Federal Civil Defense Administration. Whether or not his colleagues at the USIA knew it at the time, Goodwin maintained an equally successful career writing boys' adventure novels under the pen names of John Blaine, Hal Grodon, and Blake Savage. As the USIA's science advisor, Goodwin generated memo after memo detailing proposals on how the USIA and other government agencies might rekindle global enthusiasm for US science. Perhaps it is not surprising that the author of *The Phantom Shark* (1949) and *Rip Foster Rides the Gray Planet* (1952) proposed that the US government win global hearts and minds through science textbooks, do-it-yourself science kits, and ham radios. But Goodwin was also a serious propagandist whose ideas influenced the US effort in international science for the next decade, particularly in space. Whether recommending that the United States back programs to eradicate disease or to capture solar energy, Goodwin grasped both the promise and the limitations of scientific discovery as a storytelling device.[24]

Hired in 1958, Goodwin's primary task was to devise strategies for the USIA and, by extension, the US government to meet the promise of Eisenhower's "Waging Total Peace" speech. Goodwin encouraged the USIA to think more broadly about the psychological potential of science to win hearts and minds. With Goodwin's help, the USIA developed a set of guidelines to govern US information campaigns that distinguished between science and technology, downplayed the competition between the United States and the Soviet Union, and emphasized the potential of the products of science and technology to improve human welfare. "Science," the agency's so-called Basic Science Paper explained, "is the systematic pursuit and classification of knowledge," dedicated to establishing "general principles and natural laws that are subject to verification."

Technology applied these findings to produce "goods and systems" for agricultural and industrial use. Technological achievements might advance nationalistic goals, but science, by definition, was international and cooperative. The objective of international science is "solely the advancement of science."[25]

Goodwin so carefully distinguished between science and technology because he doubted that the United States would surpass the Soviet Union in technological achievement anytime soon, particularly in high-profile areas like rocket technology and nuclear power. Moreover, Russian science had a long tradition of achievement, and it would be foolhardy for the United States to suggest that it could consistently outperform such a distinguished community of well-funded scientists: "No nation, regardless of capability, can remain ahead of all others in all fields." For these reasons, Goodwin discouraged campaigns that focused on competition in *either* science or technology; the "ideological struggle," he advised, "exists at a level far more fundamental than that of dramatic scientific or technological achievement."[26]

Instead, the USIA should emphasize the benevolent, noncompetitive nature of US scientific research. "Scientific research is conducted for its own sake," the paper explained, "simply to increase man's knowledge of himself and his environment, in the universities, non-profit institutions, foundations, and laboratories that exist by the hundreds across the United States." Technology, in contrast, was "only the means to an end." The United States, with its free press and system of free enterprise, had a long history of "unstintingly" sharing the fruits of its scientific labor with the rest of the free world, and it would continue to do so. The Soviet Union, in contrast, jealously guarded its research, funneling its findings directly into nationalistic technology. Different economic and political systems produced very different technological results, but science transcends politics. In short, "Science is neither for nor against, moral nor immoral. It is apolitical."[27]

These were remarkable statements to be issued in the name of a government propaganda agency—all the more so because there is every indication that the paper's audience took its recommendations seriously. Universally acclaimed as "excellent" by those who read it, the USIA's Basic Science Paper received the OCB's endorsement in September 1958. Instead of treating the competition for scientific preeminence as a proxy battle for the broader struggle between capitalism and communism, the United States would fight for the title of best scientific collaborator in an "apolitical" system of international science. If the United States could claim credit for dismantling obstacles to scientific achievement, then any scientific achievement—anywhere!—would reflect well on the United States.

With a full roster of science advisors in place, the only things preventing the US government from implementing this simultaneously utopian and deeply cynical approach to defeating Communism were its own anti-Communist policies.

∷

When it came to scientific and technical cooperation, psychological warfare strategy ran ahead of US policy. Even after the downfall of red-baiting Senator Joseph McCarthy in 1954, the United States still had several anti-Communist laws on the books that limited US scientists' ability to mingle freely with scientists abroad and that complicated attempts to hold international conferences on US soil. Uneven enforcement of the McCarran Act meant that US scientists with a history of leftist associations were never sure whether they would be able to obtain a passport, and Brode's Science Office faced the same problems in arranging visas for visiting scientists that had dogged Koepfli half a decade earlier. These embarrassing restrictions on scientists' freedoms reached a crisis point when the Soviet Union and the People's Republic of China began calling the United States' bluff on international cooperation.[28]

Following Stalin's death in 1953, the United States and the Soviet Union began negotiating the possibility of small, heavily scripted exchange visits, particularly in sports. The impetus for these exchanges came from the Soviets, not the Americans. At least 101 Americans entered the Soviet Union as private citizens in 1953 and 1954, but during this same period, the United States admitted only 33 Soviet citizens. The Immigration and Nationality Act of 1952 barred "non-official" alien visitors who might be "communists or subversives" from entering the country, which meant that an entire series of cultural exchanges between youth groups, orchestras, newspaper editors, and wrestling teams got caught up in red tape. Any individual holding a nondiplomatic Soviet passport required a Proviso 9 waiver and would be subject to fingerprinting upon arrival. At least temporarily, the Soviet Union had gained the upper hand in demonstrating a commitment to openness and freedom, with foes and allies alike accusing the United States of erecting its own Iron Curtain.[29]

In 1955, the US government issued new travel restrictions on Soviet passport holders designed to mirror existing Soviet restrictions. Approximately 30 percent of each country's territory, including 30 percent of cities with populations greater than 100,000, was closed to citizens of the other. Soviet delegations to the United States could take in a Cubs game, but they couldn't sample Memphis barbecue. Most ports, coastlines, and areas housing weapons or research facilities were off-limits, as were industrial centers and several cities in the Jim Crow South. The press releases announcing the travel plans emphasized that the United States implemented such rules only as retaliation; should the Soviet Union lift its restrictions, the United States would follow. A reasonable person might have missed that message in the State Department's five-page, single-spaced, four-column list of cities closed to Soviet visitors.[30]

As early as January 1955, the NSC recognized the inherently self-defeating nature of these policies, particularly in science and technology. Severely restricting admission of Eastern bloc scientists to the United States not only undermined US claims about freedom but also cut off a potentially useful vein of scientific intelligence. Both the CIA and the FBI, however, continued to warn that the dangers of admitting foreign nationals outweighed the potential benefits.

This was the context in which Lauren Soth, editor of the *Des Moines Register*, published an editorial inviting the new Soviet leader, Nikita Khrushchev, to visit Iowa's cornfields. Khrushchev had come to power acknowledging the failures of Soviet collective farms. In a speech to the Central Committee of the Communist Party, he urged Soviet farmers to plant hybrid corn—a dramatic reversal of previous Lysenkoist policies that tarred hybrid corn with the brush of Western genetics. On February 10, 1955, Soth, acting without "diplomatic authority of any kind," invited Khrushchev to come to Iowa to "get the lowdown on raising high-quality cattle, hogs, sheep, and chickens." Soth promised not to withhold any "secrets," stating plainly that the United States should "let the Russians see how we do it."[31]

Khrushchev's surprising acceptance of this invitation found the State Department once again scurrying to find a way to admit Soviet visitors. Fortunately, most of Iowa was located outside the no-go region; the Soviets took care of the fingerprinting requirement by giving their delegation "official" credentials. Khrushchev himself wouldn't come for four more years, but in the summer of 1955, a group of twelve Soviet agricultural experts, including the first deputy minister of agriculture, spent five weeks learning about US farming techniques in Iowa and California. But while the Soviets received a crash course in the modern techniques that were filling US silos with surplus grains, the twelve US farmers sent to Ukraine and Siberia spent most of their time learning about the management of collective farms. And this, according to critics at the

State Department and the CIA, posed a threat to national security. The Soviets, according to these analysts, designed their US visit as a "concentrated course in American farm technology." The enthusiastic farmers of Iowa gave the Soviets advice on everything from the most efficient use of herbicides to the benefits of contour plowing, while the Americans returned home from the USSR without so much as recent statistics on cotton production.[32]

Had intelligence collection remained the primary goal of cultural diplomacy, scientific and technical exchange programs might have ended where they began, in an Iowa cornfield. As the Eisenhower administration developed ever more ambitious plans for ideological warfare, however, supporters of cultural exchange gained ascendance. Starting in the fall of 1955, US and Soviet authorities repeatedly expressed their desire to expand the number, kinds, and duration of exchange programs. NSC 5607, issued in June 1956, committed the United States to an "active" exchange program in keeping with the Ideological Program's emphasis on personal contacts. The policy explicitly envisioned exchanges as an offensive weapon that could combat Communism by stoking a desire for greater freedom within the Soviet Union. The Soviet Union's brutal repression of the Hungarian uprising that fall briefly threatened to scuttle the program, but by the spring of 1957, planning for exchanges was back on track. Nevertheless, modified travel restrictions remained in place, and Dulles's State Department remained skittish about granting entry visas and issuing passports, particularly to scientists, engineers, and federal employees.[33]

A medical exchange in 1957 and 1958 is a fairly typical example of how these early, somewhat informal exchanges played out. In the fall of 1957, the US government proposed a reciprocal exchange of female physicians. A group of six "outstanding Soviet women physicians," led by Dr. Ekaterina Vasyukova, spent approximately a month visiting US research institutions, public health facilities, and hospitals. In return, a group of six US doctors—almost always referred to in press releases as "women medical scientists"—spent four weeks in the Soviet Union in May 1958. Led by acclaimed Johns Hopkins cardiologist Helen Taussig, the US delegation toured medical schools, clinics, child health centers, rest homes, and research centers in Moscow, Leningrad, Kiev, Sochi, Tbilisi, Tashkent, and Samarkand. They visited an experimental monkey colony and a rayon factory and, in a special "treat," stopped by a restored version of Pavlov's laboratory.[34]

While Taussig enjoyed the trip immensely, she found it more "philosophical [sic] stimulating than medically profitable." The report she submitted on behalf of the group only briefly touched on the status of medical research in the Soviet Union or the nature of its healthcare system. Instead, like any curious tourist writing a thoughtful travelogue, she reflected on how the realities of the visit matched or departed from her expectations. She was fascinated, for example, by the idea that workers could be demoted as well as promoted, and she kept circling back to the fact that Soviet women "had" to work. Arriving in Moscow barely six months after *Sputnik* rattled US confidence, Taussig found much to admire in Soviet science. "Scientists and educators are among the most respected people in the U.S.S.R.," she wrote, and "everything is done to encourage the intelligent student to use this ability to work for the interest of the state." In her brief observations, scientists operated in relative freedom and spoke approvingly of the government—a statement remarkable for its lack of qualifiers, given the constant presence of a minder and the lack of Russian-speakers among the delegation. As either offensive propaganda or intelligence collection, this trip left much to be desired.[35]

Taussig's report also omitted the administrative hassles that plagued the organizers. This was an official visit, undertaken at the request of the US Department of State. The Soviet visitors to the United States had to be declared "official" to gain entrance to the country. The US delegates, however, only officially represented themselves. The details of their trip were arranged by the NAS rather than the State Department, and the Rockefeller Foundation—not the US government—picked up the costs of travel to and from the Soviet Union. At one point, the State Department attempted to block participation of the National Cancer Institute's Dr. Thelma Dunn, on the theory that US government employees should not accept hospitality from Soviet officials.[36]

The State Department remained deeply ambivalent about these exchanges, even as its representatives continued negotiations with their Soviet counterparts. In January 1958—in the midst of this particular exchange—State Department representative William Lacy and Soviet ambassador Georgi Zarubin signed a formal cultural exchange agreement. As the first bilateral accord signed between the two countries since the end of World War II, the Lacy-Zarubin Agreement marked a new era of US-Soviet relations. As part of the negotiations, Congress finally dropped the fingerprinting requirement, and the State Department continued to ease travel restrictions. The protocols that accompanied the agreement spelled out, in minute detail, procedures for exchanging books, films, exhibits, and people, including scientific and technical professionals. But because it specified that scientific and technical exchanges would be conducted through the countries' respective academies of science, and since the NAS was not actually a government entity, yet another round of negotiation was required before a large-scale program for scientific cooperation could begin.[37]

How many scientists would be exchanged? Who would select them? What kinds of activities might they engage in? Should the exchanges prioritize certain fields, and conversely, should sensitive fields such as nuclear physics be excluded? The task of ironing out the details fell to Bronk and Brode, who traveled to Moscow for a series of meetings with A. N. Nesmeyanov and A. V. Topchiev of the Soviet Academy of Sciences in October 1958. The eventual agreement, signed in July 1959, combined vague objectives with detailed recordkeeping. The two countries agreed to exchange "approximately twenty persons," each of whom should be "prominent," to "deliver lectures and conduct seminars on various problems of science and technology." An undefined number of researchers could be exchanged for both short- and long-term research visits. But most important, from the State Department's perspective, was the agreement's commitment to strict reciprocity. If the Soviet Union sent an applied mathematician for two weeks, the United States would send an applied mathematician for two weeks. The agreement would be subject to renewal after two years.[38]

The NAS clearly stood to benefit from this arrangement. Besides the obvious boost to the academy's prestige, the organization received a two-year NSF grant of almost half a million dollars to underwrite the associated costs. At the same time, however, both the academy's leadership and the broader scientific community questioned whether the exchange arrangement conveyed the right messages about US science. The NAS chafed at the State Department's reciprocity requirement and complained that ongoing travel restrictions in the United States wrecked the itineraries of Soviet visitors. Bronk wanted both the academy's members and his contacts at State to understand that the very concept of a controlled scientific exchange violated the precepts of scientific freedom that the Americans hoped to promote. As he put it in a letter to members, the Interacademy Exchange Agreement was "only a stimulus to the furtherance of normal, unorganized visits of scientists." PSAC members, too, questioned the wisdom of these restrictions in a plan designed to showcase US openness: "If a visiting Russian finds in the U.S. restrictions patterned after those he suffers at home, and the same tendencies towards political involvement of non-political matters, he may not carry back with him the very message we are attempting to get across." As of April 1960, not a single American had traveled to Moscow under the academy exchange program.[39]

These problems, however, were mere hiccups compared with the barriers to international scientific cooperation raised by the United States' stance toward the People's Republic of China. Since the defeat of Chinese nationalists on the mainland in 1949, the US government had maintained diplomatic relations with the Republic of China—Taiwan—but not with the People's Republic of China. The State Department's policy of nonrecognition not only barred official diplomatic contact but also prohibited US officials from participating in events or organizations that admitted Communist Chinese participants. This policy caused headaches both large and small when the People's Republic of China began seeking membership in international scientific organizations in the late 1950s.[40]

On a personal level, the policy made it nearly impossible for US government employees to attend international scientific events in an official capacity. In August 1959, for example, Alan Waterman urgently petitioned the undersecretary of state on behalf of an NSF employee who had planned to attend a meeting of the International Union of the History and Philosophy of Science in Madrid. The NSF was funding the US delegation to this meeting, and Dr. Raymond Seeger, the relevant program manager, planned to attend as an observer. The day before Seeger's flight, however, the State Department notified him that the potential presence of Communist Chinese rendered his trip "inappropriate." Waterman's appeal carefully parsed the different ways that Seeger's participation could be categorized, arguing that he should be considered "an observer named by the scientific community as represented by the Academy-Council." This was "quite different," he wrote, than if "this were a Government-sponsored conference and the individual attended as a Government representative." The ultimate fate of Seeger's trip is unclear, but the need for Waterman's intervention suggests deep and ongoing reservations about a higher international profile for US science well after the launch of *Sputnik*.[41]

Federal agencies faced an even grimmer calculus when it came to institutional participation. When the People's Republic of China, East Germany, and North Korea applied for membership in the ICSU in 1957, the State Department recommended one of two courses of action. Either the United States could actively oppose their membership, or it could stop accrediting US delegations to ICSU-affiliated meetings. Both moves seemed obviously counter to US interests. H. S. Smyth, chairman of a recent US delegation to an ICSU-sponsored meeting in Rome, explained just how such a position enhanced Soviet prestige. Whether or not the Soviets were encouraging Chinese membership for political reasons, the "avowed purpose and actual work" of international scientific unions were "non-political." If the United States opposed Chinese membership, it, not the Soviet Union, deserved blame for politicizing a scientific meeting. A member of the NAS voiced similar concerns to Bronk, arguing that the State Department's policy threatened to "bring about a dangerous rift" between scientists and the government. The policy "aids and abets the Soviet Union in its campaign to be recognized as the world leader in science . . . [and] helps Moscow achieve its goal of being proclaimed as the scientific capital of the world." No good could possibly come of it.[42]

The State Department stood firm. In a pointed letter, Brode acknowledged the desire for scientific cooperation but underscored the department's broader strategic goals. The State Department maintained that merely admitting representatives from the People's Republic of China to an international gathering of scientists furthered Communist goals by conferring legitimacy on Mao's regime. The Communist Chinese hoped to score allies not through normal diplomatic channels but by "posing as a respectable member of the world body and seeking to build up informal relationships with private groups in free world countries." It was the Communist Chinese—not the Americans, or even the Soviets—who politicized international scientific bodies by walking out when representatives from the Republic of China were seated. The State Department would not interfere if nongovernmental agencies like the NAS participated in meetings funded by private bodies, but it simply could not endorse any actions that undermined US support for the Republic of China. Brode later complained to a colleague at State that US scientists lacked an appreciation for their "responsibility" to "support their nation's policies."[43]

The problem of how to turn international scientific cooperation into a vehicle for the national interest remained an unsquareable circle for government officials. Lacking a better plan, the State Department stopped accrediting US delegations to international meetings, on the supposition that an "official" delegation sponsored by the adhering body—usually the NAS's National Research Council—would be sufficient to meet American obligations. The State Department also stopped paying dues and other fees to international scientific unions, handing over that responsibility to the NSF (a choice ostensibly justified by the NSF's ability to fund both private and public projects). Foggy Bottom took refuge in the idea that accrediting scientists might actually damage their international reputation, exposing them to charges of being used as "political tools of our government." Constrained by a logic that rejected the possibility of real cooperation with scientists from Communist countries, the government's advisors saw few alternatives other than turning the government's scientific activities over to a nongovernmental partner. And over and over again, that partner turned out to be the NAS.[44]

: :

An employee of the NSF, traveling on the NSF's time and dime, magically transforms into a private citizen by virtue of his having been issued an invitation by a nongovernmental body. Lack of paperwork somehow shields a group of US scientists, traveling as an official but unaccredited delegation, from the taint of politics. Scientists and science attachés submit reports on their international contacts but, claiming ignorance of the memo distribution list, deny having anything to do with intelligence collection. There is nothing inherently wrong with a state conducting its affairs via intermediaries, but there is definitely something peculiar about a state insisting that its chosen partners have no relationship to the government.

The architects of US policies on science in foreign relations actively chose this path. In the years following *Sputnik*, panels of scientific advisors in and out of government spent day after day in meetings actively theorizing the possibilities and limits of an arm's-length approach to science diplomacy. They collected data on scientific exchanges, attempted to tabulate public and private investment in research abroad, and argued over the limits of scientific independence from government oversight.

These seemingly endless discussions—one NSF official claimed to have worked on the issue nonstop for four months—culminated in two reports presented to the NSC in the waning days of the Eisenhower administration. The first, "Strengthening the Free World Position in Science and Technology," advocated dramatic increases in US government funding for basic research, science education, and science communication as necessary investments in national security and economic competitiveness. The second, "International Scientific Activities," depicted scientific cooperation with friends and foes alike, primarily conducted by scientists "essentially independent of government support," as integral to US science.[45]

Both documents traveled over the signature of George Kistiakowsky, who had replaced Killian as the president's science advisor in July 1959, but they represented distinct constituencies with different visions of the relationship between science and politics in foreign affairs. As a product of the Federal Council on Science and Technology, which represented the views of US government agencies, "Strengthening the Free World" assumed an overt role for those same agencies in conducting US affairs, whether at home or abroad. The heads of NASA, the AEC, the NSF, the Department of Defense, and several other cabinet-level departments focused on domestic science policy, not only because it represented their jurisdictions, but because, as the report put it, "Free World science and technology are heavily influenced by the scientific policies of the United States." The only thing remotely controversial about the text of this unclassified report was its suggestion that effective science policy, including cooperation with other free world nations, required more coordinated planning.[46]

The real tension in "Strengthening the Free World" lay in its unspoken assumption that strength in "science" meant hard power: new weapons, a strong economy, and uncontested leadership among allies. The PSAC-authored "International Scientific Activities" rejected this premise. Drafted by a panel chaired by the ubiquitous Detlev Bronk, this SECRET document embraced the role of science—particularly nongovernmental science—as a tool for political and psychological warfare. The "objective nature of science and scientific research—its independence of political belief"—made science the perfect olive branch to both Communist bloc scientists and ambitious leaders in newly independent countries. And, as Bronk and his co-panelists Lloyd Berkner and Joseph Koepfli knew well, nongovernmental scientific contacts also offered "probably the most effective way of obtaining really useful scientific intelligence." The report urged the NSC to recognize that international scientific activities, broadly defined, directly contributed to US national security objectives.[47]

On December 1, 1960, the NSC accepted the PSAC Panel's report. It did so largely over the objections of the science officers at the State Department. Throughout the spring and summer, Wallace Brode had criticized earlier drafts of the proposal, warning that the document's insistence on apolitical, international science was naive and counterproductive. His criticisms boiled down to a critique of scientific exceptionalism. It was a mistake, he argued, to see science as either "a tool" of politics or separate from it. Instead, he argued, the US government should adopt the position that "science *joins with* the economic, social, and political elements to form an integrated culture which promotes U.S. national policy and objectives." Brode's insistence on this point led him to advocate for a cabinet-level department of science well after most of the other, nongovernmental members of the PSAC had rejected it, and it contributed to his departure from the State Department that September.[48]

For Brode, a veteran of scientific intelligence and diplomacy, the idea that international scientific affairs could be separated from foreign policy was patently ridiculous. The margins of an earlier draft of the document in his files bear a taciturn "no" next to a line that claimed, without qualifications, that "science has developed over many years a tradition of separation of international political and purely scientific matters." It must have rankled, as well, to see the charge for the alternative view led by Bronk, whose NAS so consistently benefited from policies—in other words, lowercase politics—that claimed to transcend politics. Bronk's biographer tactfully refers to the biophysicist's "style of achieving common goals through many overlapping channels of management." Bronk not only presided over the NAS and chaired the PSAC's Panel on Foreign Affairs; he also chaired the governing body of the NSF at the moment that it made its half-million-dollar grant to the NAS to oversee the Interacademy Exchange programs. The NAS would continue to benefit from policies that treated it as the government's default contractor for scientific cooperation in the years to come. In 1950, it had one full-time employee working on international programs; by 1962, it had a hundred.[49]

Upon occupying the White House in late January 1961, President John F. Kennedy elevated the battle of scientific prestige to levels Eisenhower never dreamed of. But even as Kennedy embraced scientific competition, pledging to send a man to the moon, he eagerly sought opportunities for international scientific cooperation, particularly in space. More to the point, his administration identified a compromise that had eluded his predecessors. On the advice of Eisenhower's NSC, Kennedy handed over the task of identifying a strategy for international scientific cooperation to an ad hoc group within the Federal Council headed by Walter Whitman, Brode's improbably named successor at State. A more skillful negotiator than Brode, Whitman ensured that goals for international scientific cooperation were always discussed in the context of US foreign policy. The final report, released in the summer of 1961, invoked the PSAC's favored language of apolitical science—"an objectivity which transcends differences in political and social systems"—but did so explicitly in service to national goals, with guidance from federal agencies. In practice, however, the report acknowledged that most of these programs would be carried out by nongovernmental entities, including the NAS.[50]

The heady dream of a scientific internationalism freed from politics drove US cultural diplomacy involving science for most of the rest of the decade. Federal government agencies—some overt, some covert—sought partnerships with unlikely advocates for their preferred vision of science the world over. They encouraged historians of technology to highlight how US science had prospered through its unique dedication to free scientific exchange, for example, and they recruited scientists and engineers to develop journal exchange programs that the USIA could then promote as examples of US generosity. By design, very few of these programs bore the explicit stamp of the US government—something that has obscured the existence of cultural diplomacy in science for most of the past fifty years. In the next two chapters, I explore two areas where the hidden hand of government has been particularly difficult to see: nuclear disarmament and international science education.[51]

6 : Science for Diplomacy

inus Pauling should have had better things to do than sue the geneticist Bentley Glass for libel. By the summer of 1962, the US chemist had already won one Nobel Prize, for his work on chemical bonds, and was well on his way to picking up a rare second Nobel for his work on behalf of nuclear disarmament. From May 1957 to July 1958, Pauling collected the signatures of more than 11,000 scientists from around the world in support of a nuclear test ban. Over the next five years, Pauling wrote books, gave hundreds of lectures, and engaged in verbal fisticuffs with physicist Edward Teller, director of Livermore National Laboratory, the loudest advocate for ever-bigger bombs. Pauling also faced repeated investigations by the US Senate for his alleged connections to Communist groups. Somehow, in the middle of all this, he found time to mount a relentless personal and legal campaign against Glass and Glass's editors.[1]

The two men made for odd antagonists. At the time, the Johns Hopkins–based Glass was one of the United States' most famous geneticists, his profile second only to that of his former graduate advisor, H. J. Muller. In the late 1950s, Glass, like Pauling, had warned government agencies and the general public about the biological dangers of atomic radiation. But this wasn't the source of Glass's fame. During a period when many US scientists retreated to their laboratories out of fear of McCarthyist reprisals, Glass staunchly defended individual liberties and civil rights. The two men's political profiles were close enough that, when Glass refused to sign a loyalty oath to the State of Maryland in 1960, an editorialist for the *Washington Post* linked Glass and Pauling together as "men of courage" who resisted "arbitrary and foolish restrictive official actions."[2]

The comparison chafed, for each of them. A mild-mannered Baptist Sunday school teacher, Glass made his career and reputation working inside the system. Pauling, on the other hand, converted his scientific concerns about the dangers of atomic fallout into a mass political movement. The differences in their theories of political change burst into public view in the summer of 1962, when Glass published an essay on science and politics in the *Bulletin of the Atomic*

Scientists. The role of science in contemporary warfare and in the federal budget, Glass wrote, had "stir[red] in many scientists a feeling of great political responsibility," but scientists must be careful to stay within the bounds of their expertise, so as not to "abandon their scientific reserve and suspended judgment." He implied that both Pauling and Teller had let their emotions get the better of them when it came to nuclear weapons testing. Pauling called his lawyer.[3]

If Pauling's response seems out of proportion to Glass's comments, he may have been reacting to what must have seemed like Glass's blatant hypocrisy. Whatever distinctions Glass hoped to make in print, by the late 1950s both Glass *and* Pauling were involved in an international fight for the future of humanity, a fight that threw them both smack into the middle of international politics and high diplomacy. Since 1958, both men had been involved with the so-called Pugwash Conferences, an international scientists' movement on behalf of nuclear disarmament. Pugwash's leaders cultivated (and continue to cultivate) the organization's reputation as a model of nonpartisan scientific activism, a shining example of what scientists could accomplish if they worked as unaffiliated individuals, beyond the constraints of politics. In 1995, the Pugwash organization and Joseph Rotblat, one of the movement's founders, received the Nobel Peace Prize for their roles in reducing nuclear tensions at the height of the Cold War.[4]

Pugwash derived its authority from the idea that scientists could most effectively advocate for scientific internationalism (and, ultimately, world peace) when they acted as individuals affiliated with nothing but the world of knowledge. But in fairly short order, Pugwash morphed from outsider scientific critique to insider tool. Some scientists, including Glass, endorsed this closer relationship with the US government, while more radical scientists, like Pauling, thought Pugwash's effectiveness depended on its separation from the state. In the arms control debate of the late 1950s and early 1960s, the stakes for the nature of scientists' participation in the debate—as scientists, as citizens, or as diplomats—were nearly as high as those for arms control itself.

::

Even before the world's first atomic bomb lit up the New Mexico desert sky in July 1945, the Manhattan Project physicists foretold the possibility of an arms race. Once the secret of nuclear fission at the bomb's core was out, scientists warned, nothing could stop every major power in the world from building one. After the war, the scale of the devastation in Hiroshima and Nagasaki convinced other US physicists to join Leo Szilard, the Hungarian-born scientist who had first suggested the idea of the bomb in a letter to President Franklin Roosevelt, in calling for the international control of atomic energy. Although the members of the so-called atomic scientists' movement disagreed on the best route to peace—a world government, a weapons moratorium, or an international uranium authority—they all agreed that some form of action was necessary.[5]

Initially, US atomic scientists looked abroad for support for their cause. In 1946, the Federation of Atomic Scientists' Committee for Foreign Correspondence solicited the views of 3,000 scientists, including those in the Soviet Union, on what actions scientists could take to prevent atomic war. For a brief moment in 1946, it appeared that scientists' advice might compel the United States to participate in some sort of international uranium authority, with control of atomic energy placed squarely in civilian hands. By late 1946, however, the rise of anti-Communist sentiment and revelations of atomic espionage in the United States had diminished scientists' influence over atomic policy. The version of the Atomic Energy Act that Congress passed in 1946 emphasized national sovereignty over international control. The newly created Atomic Energy Commission could turn to a board of scientists for advice, but its policies and programs—and therefore the country's—would be set by political appointees. Even so, the atomic scientists' movement set a precedent for scientists' involvement in nuclear politics. Their journal, the *Bulletin of the Atomic Scientists*, would remain a platform for scientific internationalism.[6]

With no international authority limiting proliferation, and with the Cold War heating up, the nuclear arms race took off. Five months after the Soviet Union ended the United States' atomic monopoly by testing its own weapon in August 1949, President Truman raised the stakes by authorizing a hydrogen bomb program. He did so over the objections of leading US scientific advisors involved in the project, including Manhattan Project veterans J. Robert Oppenheimer, Enrico Fermi, and I. I. Rabi. By 1953, both the United States and the Soviet Union had tested massive hydrogen bombs, and the United Kingdom had assembled its first atomic weapon.[7]

Dropped over New York, London, or Moscow, a single 1-megaton hydrogen bomb could kill at least 1 million people and was likely to injure 3 million more. And beyond the immediate devastation of a nuclear attack, by the mid-1950s scientists had begun to warn of a less dramatic but equally chilling consequence of the arms race: radioactive fallout from nuclear weapons testing. Military experts in the United States, the Soviet Union, and the United Kingdom were concerned enough about fallout that they consistently conducted their tests in remote locations.[8]

Fallout became a more pressing political issue after the US Bravo test in the Marshall Islands on March 1, 1954, obliterated three atolls and sickened military personnel, weather researchers, Marshall Islanders, and Japanese fishermen aboard a small vessel, the *Daigo Fukuryu Maru*, in the area. The explosion was larger than the AEC had anticipated, forcing the agency's head, former Rear Admiral Lewis Strauss, to address public concerns. In a prepared statement delivered to the press at the end of the month, Chairman Strauss denied or downplayed every allegation leveled against the Bravo test. "At no time," he assured the press, "was the testing out of control." No islands or atolls were destroyed; the damage was limited to "a large sandspit or reef." Most importantly, Strauss emphasized, "I can state that any radioactivity falling into the test area would become harmless within a few miles." Any increase in background radiation would be "far below the levels which could be harmful in any way to human beings, animals, or crops." The death of the fishing boat's radio operator, Aikichi Kuboyama, in September exposed the wishful thinking in Strauss's insistence that nuclear testing posed no harm to the global public.[9]

The Bravo test—and the AEC's off-key response—sparked international outrage. On April 2, Indian Prime Minister Jawaharlal Nehru called for a "standstill agreement" on nuclear weapons testing, with the ultimate goal of eliminating nuclear weapons altogether. In the United States, letters demanding a nuclear test ban started pouring into the White House at a rate of more than a hundred a day. Within a year, Nikita Khrushchev, having noted the growing global popularity of a nuclear test ban, seized the moral high ground for the Soviet Union by calling for test ban talks.[10]

With calls for a nuclear test ban gaining traction among peace activists, Communists, and leaders of nonaligned nations, some scientists saw an opening to reinvigorate the atomic scientists' movement. In the United States, however, the threat of reprisals from anti-Communists tempered scientists' willingness to speak out. On April 12, 1954, just a week-and-a-half after Nehru's call for a test ban, the AEC opened a security hearing on Oppenheimer. Both during and after his time at the Manhattan Project, the chain-smoking physicist in the porkpie hat had made many enemies in the government. A man with better political skills might have gotten away with opposing the development of the hydrogen bomb, but politics was never Oppenheimer's strong suit. After a month of testimony from fellow Manhattan Project veterans who defended Oppenheimer's loyalty but not necessarily his judgment, the AEC officially revoked his security clearance in May 1954.[11]

Oppenheimer's fate spooked the older generation of US nuclear physicists, many of whom had become accustomed to being treated as trusted government advisors. They had also (despite the rhetoric of the US government's hands-off relationship to science) come to enjoy a steady stream of lucrative government contracts. If Oppenheimer, the father of the atomic bomb, could be stripped of his clearance, what fate awaited them? In contrast, scientists who operated on the margins of the federal science apparatus started to speak out, emboldened by the combination of the Oppenheimer hearings and the AEC's bald lies on the nature of fallout. The May 1954 issue of the *Bulletin of the Atomic Scientists* put the two topics in stark relief, pairing a ten-page account of the Oppenheimer hearings with Strauss's description of the Bravo tests as "harmless."[12]

The *Bulletin*'s editor, Eugene Rabinowitch, a Russian-born physicist with a penchant for poetry and bow ties, staunchly advocated for scientific internationalism as a tool for fighting Communism. It was no coincidence that the same May issue of the *Bulletin* also contained a report on the Hamburg Conference on Science and Freedom and updates on Eisenhower's plans for Atoms for Peace. In July, Rabinowitch proposed an international meeting of atomic scientists to discuss the new threats raised by the hydrogen bomb. He hoped that the assembled group might produce a "statement of facts as scientists see them," as a way to establish that scientists could agree on the dangers of fallout, even if their governments could not. To Rabinowitch's surprise, however, several chapters of the Federation of American Scientists (the successor organization to the atomic scientists' movement) largely rejected his plan. While many US scientists thought such a meeting was "desirable," it was not necessarily "feasible" or "politically advisable" in the wake of the Oppenheimer hearings.[13]

Rabinowitch's proposal received a warmer welcome from the membership and leaders of the British Atomic Scientists' Association. Rabinowitch had solicited the views of the executive director of the British organization, Joseph Rotblat, a Polish-born British citizen hailed in some quarters (and reviled in others) as the only scientist to leave the Manhattan Project because of moral opposition to the bomb. Rotblat had thrown himself into public debate on fallout after the Bravo explosion, appearing on the current affairs program *Panorama* and publishing a paper on the radiation effects of a hydrogen bomb.

When the British Medical Research Council initiated a study on the possible dangers of atomic radiation in 1955, it did so in part because of Rotblat's work on publicizing the issue.[14]

Rotblat continued organizing an international conference even after it became clear that the US effort had stalled—in part because other scientists were also seeking his cooperation. In January 1955, Rotblat received a letter from Frédéric Joliot-Curie, informing Rotblat of his intent to organize a similar international conference in Paris. At the time, Joliot-Curie (whose wife, Irène, was Marie and Pierre Curie's daughter) was the president of both the World Peace Council and the World Federation of Scientific Workers, both groups associated with the international Communist Party. Joliot-Curie's Communist ties were hardly secret; he had been removed from his role as French commissioner for atomic energy in 1950 because of his membership in the party. Rotblat and Rabinowitch hesitated, unsure of how to seize the momentum of a Communist-led initiative without ceding control of their own event.[15]

Joliot-Curie's campaign, meanwhile, had captured the imagination of the British mathematician, philosopher, and aristocrat Bertrand Russell. Russell, who since 1931 had held the title of the Third Earl Russell, had been jailed during World War I for his pacifism. He was awarded the Nobel Prize for Literature in 1950 for his vast literary output, most of which focused on the theme of humanism. In Joliot-Curie's invitation, Russell saw an opportunity to unite scientists across the political spectrum in opposition to nuclear war. The day before Christmas Eve, 1954, Russell delivered a British radio broadcast, "Man's Peril," on the theme of nuclear annihilation. In the spring of 1955, he developed the broadcast into a manifesto against the development, testing, and use of nuclear weapons. The resulting document quickly became known as the "Russell-Einstein Manifesto," not because Einstein had anything to do with writing it, but because signing it was one of his last acts on earth—Russell learned of Einstein's death before the famous scientist's signature arrived in the mail.[16]

Russell delivered the manifesto, with its eleven signatories from Britain, France, Germany, Japan, Poland, and the United States, including nine Nobel laureates (himself, Joliot-Curie, Pauling, and Muller among them), at a press conference on July 1, 1955, with Rotblat at his side. The manifesto invoked scientists' identity as members of an international community, free from national or ideological distinctions: "We are speaking on this occasion, not as members of this or that nation, continent, or creed, but as human beings, members of the species Man, whose continued existence is in doubt." He asked his listeners to "set aside" political commitments to capitalism or communism and instead focus on their identities as "members of a biological species . . . whose disappearance none of us can desire." Russell posed the problem starkly. "Shall we put an end to the human race; or shall mankind renounce war?" The manifesto received significant press coverage (no doubt driven by the pathos of Einstein's deathbed signature).[17]

The Russell-Einstein Manifesto stressed the shared fate of humanity over a special role for scientists. Even so, its opening line called for an international meeting of scientists to consider the problem of the hydrogen bomb. As it happened, the summer of 1955 featured three international meetings related to nuclear weapons. From July 2 to July 4 (immediately following Russell and Rotblat's press conference), French scientists gathered for a national Conference on Atomic Dangers in Paris. Rabinowitch was one of only two international scientists in attendance. A month later, more than 1,400 scientists and government officials from more than seventy nations gathered at the vast Geneva Conference on the Peaceful Uses of the Atom, the Atoms for Peace event where Muller's floor speech on fallout caused such a ruckus.[18]

The most curious of the three, however, was a last-minute meeting in London, inspired by Russell, organized with Rotblat's assistance, and sponsored by a British group advocating for world government. The inelegantly named World Association of Parliamentarians for World Government, led by British members of Parliament, had chapters around the world. In the years immediately following World War II, world federalism had become a popular cause among liberals who doubted the ability of the United Nations to foster global peace; it had particularly captured the imagination of many US members of the atomic scientists' movement. By the mid-1950s, the movement's popularity had declined in the United States, a victim of anti-Communist suspicion. Communist leaders, too, had by then rejected the movement as a capitalist ploy to contain the socialist revolution. Nevertheless, world federalism retained appeal throughout most of the rest of the non-Communist world as a relatively mainstream pacifist movement.[19]

In February 1955, Russell urged British members of Parliament who were associated with the World Association of Parliamentarians to host some sort of meeting of international scientists to promote the disarmament agenda. Four months later, at Russell's request, Rotblat and Rabinowitch began a more serious organizing effort on behalf of the group, issuing invitations to the rectors of universities around the world. They scheduled the meeting for early August with the hope of attracting scientists on their way to the UN meeting in Geneva. In the end, they assembled 130 representatives from twenty countries—including, somewhat to their surprise, a delegation of four Soviets that included Professor A. V. Topchiev, permanent secretary of the Soviet Academy of Sciences.

At first, none of the Westerners gathered in the ornate chambers of the London County Council knew what to make of the Soviet functionaries in their midst. Topchiev, speaking through his translator (and presumed minder), initially insisted that the conference focus on the peaceful uses of atomic energy. Soon enough, however, he dropped his demands. By the time the meeting had ended, the Soviet delegation had agreed to cosponsor a slightly altered version of Russell's resolution calling for nuclear disarmament. Afterward, Topchiev's delegation continued on to the UN meeting in Geneva.[20]

By all accounts, scientists from East and West found their encounters in both London and Geneva informative and inspiring. Under Russell's nominal leadership, scientists passionate about disarmament made plans for a much larger meeting where they could bring their unique, objective perspective to bear on the world's problems. In an attempt to keep planning for the meeting independent—by which he meant "not Communist"—Russell distanced the movement from Joliot-Curie and the World Federation of Scientific Workers, putting Rotblat in charge. After a series of delays, including scuttled plans to hold a meeting in New Delhi at Nehru's invitation, the first international Conference on Science and World Affairs finally took place at Pugwash, Nova Scotia, from July 6 to July 10, 1957. Despite sporadic later attempts to distance the organization from Cyrus Eaton, a left-wing industrialist who sponsored the first meeting in his beloved fishing village, the movement would henceforth be known as Pugwash.

: :

From the beginning, the Western participants in Pugwash, particularly Rabinowitch, emphasized what they saw as an asymmetry in the character of Pugwash participants. While US, British, and Canadian representatives described themselves as self-selected scientists who attended because of their own personal commitment to international arms control, they portrayed the Soviet participants as either apparatchiks or, more generously, well-intentioned but tightly controlled subjects of the Soviet system. Of course, Soviet authorities said much the same thing about the Americans. Both allegations held some truth, but only the Americans used the idea of scientific freedom as a bargaining chip in the court of world opinion.

Rabinowitch commented on this supposed asymmetry in almost every piece he wrote on Pugwash for the *Bulletin of the Atomic Scientists*—which was, not coincidentally, received at Soviet nuclear installations. In his wrap-up of the 1955 meeting in London, he wrote, "It has been obvious that the Soviet scientists were not free agents to the same extent as their Western colleagues, but were bound by instructions . . . Nevertheless, it was a new and promising experience to be able at least to *discuss* the matters . . . with Russian scientists as individuals." His first reflection on Pugwash, published just a few months after the first meeting in Nova Scotia, sounded similar themes, noting that the movement "must be carried on by the scientists themselves independently of all outside organizations, official or private," though, he acknowledged, "This is easier said than done."[21]

In spite of or perhaps because of these perceived restrictions, Rabinowitch and Rotblat did everything they could to ensure opportunities for informal interaction. The twenty-two participants from ten nations who attended the first Pugwash meeting shared accommodations in old railroad sleeping cars, played croquet, and strolled along the rocky shore in the long twilights of a Nova Scotia summer. Cyrus Eaton's hospitality, which included everything from arranging transportation to providing participants with carefully chosen neckties, ensured that the scientists could focus on an energizing give-and-take of ideas instead of logistics. The result, Muller wrote in a thank-you letter to Eaton, was an "extraordinary experience" that offered the potential for "mutual understanding and reconciliation."[22]

At least early on in the movement, there was some accuracy to these claims of Western scientists' independence. Although the seven US participants could hardly be considered outsiders, they came to this first Pugwash meeting representing themselves, without official credentials. The three Soviets in attendance, in contrast, had to be approved by the Central Committee of the Communist Party. Whether because of the presence of these Soviet scientists or because of the prior London meeting's association with the World Federation of Scientific Workers, the CIA regarded these early gatherings as Communist propaganda exercises. And yet, whatever goals the Politburo had in mind, the Pugwash meetings offered Soviet scientists an opportunity to mingle and exchange views with their foreign colleagues, even if—as the participants were all too well aware—

the minders were never far off. As Victor Weisskopf, a theoretical physicist and Pugwash stalwart, recalled in his memoir, "the USSR's restrictive policies worked in our favor by putting us in contact with very highly placed people."[23]

This aspect of Pugwash soon began to attract the attention of US foreign policy officials. A few months before the third Pugwash meeting, scheduled for Kitzbühel, Austria, in September 1958, to coincide with the Second UN Conference on Peaceful Uses of the Atom, State Department staffer Halvor Ekern asked his superiors whether State needed to "establish any particular liaison, surveillance, or other association with the group." Given Pugwash members' stated independence from government connections, Ekern feared that any official State Department outreach might be "counter-productive and potentially misunderstood," but he wondered whether an "informal approach [from State] by a fellow scientist" might nevertheless advance agency goals. After all, he noted, the State Department's aims—in this case, a solid footing for its own International Agency for Atomic Energy—were "not inconsistent with their [Pugwash's] objectives."[24]

Nothing seems to have come of Ekern's suggestion in 1958—or at least, nothing that made its way to the open record—but from that point on, Pugwash makes increasingly frequent appearances in State Department and CIA documents (and not just as a probable Communist front). For their part, some of the Americans associated with Pugwash began actively seeking government recognition for their movement, as nuclear test ban negotiations between the United States and the Soviet Union heated up. Since 1956, the Ban-the-Bomb effort had only grown stronger, with grassroots organizations popping up around the globe. Pauling's petitions, delivered to the UN in January 1958, represented only one strand of an activist movement that included mass protests and civil disobedience. Meanwhile, fitful bilateral, multilateral, and UN-led disarmament meetings kept the issue on the international diplomatic agenda.[25]

In March 1958, Khrushchev scored a global propaganda victory—and surprised his own nuclear scientists—by issuing a unilateral nuclear testing moratorium. Historians who have reviewed the archival evidence suggest that Khrushchev, having recently consolidated his power as head of the Soviet government, sincerely believed that a nuclear test ban could increase global security without damaging the Soviet Union's military position. He also, of course, recognized the propaganda potential of a unilateral test ban. At that point, President Eisenhower had little choice but to take the prospect of nuclear test ban talks more seriously.[26]

Test ban negotiations raised any number of international political issues, from the strength of the US commitment to NATO, to French aspirations for a nuclear weapon, to the widening split between the Soviet Union and the People's Republic of China. Eisenhower and his advisors, however, insisted on defining obstacles to a test ban in technical terms. For the Americans, who blithely assumed that the Soviets could not be trusted, the question turned on the issue of enforcement. What if one country kept its word and the other did not? Existing monitoring techniques could (and did) detect nuclear explosions in the atmosphere and in the oceans, but nuclear hawks warned that it would be nearly impossible to distinguish certain types of underground nuclear explosions from earthquakes. US foreign policy officials therefore declared that the most important obstacle to a comprehensive test ban was technical—namely, the difficulty of detecting clandestine underground nuclear explosions. In other words: don't trust; verify.[27]

The Americans' insistence on discussing technological questions as if they had no politics explains why the first multilateral talks specifically on a comprehensive test ban, also held in Geneva in the summer of 1958, focused on detection. It also explains why the United States primarily sent scientists, while the Soviet Union sent senior diplomats. As Herbert (Pete) Scoville, the CIA's assistant director for scientific intelligence, wrote in his analysis of the event, the Western delegation (which also included representatives from France, Canada, and the United Kingdom) "succeeded in maintaining the position that the conference was technical, not political." This position was, of course, political. But this obsession with defining obstacles to a test ban agreement as primarily technical helps explain why the US scientists involved in Pugwash saw disarmament as their problem to solve, and why foreign policy officials were willing to work with them—but quietly and, as always, only at arm's length.[28]

∷

By 1960, Pugwash had settled into an organizational routine. An international Pugwash Continuing Committee set meeting locations and agendas and issued formal statements; national continuing committees recruited participants and issued invitations. As head of the US Pugwash Continuing Committee, Rabinowitch was responsible for finding a more stable source of funding to get US Pugwash participants to their destinations. As head of a nominally nongovernmental group that the CIA once suspected of being a Communist front, Rabinowitch made some surprising choices.

One solution, already in evidence at the Kitzbühel meeting, was to schedule events to take advantage of attendees' dual roles as critics and government advisors. Bentley Glass, for instance, assured Rabinowitch that the AEC would cover his travel costs to and from Geneva. Glass would be attending the UN meeting on the AEC's behalf, as a member of that agency's Advisory Committee for Biology and Medicine, but he would attend Pugwash to address the consequences of fallout produced by the AEC's own weapons. Few Pugwash participants—least of all Glass and Rabinowitch—seem to have been troubled by what, in other circumstances, might easily be interpreted as a conflict of interest, presumably because they saw their work as objective, transparent, scientific, and in service to the broader good. Like the National Science Foundation's scientific staff, Pugwash's scientists easily compartmentalized their roles during this time period.[29]

Starting in 1959, Rabinowitch began recruiting US scientists with ever-closer ties to government agencies and the Eisenhower administration, including several members of the President's Science Advisory Committee, to join the Pugwash movement. His efforts produced mixed results. When the State Department declined his request to formally endorse the organization and fund it to the tune of $25,000, Rabinowitch asked his contacts at State what he could do to reverse the decision, from renaming the conference to inviting different participants. No funds were forthcoming, but by the summer of 1960, Rabinowitch was citing State Department "support" and "encouragement" in his invitations to participants.[30]

The organization's finances stabilized in the summer of 1960, in part because of a successful appeal to like-minded small donors and in part because of a new partnership with the American Academy of Arts and Sciences (AAAS). Established in 1780, the Boston-based honorary society traditionally fostered work in the natural sciences and the humanities. During the Cold War, however, the AAAS's relationship with defense intellectuals at MIT and Harvard pulled it into the fray of national and international politics. Starting in 1960, the AAAS received publication funds for its journal, *Daedalus*, from the Congress for Cultural Freedom and conference support funds from the Farfield Foundation, the CIA's pass-through organization for the congress. Around that same time, the AAAS took on administrative oversight responsibilities for a number of quasi-governmental arms control studies, as well as financial responsibility for the US Continuing Committee for Pugwash.[31]

The US Continuing Committee itself does not appear to have received CIA funds via the AAAS—or if it did, intelligence analysts at the CIA didn't know about it. A 1961 CIA account of a Pugwash meeting, for example, described it as "an unofficial, privately supported, association of scientists." That same document, however, also approvingly noted that reports on a recent meeting had been received from more than half of the US Pugwash delegation. Even without a direct transfer of funds, it is clear that Rabinowitch steered the US delegation toward cooperation with the government in test ban negotiations.[32]

This relationship was closest from 1960 to 1962, a period of tense negotiations that spanned two presidential administrations, reaching its zenith at the sixth Pugwash meeting in December 1960. The conference had originally been scheduled for September, in Moscow, but the deep freeze in US-Soviet relations that summer raised doubts about the meeting's prospects. In May, two weeks before a scheduled Paris Peace Summit between the United States, the Soviet Union, the United Kingdom, and France, Soviet surface-to-air missiles had forced down a top secret high-altitude American U-2 plane. At first, NASA claimed responsibility for the incident, calling the flight an errant "weather mission," but when Khrushchev produced the plane, the pilot, and some of its aerial photographs, Eisenhower was forced to admit that the United States had been conducting covert high-altitude aerial surveillance flights in Soviet airspace. Khrushchev accused the United States of spying, Eisenhower refused to apologize, and the Peace Summit collapsed.[33]

Given the breakdown in negotiations between the United States and the Soviet Union, as well as the upcoming US presidential election, Rabinowitch and the other Americans wanted to postpone the Moscow meeting. Rotblat objected. In August, Rabinowitch preempted Rotblat, cabling Topchiev that the meeting must be rescheduled. "The only people who would be certain to go to Moscow in September," Rabinowitch wrote, "would be those without close relation to political leaders in either party." Rotblat was furious. For Rotblat and Russell, the founders of Pugwash, the entire point of the organization was to bring together scientists outside formal political channels. Rabinowitch, in contrast, now insisted that Pugwash depended on official connections for political effectiveness.[34]

On a chilly November day, less than two weeks after Democrat John F. Kennedy defeated Eisenhower's vice president Richard Nixon at the polls, US Pugwashites gathered for a briefing with State Department officials. At the head of the table sat Walter Whitman, the State Department's new science advisor. Whitman knew the twelve Pugwash attendees well, having chaired the AAAS committee that oversaw the US Continuing Committee's finances. Though the strategic justification for the meeting went unsaid, Whitman had reached out to the Pugwash movement to prepare the scientists for their role as backchannel diplomats. While influential citizens have always engaged in freelance diplomacy, in the early 1960s the State Department was beginning to embrace a more explicit role for private citizens as government negotiators, a technique now known as Track Two diplomacy.[35]

Whitman opened the gathering with a reminder that the meeting was informal, so as to maintain the "independent character" of Pugwash. This was, however, a somewhat constrained definition of independence. The discussion at Foggy Bottom that day covered everything from the State Department's preference that Pugwash not issue consensus statements, to the caliber and political independence of the Soviet scientists in attendance, to how the Americans should conduct themselves during toasts. Charles "Chip" Bohlen, a former ambassador to the Soviet Union, counseled the scientists to "avoid statements which could be classified as an attack on government policy." Paul Doty, a Harvard biochemist and outspoken advocate of disarmament, assured Bohlen that he would "minimize any basic conflict" between his own position and that of the US government.[36]

Rabinowitch occasionally stumbled along this precarious tightrope, but the delegation as a whole kept its balance. Just in case, the US ambassador to the Soviet Union, Llewellyn "Tommy" Thompson, kept close tabs on the US participants. Each day, the Americans took their places among the seventy-five seats at the U-shaped table in the Soviet House of Friendship; each night, Thompson debriefed a select group and prepared them for the next day's negotiations. Officially, the Pugwash meeting focused on technical issues of disarmament and the problem of surprise attack. But between performances at the Bolshoi and a reception at the US Embassy, Rabinowitch, Doty, Jerome Wiesner, and Walt Rostow (the last two soon to be appointed to the Kennedy administration as scientific advisors) pushed their Soviet colleagues on Cuba, release of the captured U-2 pilot, and Soviet propaganda campaigns—issues notably devoid of scientific content.[37]

Government officials in the United States clearly saw the same opportunity in Pugwash's "informal character" that Victor Weisskopf had commented on in 1957. Even if the Soviet scientists in attendance lacked formal political power, they obviously had *some* political pull, or they would not have been allowed to mingle with Western scientists in Moscow or anywhere else. Rostow, for instance, described the meeting as an "extraordinary experience" in which the Americans "met and lived with a 'top drawer' group of Soviet academicians and military people." Wiesner, for his part, reported that the "Soviets appeared to talk more freely" than they had at previous meetings. In its review of intelligence gathered at the meeting, a CIA analyst remarked that, to a man, the US participants came away believing that Soviet scientists wanted complete and total disarmament. He also noted that the Americans believed that the Soviet scientists' views "probably reflected official thinking." On that point, the CIA remained agnostic.[38]

::

For advocates of disarmament inclined to work through the state, including Rabinowitch, Glass, Doty, Wiesner, and Rostow, these ambiguous relationships offered an unprecedented opportunity to make the world a safer place. But the US position on arms control, in the form of a test ban and a limited supply of weapons to facilitate deterrence, fell far short of the original goal of the Einstein-Russell Manifesto: full and total disarmament. An exchange between Russell, Rabinowitch, and Glass in 1959 illustrates the growing distance between these two objectives. Pugwash's erstwhile patron Cyrus Eaton had cabled Russell, asking the group to endorse Khrushchev's most recent call for disarmament. Rabinowitch and Glass, speaking on behalf of the US delegation, refused, citing the need for Pugwash to remain "independent" of official government positions. As Rabinowitch put it, "Our movement would be immediately split if we tried to use it for public support of the politics of one of the great powers, however desirable they may appear to most of us."[39]

Such attitudes proved particularly frustrating to more radical US disarmament advocates. Even while officials at the State Department, the CIA, and the White House were exploiting Pugwash as an effective backchannel for negotiations, the Senate was harassing one of the original signers of the Russell-Einstein Manifesto. Linus Pauling, fearless to the point of recklessness, had taken an aggressive stance against the US government from the earliest days of the fallout debate. In April 1958, Pauling joined Russell and sixteen others in a lawsuit against the AEC for the damage its nuclear tests were inflicting on the world's present and future children. At one press conference, Pauling estimated that genetic damage from radioactive fallout would cause stillbirths and birth defects in 5 million children. He resigned his position as head of Caltech's Division of Chemistry and Chemical Engineering to free up time to deliver antiwar speeches to audiences from London to Hiroshima. And, crucially, he circulated antinuclear petitions among scientists the world over.[40]

Thomas Dodd, the junior senator from Connecticut, found it inconceivable that Pauling had gathered 11,000 scientists' signatures without outsider—presumably Communist—help. In June 1960, he ordered Pauling to appear before the Senate Internal Security Subcommittee to explain himself. Following Senator Joe McCarthy's downfall in 1954, red-baiting hearings had become relatively rare occurrences on Capitol Hill, but Dodd found Pauling's persistent flouting of the mores of apolitical science a tempting target. Dodd had miscalculated. Instead of pleading the Fifth Amendment—the standard defense against self-incrimination for those unwilling to name names—Pauling invoked his First Amendment rights of freedom of speech and petition. This turned out to be a winning strategy: the press lauded Pauling's steadfastness and chided Dodd for resurrecting the spirit of McCarthy.

Pauling emerged from the first hearing strengthened instead of humiliated. Dodd's committee nevertheless ordered the scientist to return in October, this time with a list of names of those who had helped him rally antinuclear support. Pauling, who had no intention of revealing those names, steeled himself for a contempt-of-Congress charge. And yet, at the critical moment, Dodd backed down. Instead of having Pauling arrested, he read a long list of Pauling's affiliations with left-wing organizations into the record. The hearing ended with a whimper, with the media mostly ignoring Dodd's charges. This outcome only emboldened Pauling, who mounted a media campaign against McCarthyism in general and the Senate committee in particular. In print and on television, he described Dodd as "evil."[41]

Dodd retaliated by investigating disarmament groups, including Pugwash, in the spring of 1961. His committee's report on the movement described it as a vehicle for Communist propaganda, primarily based on its prior—and, unbeknownst to Dodd, deteriorating—relationship with Pauling. The report rattled Pugwash, but, given the continued assurances of the State Department's support and the press's studied indifference to Dodd's crusade, its leaders carried on.[42]

Pauling, meanwhile, continued his antinuclear campaign undeterred. When first the Soviet Union and then the United States resumed nuclear testing in September 1961 after yet another series of peace negotiations yielded little progress, he telegrammed Khrushchev in the name of "the whole human race and unborn children who will be damaged by radioactive fallout." In an accompanying press release, Pauling claimed to speak "for scientists all over the world." Given the timing only days before two upcoming Pugwash meetings in Stowe, Vermont, some members of the press inevitably assumed that Pauling—in attendance for the first time since Kitzbühel—spoke on behalf of Pugwash. The US Continuing Committee hired a publicist to clarify the distinction, but the rift between Pauling and the organization men was growing.[43]

Pauling's continued advocacy on behalf of total nuclear disarmament forced the US leaders of Pugwash to reckon with the hypocrisy that threatened to derail their movement. After the Stowe conference, Pauling chastised the US Continuing Committee (now consisting of Rabinowitch, Glass, and Harrison Brown, foreign secretary of the National Academy of Sciences) for inviting persons with military rather than scientific expertise. He particularly objected to the presence of such national security figures as Donald G. Brennan, founder of the Hudson Institute, a defense think tank, and Henry Kissinger, then a young professor at Harvard with close ties to the National Security Council, the CIA, and the Department of Defense. The organization men—Rabinowitch, Glass, Brown, and Doty—held firm, pointing to differences of opinion as synonymous with American liberty. As Brown put it, "We are not monolithic and I see little reason to pretend that we are." And yet, Pauling's hunch that Brennan couldn't be trusted was on the mark. Brennan regularly submitted reports on his experiences at Pugwash to officials at the State Department and the CIA.[44]

Such was the context in which Bentley Glass published "Scientists in Politics" in the *Bulletin of the Atomic Scientists* in May 1962. Writing at a moment when scientists seemed to have unprecedented sway over national policies and politics, Glass attempted to differentiate between acceptable and unacceptable modes of political involvement. As his primary example, he chose the public debate over fallout, contrasting Pauling and Teller's heated exchanges with the cool, "carefully chosen words" of studies produced by the NAS and the British Medical Research Council. The differences, he wrote, revealed the "all-too-common tendency" for scientists to yield to the temptations of political passion.

Glass warned that a similar dynamic threatened to undermine arms control negotiations because detection depended on scientific and technical expertise. "The trouble," he wrote, is that the scientists involved "have forsaken the scientific approach to the problem" in favor of political machinations and table debates. Both Pauling and Teller, he argued, had been "carried away by the conviction of the rightness of their opinions." They made "extrapolations from the existing facts . . . or are even lured into flat misstatements." The solution? Pugwash, which offered an attempt to reach agreement through "the international character of science itself and the objectivity of its practitioners." At the Pugwash meetings, Glass explained, scientists could overcome feelings of mutual distrust to discuss the technical problems of arms control "objectively."[45]

Glass listed several examples of Teller's speaking beyond his expertise, but of Pauling he said nothing further. But this was enough for Pauling, who never shied away from a lawsuit, to threaten Glass, the *Bulletin*, Rabinowitch, and another publication that reprinted the article with libel suits. The specifics of his threatened defamation suit rested on Glass's claims that Pauling had made "flat misstatements" and worked beyond his areas of expertise, but Pauling seems to have been equally infuriated by Glass's description of arms control as a technical problem and Pugwash as an objective, apolitical organization. What gave Glass, a geneticist, the expertise to advise the US government on the feasibility of detecting underground nuclear explosions?[46]

In the end, Pauling dropped his suit. He eventually resigned from the *Bulletin*'s board of sponsors, but only after subjecting Glass and the *Bulletin* to a year's worth of demands. He also announced that he would henceforth boycott Pugwash so long as the delegation included Glass, Rabinowitch, and other "low-calibre" scientists. When Russell took Pauling's side, and as the volley of letters from Pauling continued, Glass offered to resign from the US Continuing Committee. At Rabinowitch's insistence, Glass remained, but from that point on, Pauling had little use for the *Bulletin*, Glass, Rabinowitch, or Pugwash itself.[47]

∷

How could Pugwash simultaneously be an organization dedicated to harnessing the objective powers of outsider scientists to solving the world's problems and an establishment venue for informal discussion? It couldn't, which is why in 1962 Pugwash split into separate factions. After several months of hand-wringing, the Pugwash leadership agreed to reorganize. Larger, more public meetings would bring together a wide variety of scientists from the developing and developed world to discuss an open-ended set of questions concerning science, technology, and scientific responsibility. A separate track of smaller, more elite study groups, often with close ties to government, would address specific topics—notably, arms control.[48]

The most influential of these smaller groups would be the Soviet-American Disarmament Study Group (SADS), headed by Harvard biochemist Paul Doty. The group maintained a purposefully ambiguous relationship with Pugwash. While Pugwash continued to scrape by on small donations, SADS received $325,000 in grants from the Ford Foundation. In theory, SADS and Pugwash were separate organizations. In practice, it is clear that Doty, his Harvard colleague Bernard Feld (the new chair of the AAAS committee overseeing the US Continuing Committee), and the CIA hoped that different audiences might interpret the situation differently. To Pugwash stalwarts concerned that the organization had grown too close to the US government, the existence of SADS offered a way to separate state and non-state efforts. To US government officials who viewed Pugwash with suspicion, the separation offered reassurance that scientific advisors to the departments of State and Defense were keeping their distance from radicals and peaceniks. But for key Soviet scientists who had received blanket clearance to attend Pugwash events, colocated meetings and overlapping personnel removed barriers to participation.[49]

For a variety of reasons ranging from political obstacles to the deaths of key participants, the SADS model proved difficult to implement. Nevertheless, a CIA memo on the only SADS meeting to actually take place made these various motivations explicit. Filed in May 1964, just weeks before the meeting, the document relied on information provided by Paul Doty and described the group as "Associated with Pugwash." "This Group," the author explained, "was formed" to assemble "a small number of carefully selected participants" in a setting that would allow "much more continuity and privacy than is possible under the Pugwash arrangement." Note the passive voice.[50]

By this point, however, it was too late for the broader Pugwash movement to repair its relationship with State. Ever since the 1960 Moscow meeting, State Department officials had warned US Pugwashites, particularly Rabinowitch, that they risked being used as tools for Soviet propaganda. While Pauling's continued activism certainly didn't help, the real problem, from the State Department's perspective, was the organization's insistence on issuing consensus statements that inevitably endorsed disarmament rather than arms control. By October 1963, an unnamed diplomatic official remarked that "the value in Pugwash of U.S.–Soviet conclaves is diminishing," particularly as discussions shifted from "scientific to scientific-political." The following year, Ambassador Thompson, who had earlier expressed such optimism about the power of Pugwash to further US goals, informed Feld that he had become "extremely dubious about the usefulness of the entire exercise."[51]

Pugwash had never been "nonpolitical," but the changing character of the arms control debate eliminated the purpose of that once useful fiction. The on-again, off-again nuclear test ban talks continued into 1963. Khrushchev maintained a consistent commitment to a comprehensive test ban throughout these discussions. The US foreign policy establishment's opinion on this issue vacillated wildly but continued to run aground on the supposedly technical issue of verifying underground nuclear explosions. It took the near-apocalypse of the Cuban Missile Crisis in October 1962, when the Soviet deployment of nuclear missiles just 90 miles off the coast of Florida nearly sparked a nuclear war, to startle both Kennedy and Khrushchev into more flexible negotiating positions.

In July 1963, Khrushchev signaled that he would accept a more limited treaty banning nuclear explosions in water, air, and outer space. Since existing technologies could easily detect these sorts of explosions, negotiations proceeded quickly. Allowing continued nuclear testing underground rendered the technical problems of seismic detection moot. On August 5, 1963, representatives from the United States, the United Kingdom, and the Soviet Union signed the Limited Test Ban Treaty, prohibiting participating nations from testing atomic or nuclear weapons "in the atmosphere; beyond its limits, including outer space; or under water, including territorial waters or high seas."[52]

By forcing nuclear explosions underground, the Limited Test Ban Treaty (mostly) solved the original motivating problem for many Pugwash scientists: fallout. What the Limited Test Ban Treaty most definitely did *not* do was remove the threat of nuclear annihilation. The nuclear powers would keep their existing weapons; they could even develop new and more powerful ones, so long as they could figure out how to do it underground. Nor did the treaty solve the intractable problem of nonproliferation. France joined the nuclear club in 1960, even before the treaty was signed; China would do so in 1964. Israel, India, and South Africa continued to pursue their nuclear research programs.[53]

Nothing illustrated the gap between those two goals—a test ban and disarmament—better than the official reaction to the news that Linus Pauling had been awarded the Nobel Peace Prize. Having declined to name a recipient in 1962, the prize committee announced Pauling's selection as the previous year's winner on October 10, 1963, the day the treaty took effect. Pauling, of course, was honored and delighted, but administrators at Caltech offered only lukewarm congratulations to their most famous employee. *Time* rehashed Pauling's Communist associations, and *Life* described the award as an "insult to America." While the test ban treaty elicited cheers and sighs of relief, Pauling's pathbreaking work on behalf of disarmament catalyzed another round of red-baiting.

Pauling's approach to science and politics, so different from that advocated by Rabinowitch and Glass, played better in Oslo than in Washington.[54]

::

Pugwash, like so many aspects of science in foreign affairs, was shot through with contradictions. Famously described as a group of independent scientists unaffiliated with any government, Pugwash offered a reliable backchannel for diplomats and intelligence officers in both the United States and the Soviet Union. Some US scientists, like Glass and Rabinowitch, endorsed this closer relationship with the government, while more radical scientists, like Pauling, thought Pugwash's effectiveness depended on its separation from the state. The willing cooperation of sympathetic US scientists gave the State Department and the CIA access to privileged information about the Soviet stance in negotiations, but those agencies proved resistant to the content of that intelligence— namely, that Khrushchev sincerely desired a comprehensive test ban and possibly even greater disarmament measures. Most members of the US foreign policy establishment seem to have regarded Pugwash as an organization by and for credulous scientists that might nevertheless prove useful in a pinch.

Viewed through the lens of cultural diplomacy and even domestic politics, Pugwash has greater significance. Encouraging, or at least not discouraging, individual scientists' place at the table allowed the US government to simultaneously participate in and dodge disarmament negotiations. The State Department's tolerance for Pugwash, at least in the early days, reassured scientists of their vaunted position in US democracy after the Oppenheimer hearings shook their confidence. And at least some foreign policy officials, notably State's Walt Whitman and the CIA's Pete Scoville, understood the value of keeping the government's scientific critics on a short leash.

Perhaps more to the point, Pugwash's global commitment to a supposedly apolitical objectivity, as stated in countless press releases, consensus statements, and articles in the *Bulletin of the Atomic Scientists*, demonstrated to all who would listen a peculiarly Western way of treating scientific and technical knowledge. Just as the US Information Agency's Hal Goodwin recommended, the US government's support of Pugwash demonstrated to its free world allies that the United States respected science and scientists. And at least some of the physicists working at the Soviet nuclear installations heard this message loud and clear. No amount of cynicism can negate the fact that Pugwash was one of the only places where a select group of Soviet scientists could gather with their international peers for relatively free and open discussions, an experience that deeply shaped their future expectations of civil society.[55]

In this light, Glass's article on science and politics can be read not so much as a criticism of Pauling and Teller but more as reassurance: to the Soviet scientists he probably knew were reading his words; to the US scientists, like himself, who thought they could achieve more change from within the system than from without; and to US government officials, who anxiously watched over their unofficial emissaries. Always aspirational, scientific neutrality nevertheless remained a key value of US cultural diplomacy through most of the Cold War.

7 : Developing Scientific Minds

Lyman Hoover, the Asia Foundation's representative in the Republic of China, needed his bosses in San Francisco to make a decision. Since January 1966, he had been urging the foundation to fund the printing of a biology textbook prepared by Dr. Koh Ting-pong, one of the leading biology educators in Taiwan. As chairman of the Biology Subcommittee of the Ministry of Education's Science Curricula and Textbook Revision Committee, Koh had urged the government to adapt one of the US Biological Sciences Curriculum Study's textbooks for use in Taiwanese classrooms. With the blessing of the minister of education, Koh and an illustrator from his university, Lee Sheng-hing, spent a year at BSCS headquarters in Boulder, Colorado, adapting the text and illustrations for the Chinese market. In the meantime, though, a pirated translation of the book, stolen from the revision committee's earlier work, had begun to circulate on the island. The Asia Foundation faced a choice: invest up to $10,000 in underwriting Koh's authorized adaptation, or lose the nearly $35,000 that it, along with the National Science Foundation, had invested in the project thus far.[1]

The BSCS textbooks were no ordinary textbooks, and the Asia Foundation was no ordinary foundation. In the years immediately following the launch of *Sputnik*, the NSF spent millions of dollars developing high school science curricula designed to recapture the US lead in science. The BSCS was arguably the most successful of these programs, with its textbooks used in approximately 50 percent of US high school biology classrooms within a few years of their introduction. More so than the post-*Sputnik* books in physics, chemistry, or mathematics, the BSCS's curriculum promoted a philosophy of science alongside discussions of modern genetics, evolutionary biology, and forest ecology. The program's approach to learning stressed the nature of inquiry, with both students and teachers urged to ask questions, consider evidence, and draw their own conclusions based on laboratory observation.[2]

The BSCS's emphasis on inquiry, hypothesis, and rationality made it an excellent choice to advance the Asia Foundation's goals, not only in Taiwan but throughout Asia. Nominally a nonprofit foundation dedicated to fostering the development of democratically oriented voluntary organizations in Asia, the Asia Foundation was in reality one of the CIA's largest proprietary organizations. Unlike the Congress for Cultural Freedom, which was overseen by a CIA officer but largely run by freelance intellectuals, the Asia Foundation's operations were almost entirely under the CIA's control from the time of its creation in 1954 until its cover was blown in 1967. Operating under codename DTPILLAR, the foundation spent approximately $8 million a year backing Asian-led organizations and their US collaborators. Its representatives formed partnerships with Asian elites who were seeking sponsors for projects that would benefit both partners. Hence the urgency, for Hoover, of supporting Koh's efforts.[3]

Understanding how and why the CIA came to support textbook translations across Asia requires a closer look at the broader phenomenon of post-*Sputnik* curriculum reforms and why the BSCS, in particular, attempted to internationalize its programs. Though biological education might not be as dramatic as solar power, desalination plants, or nuclear reactors, US officials embraced the new curricular approaches as just the kind of scientific breakthrough that might restore the United States' reputation for scientific leadership, especially in the developing world. The BSCS's goals and those of the Asia Foundation didn't entirely overlap, but the leaders of the two organizations shared a belief that scientific thinking, cultivated through inquiry-based learning, offered the key to a democratic future, both at home and abroad.[4]

∷

In an ideological war, everything is a competition. A presentation delivered to the National Security Council in 1956 warned that the Soviet Union was outpacing the United States when it came to scientific and technical training, with a series of ominous charts and graphs pointing to a "widening gap" in the number of scientific and engineering doctoral degrees. The Soviet Union's supposed ability to "select and train" any student with scientific aptitude had given it an edge over the United States, where students were free to pursue whatever topics interested them. Science, it seemed, did not appeal to the typical US teenager. According to NSF director Alan Waterman, only one in four US high school students took any science classes at all.[5]

This anxiety manifested at a moment when educators were struggling to find ways to match the high school curriculum with the diversity of US high school students. By the 1950s, US high schools, once reserved for the children of elites, had become multifunctional institutions idealized as training grounds for college students, homemakers, and industrial workers, all of whom should become upstanding citizens. Even as legal and de facto segregation created vast disparities in the resources available to white, black, and Latino children, attending public high school, at least for a few years, was becoming a standard and accepted part of the transition to adulthood in the United States. By the end of the 1950s, the vast majority—90 percent—of high-school-aged children in the United States attended high schools. What should they be taught?[6]

Two competing visions for the US high school had emerged in the years immediately following World War II. In *General Education in a Free Society* (1945), a committee of Harvard faculty members, chaired by James Conant, beat the familiar drum of high schools as liberal arts training. The US Department of Education, however, backed a new plan of "life adjustment"—a less academically rigorous plan of general education geared toward student interest and functional relevancy. In the words of the Commission on Life Adjustment Education for Youth, life adjustment was "that which better equips all U.S. youth to live democratically with satisfaction to themselves and profit to society as home members, workers, and citizens." Courses on home life, social interaction, and physical education played increasingly prominent roles in the high school curriculum. But as the life adjustment approach became more popular, many intellectual leaders, including Conant, blamed students' focus on individual and social development for the country's perceived shortfall of scientists and engineers. Caught off guard by the Soviet nuclear test in August 1949, critics feared that Americans failed to appreciate scientific and technological achievement.[7]

Over the course of the 1950s, Conant attempted to reconcile these two approaches through attention to the methods of science instruction rather than its content. His numerous books, including *On Understanding Science* (1947), *Education in a Divided World* (1948), *Science and Common Sense* (1951), *Modern Science and Modern Man* (1952), and *Education and Liberty* (1953), consistently argued that all educated Americans needed a working familiarity with the nature of science. Conant, who had resigned the presidency of Harvard in 1953 to become US high commissioner for Occupied Germany, did not advocate this approach because he thought all students should become scientists. As he wrote in his Carnegie Corporation–sponsored study, *The American High School Today* (1959), the effective US school could "at one and the same time provide a good general education for *all* the pupils as future citizens of a democracy, provide elective programs for the majority to develop useful skills, and ed-

ucate adequately those with a talent for handling advanced academic subjects."
As he had since the late 1940s, Conant praised the potential of instruction in
science—particularly the history of science—to help students learn to think for
themselves. By the mid-1950s, in large part through Conant's influence, "science" and "democracy" had become increasingly entwined in US education.[8]

Scientific curriculum studies proliferated in the 1950s, beginning with the
University of Illinois Committee on School Mathematics in 1952 and MIT's
Physical Science Study Committee (PSSC) in 1956. While each of these programs had its own particular pedagogical slant, their leaders—almost all university scientists—shared two key assumptions: first, that scientific education
was too important to be left to educators; and second, that the pace of scientific
change required an emphasis on science as a way of knowing rather than on
scientific fact. All of the "new" curricular programs therefore emphasized the
concept of science as inquiry and stressed lab work, student-driven experiments,
and open-ended questions rather than rote memorization. As cognitive psychologist and Project Troy veteran Jerome Bruner put it, "What a scientist does
at his desk or in his laboratory, what a literary critic does in reading a poem, are
of the same order as what anybody else does when he is engaged in like activities—if he is to achieve understanding." By this logic, "the schoolboy learning
physics *is* a physicist, and it is easier for him to learn physics behaving like a physicist than doing anything else."[9]

Sputnik catalyzed the transformation of this nascent scientist-driven educational reform movement into a national cause. The 1958 National Defense Education Act included funding for education programs, administered through
the NSF and the Department of Health, Education, and Welfare. The majority
of the $1 billion earmarked for Health, Education, and Welfare funded 40,000
merit scholarships for students pursuing college science education, with the re-
mainder supporting graduate fellowships, programs to improve science and math
teaching in high schools (including funds to purchase laboratory equipment and
supplies), and special centers for foreign language instruction. Virtually all the
NSF money, in contrast, was earmarked for revising high school science curricula, including course outlines, textbook projects, and summer institutes for high
school teachers.[10]

By 1962, university scientists had created NSF-funded curricular programs
for high schools in mathematics (the School Mathematics Group), chemistry
(the Chemical Bond Approach and CHEMStudy), physics (the PSSC), biol-

ogy (the BSCS), and the earth sciences (the Earth Sciences Curriculum Project). Most of these programs were modeled on the PSSC, the MIT-based physics program created by Jerrold Zacharias. Under Zacharias's leadership, the PSSC created instructional films, laboratory modules, and textbooks. Its work successfully infiltrated US high school physics classrooms, at its peak supplying the materials for half of US physics courses. But while school corporations readily adopted the program, many teachers resisted it. The course was notoriously difficult for students as well as teachers, and the innovative instruction failed to attract new students to physics courses.[11]

The teachers had a better time with the biology textbooks being developed by the BSCS, a project of the American Institute of Biological Sciences. The BSCS program was intended to completely overhaul the biology curriculum, beginning at the secondary level. Within a year of its proposal by the institute's executive director, Hiden Cox, the BSCS had received an NSF grant for $143,000; soon the NSF granted an additional $595,000. The BSCS eventually prepared three sets of sophomore-level textbooks, laboratory manuals, and teachers' guides as well as numerous instructional films. The materials were identified simply by color—yellow, blue, and green—with each version focused on a particular aspect of general biology. Yellow took a cellular or evolutionary approach; blue stressed biochemistry and physiology; green emphasized ecology. During the first few years of the project, selected teachers used draft materials in typical tenth-grade classrooms. By the mid-1960s, any interested school corporation could purchase the materials from commercial publishers.[12]

Arnold Grobman, the BSCS's first director, recruited top-notch biologists to the task of reimagining the biology classroom. In an article in the *AIBS Bulletin*, he urged his colleagues to "somehow bring themselves together and to agree upon what things in biological science should be known by an educated person and how these things can best be taught." Grobman suggested that the biological community was responsible for the production of an educated citizenry. A number of leading US biologists, including H. J. Muller and Bentley Glass, answered the call. Muller served on the early advisory boards, while Glass chaired the BSCS Steering Committee from 1958 to 1965. The early records of the BSCS, scattered across the personal papers of scientists involved with the project, indicate that this was more than a plan to improve the scientific content of high school classrooms. For Glass, Grobman, and their colleagues, the BSCS offered biologists a critical opportunity to preserve the future of democracy.[13]

∷

Imagine that you're a fifteen-year-old high school sophomore. It's late August or early September 1962, and you and your parents have been told that your suburban high school is participating in a pilot program to try out a new biology curriculum. Your teacher has already handed out mimeographed copies of your experimental "blue" BSCS textbook. But before she starts the first lesson, she announces that you'll be taking a test to gauge your baseline knowledge of scientific concepts.

Pencil ready, you sit up in your seat. The test begins:

(1) The chief purpose of the science of botany is to:

(A) Teach farmers how to produce more food.
(B) Develop new drugs and medicines from plants.
(C) Provide explanations of how plants grow and reproduce.
(D) Tell us what kinds of plants will grow best in various kinds of soils.

Since "all of the above" isn't an option, you select (A). Your parents were raised on farms, after all. The questions get more confusing from there:

(7) It has often been said that published reports of scientific research are generally very accurate and honest.

(A) This is true because scientists are very accurate and honest people.
(B) This is true because only one answer can be correct in science.
(C) This is true because reported results can be checked by other scientists.
(D) There is little basis for this claim.

You respect your science teacher, so you're tempted to pick (A). On the other hand, this is a multiple-choice test, suggesting only one right answer is possible. You pick (B).

(26) In discussing our country's disarmament policy, a famous scientist declared that we must continue our experimenting with nuclear bombs. What is the best evaluation we can give to this scientist's statement?

(A) His conclusion is probably right, since he approaches the problem with a scientific attitude.

(B) His conclusion is probably wrong, because scientists seem to be destroying the world.

(C) His conclusions and reasoning are probably correct, because scientific results are the most reliable kind.

(D) His conclusions and reasons should be weighed according to his knowledge of international affairs.

Is this biology class or social studies? Baffled, you pick (C).

The last ten questions ask you to indicate your level of agreement with various statements describing scientists. You're slightly taken aback by their dim view of the scientific life:

(36) The scientist is not able to have a normal family life.

(37) Scientists are more likely to be mentally ill than people engaged in other types of work.

(38) The scientist is more likely to be unpatriotic than other people.

(39) All scientists should be employed by the government so that control can be kept over their findings.

(40) Most scientists are not religious.

(41) Most scientists are geniuses.

(42) The scientist seeks to find out the truth with no thought to the consequences of his work.

(43) Scientists are usually impractical in the way they try to solve the problems of everyday living.

(44) Most scientists are more than a little bit "odd."

(45) Things like the development of the atom bomb indicate that scientists have little regard for humanity.

Your teacher assures you that there are no right or wrong answers and tells you that you'll be asked to complete the test again at the end of the year to see what you learned from the course. You've taken lots of tests before but find this one unsettling. What kind of biology course is this? What, exactly, were its designers hoping you'd learn?[14]

Like the authors of civic biology textbooks at the beginning of the twentieth century, the BSCS's planners sought to teach the elements of biology that were necessary for the creation of sound, responsible, and productive citizens. But what might that information be, and how specific should it be to biology? Judging from the above examples—real questions from an impact test in 1962— the BSCS interpreted its mandate broadly. Its original list of twenty-eight objectives to be measured by the impact test is remarkably devoid of biological content. Instead, the impact test—an instrument specifically designed to measure how a student's attitudes changed by exposure to a course—tracked more general ideas about the nature of science. For example, a scientist "defines a problem in researchable terms," "uses some specialized tools," "bases his interpretations on the evidence at hand," and "does not base judgments on authority." An additional nine objectives, added a few months later, clarified that students should gain "an understanding of and appreciation for the scientific enterprise" by, for instance, recognizing that "people create the ideas in science" and distinguishing between science and technology.[15]

If these objectives sound remarkably general for a course in biology, this was by intent. As the BSCS's publicity materials so often pointed out, biology was the only science course that eight out of ten US high school students would ever take. Grobman, Glass, and the rest of the BSCS leadership interpreted this statistic as proof that they had a special responsibility to correct misconceptions about science and scientists. A student who successfully completed the course would be open-minded, inquisitive, and capable of distinguishing between fact and opinion, and she would accept scientists as part of the fabric of society.[16]

The BSCS's content and writing teams faced the hard work of incorporating these objectives into something resembling a high school biology course. They found the answer in the concept of nine "pervading biological themes" that would thread through all three books, regardless of their specific focus on ecology, evolution, or molecular biology: the nature of scientific inquiry, the intellectual history of biological concepts, genetic continuity, regulation and homeostasis, the relationship between structure and function, the biological roots of behavior, regulation and environment, biological diversity, and change over time (evolution). As Glass explained in a memo to the drafting teams, all of the biological themes ultimately contributed to students' understanding of evolution, "the most pervasive, most significant biological theme of all." But in discussing evolution, as in discussing contemporary genetics, the books should above all stress the empirical basis of science. As Glass put it, "Let us avoid presenting shovelfuls of authoritative facts and doses of scientific dogma. Rather, let us indicate clearly and frequently that scientific concepts and theories are based on *evidence*."[17]

The experimental blue BSCS textbook demonstrates both the strengths and limitations of this approach for the high school classroom. The thirty-three-chapter book, issued in three parts over the course of the 1960–1961 school year, was arranged according to scientific concepts rather than practical applications. For example, section seven, on genetics, contained chapters on the cellular and chemical basis of heredity, chromosomes, patterns of heredity, population genetics, and gene expression, but mentioned plant and animal breeding only as part of a historical narrative on the origins of modern genetics. While the text's authors claimed that they wanted to encourage all students to "think like biologists," in practice this translated into heavy-handed descriptions of "what biologists think." Typical divisions in the introduction, for instance, were titled, "The biologist sees similarity among living things," "How the biologist works," and "The biologist studies the effects of radioactivity on life."[18]

In keeping with both Conant's and Glass's admonishments on scientific "dogma," the blue version repeatedly stressed that today's theories might be superseded by future findings. The text explained scientific theories as part of the long history of ideas and downplayed the role of individual genius. The first chapter, on evolution, contains the following text:

> You may ask the question: Do biologists really believe the theory of evolution? The answer is that "believe" is hardly the right word. Scientists *accept* the theory of evolution because all the available evidence points to its being true. Evidence from fossils, evidence from the development of organisms, evidence from their comparative structure, evidence from ecology and from the fields of taxonomy, . . . genetics, and other biological fields all indicate the soundness of the theory.[19]

In case the student missed the point that evolutionary theory was not based on a personality cult, the following page remarked that "the theory of evolution was not composed by Darwin 'out of thin air.' The clues which led to his theory were gleaned from the discoveries of other men of science, and from his own studies and experiences as well." This is not to say that the textbook elided the contributions of individual scientists. The chapter on the mechanisms of heredity included lavish portraits of Gregor Mendel and Thomas Hunt Morgan. The text presented "evidence" for the chromosome theory of heredity and reminded students that scientific "proof" meant only that a theory continued to account for experimental evidence.[20]

The teachers' guide reinforced the point that students should gain a realistic idea of how modern science progresses. Echoing Vannevar Bush's warning from 1949, the guide proclaimed, "Science is not magic." Students needed to realize that the goal of science was understanding, not the production of useful tools or products: "[Science] is generated by inextinguishable human curiosity, and has no necessary connection with efforts to improve the circumstances of human life although such fruits do arise from scientific labor more generally than in any other way." (The correct answer to the first impact question was, apparently, [C].) Finally, and perhaps most importantly, "honesty and integrity are also of the essence of science."[21]

The BSCS's leaders envisioned laboratory exercises as the most efficient way for nature, with the assistance of high school teachers, to reveal the true character of science. The laboratory manual contained problems, classroom demonstrations, didactic exercises, and a few true experiments. The manual's introduction presented laboratory exercises as the *sine qua non* of understanding science—without them, "no matter how much you learn about the facts of science, you will never quite understand what makes science the force it is in human history." Students needed to be prepared for challenges and even failure in the laboratory. "Nature," the manual warned, "is a hard master and reveals her secrets only to those who are patient and careful, persistent to the point of obstinacy, and willing to spend time and energy without reckoning the cost."[22]

Lysenko's name is never stated explicitly in these texts, but the specter of Lysenkoism is unmistakable in the BSCS's course materials. The foil of Lysenko is evident not just in the course's emphasis on rejecting received knowledge but also in its specific content, from the defense of *Drosophila* studies to the portraits of Mendel and Morgan. Considering that Glass had trained under Muller, perhaps this is not surprising. What is startling, however, is how many people, both within and outside the BSCS community, thought it an appropriate message to export to the rest of the world.

∷

The constellation of agencies that directed US cultural diplomacy expressed an interest in the curriculum reform projects almost as soon as they started. As early as March 1958, for example, the US Information Agency's Hal Goodwin wrote to the director of the PSSC requesting more information on the project. He welcomed suggestions as to how "USIA could utilize the public aspects of your work in explaining American science and education to overseas audiences." Goodwin similarly included "textbooks" in his list of twenty-five possible focus areas for US government efforts to use science to strengthen ties throughout the so-called free world. By 1962, the BSCS had donated 176 sets of draft textbooks for the agency to distribute to its network of international libraries. By the mid-1960s, the US Agency for International Development (USAID) and the Peace Corps were also experimenting with the textbooks, particularly in India, Pakistan, and Afghanistan.[23]

The US government's private partners also supported attempts to internationalize science curriculum reform efforts. Both the Ford Foundation and the Rockefeller Foundation generously supported curriculum reform efforts in chemistry, physics, mathematics, and biology throughout the developing world, particularly in Asia and Latin America. Both foundations bankrolled teacher training programs, lab equipment purchases, and the infrastructure for curriculum reform programs. The Rockefeller Foundation played an especially important role in Latin America, partnering with the NSF and Colombian biology educators to create a teaching center that could serve as a beachhead for reform throughout Central and Latin America.[24]

But it was the CIA's Asia Foundation that most fully embraced curriculum reform as a tool to achieve US strategic objectives. The Office of Policy Coordination created the Asia Foundation's predecessor, the Committee for Free Asia, in 1951. Like its European analog, the Committee for Free Asia worked with refugees (in this case, mostly overseas Chinese) and established a short-lived broadcast service, Radio Free Asia, that each day provided ninety minutes of news and commentary in Mandarin, Cantonese, and English. Other initiatives in the first six months included bringing the Dalai Lama's brother to the United States and translating a Boy Scout manual into Tagalog, Sinhalese, and Hindi. State Department officials objected to all of this, arguing that the committee's anti-Communist programming and internal structure, which mirrored the State Department's Information and Educational Exchange program, too easily flagged it as a US government entity.[25]

By 1954, the Committee for Free Asia had retooled and reorganized as the Asia Foundation. Its new articles of incorporation, issued in California in October of that year, described it as a "non-political, non-profit organization" aimed at supporting "Asian groups and individuals working for Asian objectives which parallel the long-term interests of the free world," a strategy that US officials repeatedly described as "Asia for Asians." With a new charter and new purpose, the Asia Foundation quickly set up an infrastructure throughout non-Communist Asia, with permanent station offices established in Afghanistan, Burma, Cambodia, Ceylon, Hong Kong, Japan, Korea, Malaya, Pakistan, the Philippines, Taiwan, and Thailand by the summer of 1955. In its new form, the Asia Foundation partnered with almost any voluntary organization headed by Asians that might plausibly support the development of civil society, from 4-H clubs and seed exchanges to film societies.[26]

Educational reform programs meshed nicely with the Asia Foundation's mandate to cultivate democratic institutions, foster economic growth, and strengthen relationships between Asian and American elites. From the beginning, the Committee for Free Asia had supported book exchange programs; the foundation now redoubled the organization's involvement with libraries and began funding translations and providing publishing subsidies. In a status report submitted to the State Department in August 1954, the foundation claimed to have distributed more than 400,000 copies of nearly sixty textbooks in unspecified Asian countries, all "designed to replace Communist texts." As US educators began exploring curriculum reform, so did the Asia Foundation, which funded study groups, teacher training, textbook translations, and, eventually, regional conferences of educators. But despite the Asia Foundation's stated commitment to "Asian programming for Asians," the organization did not attempt to create or even encourage the development of indigenous curriculum reform programs. Instead, it provided various kinds of financial and material support for translating and adapting the newly revised US science textbooks for Asian markets.[27]

It was the *spirit of inquiry*, much more than any actual scientific content, that appealed to Asia Foundation officials. From their perspective, it didn't really matter which scientific discipline came first: adopting any of the post-*Sputnik* curricular programs would inject a new spirit of reform into what they regarded as the backwater of Asian secondary education. On the one hand, as one official pointed out, the programs in physics and chemistry related "more closely to TAF [the Asia Foundation] interests in the development of science."

On the other hand, as Marty Fleischmann in the Program Services Division replied, biology had some bearing on agriculture, an essential sector for economic growth. Really, Fleischmann noted, the more important thing was whether "the study of any one of these sciences would contribute more to the development of a scientific attitude" than the others, since "the primary goal" was the development of a "spirit of inquiry."[28]

At least at first, the Asia Foundation split the difference by supporting the international work of the PSSC, the Chemical Bond Approach, and the BSCS at more or less equal (and equally low) levels. Over time, however, the foundation's investments in curricular reform focused more and more on biology. In part, this was because the agency's program officers reported more enthusiasm for this topic from their local partners. But at least as important were persistent nudges from Arnold Grobman.[29]

Grobman was ecumenical in his search for partners to underwrite his dreams of an overseas empire for the BSCS. In 1961, two years before the textbooks appeared on the commercial market, the BSCS claimed in its grant applications that it had received inquiries for translation or adaption from educators in more than forty nations. When no foundation seemed interested in sponsoring this work, James Dickson, chair of the BSCS Committee on Foreign Utilization, pointed out that the "CIA has funds in abundance for international support of projects of interest to AIBS." Four days later, Dickson pushed the cultural diplomacy angle with a grants officer at the NSF, claiming that the Soviet Union and China had placed "course outlines and supplementary materials" in classrooms all over the developing world and that the United States needed to keep up.[30]

The BSCS's leaders—and Dickson in particular—seemed surprisingly clued in to why these appeals might resonate with government officials. None of the members of the BSCS Steering Committee or Foreign Utilization Committee were members of the President's Science Advisory Committee, for example, yet Dickson offered knowledgeable summaries of the PSAC's classified recommendations on the role of science in international relations, explaining that the administration preferred to support the activities of "private, nonpolitical, nongovernmental organizations of scientists in contrast to [the activities] of a political body." Dickson explained that "a recent reorientation of thought in administration circles" emphasized scientist-to-teacher interactions as a tool for advancing "international understanding and development." It's unclear where Dickson obtained this information, but he was, of course, right about all of it. Eventually, the BSCS could count the Asia Foundation, USAID, UNESCO, and the Rockefeller and Ford foundations among the backers for its work abroad, in addition to the millions it received from the NSF for its domestic work. Over time, the sheer scale of the BSCS's work became a reason to continue to fund it, with the argument being made (repeatedly and apparently persuasively) that, since the NSF had spent more than $8 million of public money supporting the organization's work, it was in the public interest to share the results with educators across the globe.[31]

But Grobman was more than just savvy and persistent. His commitment to "adaptation" made the BSCS's approach to internationalization more attractive for US development work than, say, the PSSC's. These textbooks were not "straight translations" but cultural adaptations, with local flora, fauna, and cultural examples substituted in the original texts. References to baseball might be changed to cricket; genetics examples involving hair or eye color might be changed to the ability to taste phenylthiocarbamide. The BSCS's leaders repeatedly stressed the need to avoid even the appearance of cultural imperialism, to the point that they refused to allow foreign ministries of education to license BSCS texts unless they promised to undertake a full "adaption," not just a "translation." This feature of the BSCS approach offered multiple benefits for its US patrons, from familiarizing international students with the natural resources in their own backyards to building national identity through scientific achievement. It also, of course, held more appeal for local teachers, students, and technocrats.[32]

::

The combination of Grobman's enthusiasm, government and foundation interest, and local ambition proved powerful. By November 1963, the BSCS reported adaptation programs underway in Argentina, Brazil, Colombia, Indonesia, Israel, Japan, the Philippines, Taiwan, Thailand, and Turkey. In some cases, as in the Philippines and Thailand, the US offices of the BSCS provided significant technical assistance; in others, its involvement was limited to exchanging correspondence. Just six months later, Grobman claimed to have fielded correspondence from educators in almost ninety nations. The countries represented on this list ranged the full gamut of educational infrastructure and overall development, from Laos and Liberia to Great Britain and West Germany.[33]

In 1963, the BSCS—or, more specifically, Grobman—introduced a strict policy toward adaptations. As repeatedly outlined in policy documents, newsletters, and correspondence with individuals, Grobman required educators interested in using BSCS materials to follow three steps. First, they should form a study committee to identify precisely which materials they might wish to adapt. Besides the three versions of the textbook (green, yellow, and blue), the BSCS produced teachers' preparation manuals, laboratory blocks, and student research guides. Educators needed to decide which, if any, of these materials best fit the needs of their students. Second, having identified the materials, the adaptation committee should arrange some sort of summer institute or conference that included the participation of "a person fully familiar with the BSCS program"—most likely a representative from the Boulder, Colorado, office. Funding for such a conference, Grobman typically advised, might be available from the usual suspects: the Asia Foundation, USAID, the US Department of State, or the Ford or Rockefeller foundation. Then, and only then, should the committee move on to the third step: a writing team of local biologists and educators, paired with an "experienced BSCS writer," should adapt the selected text for local conditions. This process might or might not require a full translation of a reference copy of the original materials, depending on the local committee's facility with English.[34]

Grobman's correspondence from the 1960s is filled with letters attempting to discern the intent of overseas educators who requested information on the books. If he detected any hint that the educators were planning to pursue what he referred to as a "straight translation," he might try to block their access to the books or refuse to sign a licensing agreement. When, for example, a professor of education at Tokyo's International Christian University requested permission to distribute photocopies of a "straight translation" as a pilot project to see how the books fared in Japanese classrooms, Grobman objected, arguing that doing so would undermine the entire program. (He eventually authorized the distribution of "not more than 500 mimeographed copies" for that purpose.)[35]

But try as he might, Grobman could not entirely control the books' circulation. Nor was it entirely clear that such strict regulation made pedagogical or ideological sense if the goal was to spread the BSCS approach—the US approach—as far and wide as possible. By the mid-1960s, Grobman's insistence that all international use must take the form of adaptations, ideally adaptations written with US supervision, was encountering increasing resistance.

The controversy over pirated textbooks in Taiwan highlighted the underlying tensions in the adaptation strategy. Taiwanese educators had been exploring BSCS books on their own as early as 1962. Grobman made his first visit to

Taipei (but not his first trip to Asia) in early 1964 at the request of two Taiwanese educators, Dr. Feng Kuo-Ao and Koh Ting-pong. As part of the Ministry of Education's broader plan to revise the high school science curriculum, Feng and Koh formed a committee to explore an adaptation of the BSCS yellow edition. The Asia Foundation supplied the textbooks for their review. With Grobman's help, Koh cobbled together support from the NSF, Taiwanese institutions, and the Asia Foundation to underwrite his and his illustrator's preliminary work on the adaptation in Boulder. With the encouragement of Asia Foundation program officers, in the summer of 1965 Koh submitted a grant application to establish a permanent BSCS adaptation committee, with membership drawn equally from Chinese educators living in Taiwan and abroad (the United States and Canada). The Asia Foundation's representatives were impressed with Koh, a noted paleontologist, and began treating him as a sort of ambassador for the BSCS in England, Israel, and India. They considered hiring him as a consultant to kick-start a BSCS program in South Korea. While the program officers remained skeptical about supporting a textbook program over the long term, they clearly found Koh's efforts worthwhile.[36]

Koh, the BSCS, and the Asia Foundation were therefore dismayed to learn that a pirated translation of the text had been released by a commercial publisher. William Mayer, the BSCS's new director, explained that, despite "appreciative nods and lip service," the Ministry of Education had "acquiesced in the production of the pirated materials." Mayer was sure that the "indigenous version" (by which he meant Koh's adaptation) would ultimately be "more acceptable to the biology teachers in Taiwan than the current pirated American versions," but the foundation would have to act quickly. Edith Coliver, director of the program services division and the recipient of this letter, scrawled "ouch!" in the margins alongside this news.[37]

Koh and Lyman Hoover, the Asia Foundation's new program representative in Taiwan, acted quickly to contain the damage. Koh proposed using his textbook's royalties to fund an organization of Taiwanese science educators to promote the new, inquiry-based science teaching methods. Such an organization, Hoover agreed, would support the foundation's larger objectives, which included forging closer relationships between Asian scientists and US scientific institutions and "assist[ing] the intellectual community to achieve its goals of free inquiry and constructive thought as a positive renovating force in the community." A telegram to the San Francisco office drove home the point: "Emphasis on revolutionary educational breakthrough not science itself." A day later, on April 22, 1966, the San Francisco office finally relented, approving a grant of $9,500 to Taiwan Normal University to print 10,000 copies of Koh's textbook. The grant was apparently successful. By November 1970, BSCS textbooks were in wide use in Taiwan; according to one report, Koh's version "dominates the field."[38]

The Taiwanese case highlights the peculiar politics of textbook adaptation as cultural diplomacy: the Asia Foundation, a CIA proprietary, backed the book that looked *less* American on the theory that it would more effectively convey the US approach to science instruction. It is no coincidence that the officially sanctioned versions granted authorship to local adaptation committees, while pirated versions tended to credit publication to the BSCS itself. The BSCS's approach recognized the labor of its overseas partners but nevertheless attempted to control their behavior. Grobman believed the adaptation policy to be in his local partners' best interests, and he refused to listen when local educators objected to this cumbersome process. Some observers within the US government realized the dangers. A program officer at the NSF, for instance, strongly advised Grobman against pursuing international outreach without the NSF's permission, as his actions reeked of "educational imperialism."[39]

By the mid-1960s, these tensions were playing out within the BSCS's own committees. Should the BSCS move heaven and earth to make sure that its foreign partners followed the rules to produce a text that replicated US values in foreign bindings? Or should it facilitate the broader task of worldwide curriculum reform? At a meeting of the BSCS Steering Committee in early 1964, two new members expressed shock at Grobman's adaptation policies, finding them overly restrictive and "paternalistic." The BSCS International Cooperation Committee (the successor to the Foreign Utilization Committee) took up the issue in March. There, Ralph Gerard, a neurophysiologist who had traveled extensively abroad as a consultant for the Asia Foundation and the Rockefeller Foundation, pointed out that Grobman's insistence on adaptation rather than translation might create ill will toward the United States. As he put it, "Many countries may feel that we are deliberately withholding from them materials which they could use but which are denied them and this may have unfortunate and unpleasant repercussions." Other members of the committee agreed, pointing out that the BSCS materials were in the public domain in any case. All present nevertheless thought it essential that the BSCS somehow remain involved, if for no other reason than to retain the "flavor" of inquiry that so distinguished its textbooks from others on the market.[40]

Following the International Cooperation Committee's recommendations, the BSCS loosened its policy toward adaptation in the spring of 1964. Tensions remained, however, and the BSCS leadership continued to debate ways to exercise control—but not too much control—over international editions. In the following year, an internal memo reiterated that "the BSCS is not interested in exporting its own courses. Rather, it is interested in assisting other biologists in developing their own courses indigenously for their own localities," all of which was underlined for emphasis. But what, exactly, might these foreign educators have in mind?[41]

::

Secondary science education is not and cannot be politically neutral. Science instruction teaches students not only facts but also what sorts of facts are worth knowing. The stakes are even higher for biology—the science of life—since biology textbooks inevitably advise students about their origins and reproductive choices. From the beginning, the BSCS's leaders positioned classroom empiricism as civic instruction, as made clear by the impact tests developed for US classrooms and by Grobman's story of the Hong Kong worms, recounted in this book's introduction. Its funding agencies shared the BSCS's faith in democracy but prioritized the civic education of educators—the possibility for organizational reform—over that of students. The US supporters of the BSCS steadfastly refused to acknowledge the legitimacy of other obvious uses of biology instruction that appealed to educators in the developing world: vocational training and national identity.[42]

The BSCS's approach to curriculum reform was, at best, an indirect path to technical training. All three of the editions (green, blue, and yellow) excluded "applications," leaving those to the discretion of individual classroom teachers, or perhaps something that could be taught in a laboratory. In Israel, for example, Hebrew University's experimental edition of the yellow version encountered resistance from the Ministry of Education because of its lack of agricultural content. Similarly, Dr. Valentine Basnayake, a physiologist who directed the School Biology Project in what was then known as Ceylon, commented in a report to the Asia Foundation, "It is an open question whether the balance chosen in the BSCS materials . . . is the best one for everybody. And applied biology might well loom large . . . in a country . . . striving to achieve technological development." A country dedicated to development via agriculture, forestry, or fisheries might well avoid the BSCS altogether. In Ghana, the Ministry of Education did just that, investing in an elaborate laboratory-based science education program that does not appear to have used any of the inquiry-based US materials.[43]

The green version, with its focus on ecology, offered the most plausible rebuttal to the claim that BSCS textbooks prioritized theory over practice. The ecological focus also lent itself well to the concept of "bioethnic regions," a money-saving idea cooked up by the International Cooperation Committee to allow neighboring countries that shared a common language (and colonial history) to develop shared textbooks. Accordingly, a group of educators working in Argentina developed a version of the green edition for temperate South America, while another in Colombia had reasonable success popularizing a tropical green version for Central America and equatorial South America. And yet, the versions continued to proliferate, suggesting that the biologists in Boulder struggled to understand the needs and desires of their overseas partners. Departments of education in both Peru and Puerto Rico planned to use the tropical version in their schools, but only after further adaptation. Nor, apparently, were the Brazilians interested in translating one of the Spanish-language versions, having already created their own, Portuguese-language tropical green version. Educators in Mexico and Uruguay, meanwhile, were busy preparing competing Spanish-language, temperate yellow versions, while the Venezuelans set their sights on a tropical blue version. In Latin America, at least, no one wanted a textbook adapted by someone else.[44]

The phenomenal ecological, cultural, and linguistic diversity of Asian countries made the regional approach even less likely to succeed there. But perhaps more to the point, the main sponsor of the BSCS adaptation effort in Asia, the Asia Foundation, actively encouraged countries to develop their own, specialized editions (via Grobman's elaborate adaptation structure) as a means to kickstart broader administrative reform. For the Asia Foundation, curriculum reform served as a rung on the development ladder; the content of the books was not nearly so important as the process of their development. In keeping with longstanding US stereotypes about Asians, the Asia Foundation's program officers, representatives, and senior administrators thought that investing in science curriculum reforms might be a way to break through what they saw as the overly deferential aspects of Asian cultures. In the matter of funding for Koh's textbooks, for example, a program officer explained that the problem in modernizing Taiwanese science education wasn't the universities, which were doing an admirable job in training teachers in modern scientific methods, but the administrators in charge of local high schools. In his assessment, "the old guard of teachers" was stifling young teachers' attempts at innovation. But supplying these teachers with a textbook system that stressed critical inquiry, as well as establishing a teachers' association where young, modern teachers could share insights about new methods of science teaching, might be just the thing to "break the bottleneck" of reform.[45]

For the Asia Foundation, with its emphasis on capacity building, the complicated nature of Grobman's adaptation process was precisely the point. The nearly $10,000 that the foundation spent on Taiwanese BSCS textbooks represented only about 5 percent of the organization's total investment in Taiwan's scientific and research infrastructure between 1960 and 1968. The balance of funds was directed at programs that overwhelmingly focused on developing human capital, whether in the form of the underwriting of visiting professorships, the sponsorship of research prizes, or faculty training programs. In keeping with the Asia Foundation's broader strategy, the programs were usually introduced at the request of local elites. The funding figures are similar for the other countries in which the Asia Foundation invested in educational reform, particularly the Philippines, Thailand, and Japan.[46]

The Asia Foundation's representatives saw themselves as engaged in a generation-long effort to bring the Republic of China (and other countries on the periphery of the People's Republic of China) into a Western modernity, with science as an agent for changing the spirit as well as the economy. But they could do so only with the support and enthusiasm of local leaders and preferably through mechanisms that gave the appearance of being "indigenous," not imported. Whether or not the Asia Foundation's local representatives realized that they were pursuing US foreign policy on behalf of the US government (the question of "wittingness"), it is exceedingly clear that they believed themselves to be assisting low- and mid-level Asian technocrats in achieving their own, self-identified goals.[47]

The Asia Foundation's representatives were on to something. Many leaders in the developing world hoped to use international aid programs to pursue their own goals of national advancement. In Tanzania, Julius Nyerere tied his country's future to hydroelectric power. In Egypt, Gamal Abdel Nasser staked his country's geopolitical status on a canal and a dam. In the same way, a number of Asian leaders tied their political fortunes to economic growth through investments in science and technology, including investment in secondary school education. For these ambitious leaders, foreign assistance from an obviously richer state or entity, be it the United States, the Soviet Union, or the United Nations, did not necessarily pose a threat to national identity: indeed, it might demonstrate to potential allies an enlightened self-interest.[48]

∷

International science educators, US propagandists, Asian politicians—all saw opportunities in US-funded science textbooks, even if they disagreed on what those opportunities were. By the time the Asia Foundation's ties to the CIA were uncovered in 1967, the foundation was directly underwriting BSCS textbook adaptation programs in five countries (Taiwan, the Philippines, Ceylon, Hong Kong, and South Korea) and funding teacher training programs in three others (Afghanistan, India, and Japan) where overt US agencies backed similar adaption programs. Somewhat remarkably, all of these programs continued uninterrupted, even after the Asia Foundation's ties to the CIA became known and the agency struggled to meet its financial commitments to its grantees. The Ford Foundation picked up $50,000 of the tab, with the NSF and USAID sponsoring the rest of the BSCS's international work in Asia until 1973.[49]

By 1971, more than thirty-five countries had BSCS adaptation projects underway, some for multiple versions. The ready availability of these modern, accessible, locally specific textbooks changed the way that high school biology was taught around the globe, replacing rote memorization with inquiry-based learning. Grobman's elaborate plans for country-specific adaptation programs were patronizing and expensive, but they provided substantial room for building national autonomy and educational infrastructure—albeit within parameters imposed by the United States. The flora and fauna in the BSCS's international editions changed at each national border, allowing both local sponsors and their US backers to obscure the anti-Communist message at the books' core. And when it came to the classroom, US influence could only extend so far. Teachers might not assign the entire textbook, or they might alter its message through their classroom presentation.[50]

In an extraordinarily unbalanced global political and economic system skewed by US power, foreign educators might very well advance their own agendas by accepting the assistance of US government agencies and foundations in reforming their own curricula. When PSAC, the State Department, USIA, and the Asia Foundation first considered using the textbooks to promote US ideals in the late 1950s, they intended the books to broadcast certain ideas about the nature of science in the United States. As a USIA planning paper put it, programs should strive to improve foreign peoples' attitudes toward the United States by "explaining and interpreting the place of science and technology in modern life and progress," that is, to "tell the story" of how US science and technology "improve the general welfare, not only in the U.S., but in the world over." By 1968, in contrast, the USIA was at least acknowledging the goal of self-determination: "The developing world, while admiring American science and technology, is determined to catch up." As stated in its new policy paper on science and technology, the agency now sought to "establish that U.S. knowledge and discoveries in science are shared with others to meet practical human needs, and to show that the U.S. uses its science and technology to assist developing nations in modernization and nation-building."[51]

High school science textbooks, somewhat miraculously, did all of these things at once. Through their focus on inquiry-based learning, the post-*Sputnik* generation of textbooks denounced dogma, scientific or otherwise. By partnering US scientists with overseas educators, the teacher training programs strengthened personal networks among technical elites and brought cutting-edge laboratory instruction equipment to global classrooms. By insisting that local educators do the work of translating and revising the textbooks, the BSCS's elaborate adaptation process ensured that the books not only escaped the charge of cultural imperialism but also genuinely incorporated local perspectives. Not even the taint of a CIA connection could dampen international technocrats' enthusiasm for development through science-based education. The same cannot be said for the rest of US cultural diplomacy programs, in science or otherwise.

8 : An Unscientific Reckoning

n February 12, 1967, the White House received a deeply unwelcome tip. *Ramparts*, a radical West Coast magazine, planned to expose the CIA's long-term involvement with the National Student Association (NSA), a respected mainstream organization of collegiate leaders. Rumors of secret CIA funding channels had been circulating for years, but the timing and content of the *Ramparts* article threatened to blow the lid off the CIA's entire network of private organizations. A wave of student protests against the Vietnam War at home, combined with growing resentment of US power abroad, meant that US authorities could no longer count on the discretion of students and private citizens, let alone journalists, to keep the government's secrets.[1]

Over the next month, report after report in *Ramparts*, the *New York Times*, and the *Washington Post* informed the reading public of the CIA's relationships with youth groups, organized labor, cultural organizations, and private foundations. The *Times* not only explained the CIA's pass-through funding model in excruciating detail but also noted, almost in passing, the ease of assembling this information from tax forms that reported the contributions and spending of tax-exempt nonprofit organizations. In May, Thomas Braden, the former head of the Office of Policy Coordination's International Organizations Division, eliminated any remaining mystery about the nature of the CIA's private network in a non-apology published in the *Saturday Evening Post*, provocatively titled "I'm Glad the CIA Is 'Immoral.'"[2]

The moment of reckoning for US covert psychological warfare operations had finally arrived. The news forced all Americans involved in public life, including scientists, to reconsider their roles, witting or not, in supporting US foreign policy. None of the revelations specifically mentioned scientific organizations—the National Academy of Sciences, for example, did not face the mix of outrage and sympathy showered on the NSA. But plenty of scientists had had contact with some of the organizations named, including the Asia Foundation, and many more had received funds from overt US agencies and private organizations whose goals seemed difficult to distinguish from those of the

CIA's disgraced fronts. Scientists' responses to these revelations spanned the political spectrum, but those on the left were particularly chagrined to realize how far ideas about science had wormed their way into US propaganda.

Was it possible, scientists wondered, to cultivate both a scientific mindset and sympathy toward those on the receiving end of US power? Like the senior government officials who struggled to salvage the potential of private-public partnerships, scientists who had been complicit in US foreign policy grappled with the hidden costs of their commitment to scientific freedom. Radicals, especially, attacked the concepts of objectivity and apolitical science. By the time the dust had settled, any semblance of consensus on the relation between science and politics had been destroyed.

∷

The most remarkable aspect of the 1967 revelations was that they came as late as they did. Since 1949, when Frank Wisner's OPC first began cultivating relationships with Eastern European refugees, the CIA had trusted some of its most inflammatory secrets to private individuals. In theory, the pass-through funding system provided plausible deniability to both the US government and its private partners. The Congress for Cultural Freedom, for example, could maintain that its money came from the Farfield Foundation, without acknowledging that the Farfield Foundation existed solely to launder the CIA's contributions. Private citizens who wished to remain ignorant of the true sources of the funds could simply choose not to ask too many questions about the obscure but apparently generous foundations that popped up in the 1950s. Inevitably, others displayed more curiosity.[3]

The very nature of these arrangements put hundreds, and possibly thousands, of scientists, artists, writers, philosophers, and labor organizers in positions that exposed them to knowledge that they either shouldn't have had or didn't want. Over time, the system inevitably sprang leaks, only some of which could be blamed on witting participants. The most potentially damaging came in 1964, when, in the course of a US congressional investigation into the tax status of nonprofit foundations, Texas Congressman Wright Patman disclosed that the Kaplan Fund had received CIA funding since 1959. The *New York Times*'s small report on the story identified eight foundations (the Andrew Hamilton Fund, the Beacon Fund, the Borden Trust, the Edsel Fund, the Gotham Foundation, the Kentfield Fund, the Michigan Fund, and the Price Fund) that had served as the CIA's partners in the arrangement. With the exception of an editorial in *The Nation*, however, the story died in the national press. The CIA's cozy relationship with the media, which had included everything from the right to review stories to using newspaper credentials as cover, remained warm in 1964.[4]

College students ultimately proved less compliant. The CIA's partnership with the NSA started in 1950, when the agency supported an international survey of student opinion designed to identify pockets of anti-Communist support. In the fifteen years since, the CIA had invested hundreds of thousands of dollars in the group's infrastructure and activities, especially its overseas programs. Each year, the CIA worked behind the scenes to influence the association's election process, making sure that the agency's preferred and carefully groomed candidates prevailed. Witting former leaders then shepherded the new officers into the conspiracy, after first requiring them to sign an oath of secrecy that threatened violators with twenty-year prison terms.

At first, this arrangement worked well enough; many NSA leaders, including a young Gloria Steinem, saw the CIA's help in making the organization's moderate-to-liberal work possible as "honorable" and exciting. But from the beginning, the NSA's domestic politics veered to the left of its international anti-Communism. By the mid-1960s, an outspoken contingent of NSA members was pushing the organization's leadership to condemn the war in Vietnam, in solidarity with students in the decolonizing world. Students, and youth movements in general, were rapidly rejecting the Cold War consensus on which the CIA's private partnerships depended. In 1966, the NSA's new president, Phil Sherburne, informed the CIA that the organization would no longer accept agency funds. He did this quietly, so as not to blow the CIA's cover or violate his oath of secrecy. But in the process of searching for replacement funds, an NSA staffer, Michael Wood, sniffed out the truth. Opposed to the Vietnam War and unencumbered by a legal obligation of silence, Wood unburdened his conscience to Warren Hinckle, the editor of *Ramparts* magazine.[5]

Originally founded in 1962 as a thoughtful venue for Catholic cultural criticism, by 1967 *Ramparts* had become the most radical magazine on national newsstands. Based in San Francisco, the magazine's leftist politics embraced both cultural criticism and social justice, with its alumni founding *Mother Jones* and *Rolling Stone*. The NSA story was not the first time that *Ramparts* had reported on the CIA, but in its careful tracing of front organizations, dummy foundations, and witting leadership, the story opened the floodgates to a new wave of reporting from the magazine's more establishment-oriented East Coast rivals. On February 13, *Ramparts* ran full-page ads in the *New York Times* and the *Washington Post* protecting its scoop, which would not be published for another two weeks. The strategy worked. With both papers pushing the story, the March 1967 issue of *Ramparts* sold an astonishing quarter of a million copies.[6]

News of the forthcoming bombshell put the administration of President Lyndon B. Johnson in a peculiar bind. Unaware of the specifics of the *Ramparts* article or what might be published next, the White House wasn't sure what it should be defending. A blanket defense of the CIA's use of any and all private covers risked alienating unwitting liberal allies, yet mentioning specific organizations might blow covers that remained secure. Johnson turned to Nicholas Katzenbach, the undersecretary of state and former attorney general, for a solution. No stranger to crisis, Katzenbach had spent two years in prisoner-of-war camps in Italy and Germany and famously stared down reactionary governor George Wallace during a desegregation standoff at the University of Alabama. Katzenbach advised the president to reveal as little about the CIA's cover networks as possible—in part because he doubted whether the administration had yet gotten the straight story from the CIA.[7]

Deeming the CIA's political reputation a lost cause, the State Department launched into damage-control mode. A series of diplomatic cables from Washington updated the embassies on the latest revelations, requested reports on how the news was being received abroad, and urged diplomatic posts to "volunteer nothing." "If and when groups that have received government funds do make statements," Washington advised, "Department will make case-by-case determination of how it will respond." Otherwise, inquiries should be directed toward a written statement from Katzenbach, read into the record by the president's press secretary, George Christian, at a news conference on February 15. The statement announced that the CIA "did not act on its own initiative" but rather on instructions from the National Security Council, which had in turn established these policies "in times of challenge and danger to the United States and the free world." In an attempt to defuse criticism of Johnson, the statement reminded the press that these policies had been in place under four presidents. Even so, the administration acknowledged that the situation raised "difficult and delicate problems." A three-man committee, consisting of Katzenbach, Director of Central Intelligence Richard Helms, and Health, Education, and Welfare Secretary John Gardner would investigate the problem on the president's behalf.[8]

Katzenbach's statement, we now know, was surprisingly truthful in its bare admissions of facts. NSC 4-A had first authorized the CIA to engage in covert operations under President Truman, and those programs expanded under Eisenhower and Kennedy. At the same time, the statement contained key omissions that cultivated ambiguity. It noted, for example, that the CIA acted on the approval of "senior interdepartmental review committees, including the Secretaries of State and Defense or their representatives." The statement didn't elaborate, but the most recent iteration of this group, the 303 Committee, was chaired by the president's national security advisor and included the head of the CIA, as well as representatives from State and Defense at the rank of assistant secretary or higher. A sharp reporter asked the obvious question: did this slippery wording mean that the president was or was not aware of these activities? Christian's response, "I can't draw any inference on this," exacerbated the confusion. Another reporter noticed that Katzenbach's statement referred to a controversy involving "certain private organizations," while the full-page *Ramparts* ad that ran in the *Times* and the *Post* limited its claims to educational institutions. Did this mean that the investigation had been broadened to, as a reporter put it, "labor and all of these other things that have been tossed around in the press?" Yes, Christian replied.[9]

::

Unbeknownst to the reporters, the question of what the president knew and when he knew it would drive both Soviet propaganda and histories of the CIA's covert activities for years to come. Even today, it shapes how we, as historians, curious readers, and observers of politics, grapple with private citizens' involvement in Cold War propaganda campaigns.

In April 1967, when the pace of the revelations had slowed somewhat, the State Department reported that *Pravda* and *Izvestia* had "dredged up" old charges that all US foreign operations contained elements of espionage. Soviet media particularly targeted the Peace Corps, alleging that "70 percent" of Peace Corps volunteers were "in the pay of the U.S. intelligence agency." The US Information Agency was attacked as being "largely a front for [the] CIA," as was USAID. Most ominously, the State Department reported that Soviet propaganda portrayed the CIA "not so much as the instrument but as the motive force of American activity abroad"—an interpretive line that will be familiar to anyone who's spent much time reading about the CIA.[10]

Contact paranoia is a real hazard for people who write about the history of the CIA. In the summer of 2012, I was preparing for a trip to California to examine the archives of the Asia Foundation, now housed at the Hoover Institution Library and Archives on the campus of Stanford University. I had only recently learned that the pre-1967 Asia Foundation was not, in fact, an independent nongovernmental foundation. An archivist at Hoover had sent me the finding aid for the collection, which lists the individual folders contained in more than 400 boxes' worth of materials. I spent an entire afternoon, transfixed and appalled, scrolling through page after page listing materials on every conceivable category of nonprofit organization, from the Brookings Institution to the Girl Scouts and the Suicide Prevention Center. The list included entries for most US scientific organizations, including the American Anthropological Association, the American Association for the Advancement of Science, the American Chemical Society, the American Geophysical Union, the American Institute of Biological Sciences, the American Mathematical Society, the American Medical Association, the American Psychological Association, the American Sociological Association, the National Science Teachers Association, and many smaller professional organizations, as well as several folders on the NAS and the National Science Foundation. A person browsing this list could be forgiven for wondering whether every single US voluntary group doing work abroad was, in fact, doing the CIA's bidding.[11]

But the reality of these arrangements was more complicated than this. The Asia Foundation did not necessarily interfere in the *operations* of these organizations; instead, it interfered in the sovereignty of foreign countries by easing the *possibility* of US groups' operating in places that US policymakers wanted to influence. The distinction is subtle but meaningful. The US image abroad depended deeply on the work of genuinely voluntary groups, like the American Mathematical Society or the Girl Scouts, that frequently had no idea that their overseas work was being assisted, at arm's length, by an intelligence agency. The actions of that intelligence agency had, in turn, been authorized by a series of NSC directives issued from the late 1940s through the mid-1950s. The US government's policy of supporting and, in some cases, creating voluntary organizations indistinguishable from truly independent private organizations fertilized the already rich soil from which a thousand conspiracy theories bloomed.

Based on the panic in the White House in the spring of 1967—as unbelievable as this now seems—it would appear that neither the CIA nor the State Department had prepared contingency plans for the inevitable moment that the truth emerged. Since 1947, the NSC had been creating layers of secrecy to ensure that the government would be shielded from the potentially embarrassing acts of nominally private organizations. At no point, however, did the planners take seriously the problem of protecting private organizations from the embarrassment of being associated with the government. This was the task of the Katzenbach Committee. When the committee, only a month after the initial *Ramparts* revelations that had spurred its creation, recommended that "no federal agency shall provide any covert financial assistance or support, direct or indirect, to any of the nation's private voluntary organizations," President Johnson felt he had no choice but to accept.[12]

::

Johnson's announcement set December 31, 1967, as the deadline for the CIA to end its secret subsidies to US private organizations. But practically speaking, what would such a policy mean? At the time of the revelations, the CIA had been supporting thirty-eight nominally private organizations at a cost of $15 million annually. These monies funded a wide variety of activities, including the work of anti-Communist international organizations, US participation in international events, exchange programs, and educational and technical aid. The majority of funds, estimated to make up 58 percent of the total, went to private foreign groups to "build essential political, social, and economic institutions and prevent communist control of such groups." If the new policy were to be followed to the letter, it would also spell the end of covert support for Radio Free Europe and Radio Liberty, which represented a $350 million investment since 1952. Maintaining the status quo would erode what little public trust remained in the private voluntary groups working abroad without government subsidy (estimated to possibly number as high as a thousand), but an abrupt end to the subsidies would create chaos.[13]

The fate of the Asia Foundation, the largest of the CIA's now-blown covers, crystallized the issue. On March 21, 1967, the *New York Times* ran a relatively small article, buried on page 17, announcing that the Asia Foundation, too, had received CIA funds. The Johnson administration had been expecting this. The previous summer, Cord Meyer, head of the CIA's covert action programs since 1962, had warned the 303 Committee of the foundation's "vulnerability" to a "gadfly publication like *Ramparts*." The members of the top-secret oversight group agreed at that time to establish an endowment fund to allow the organization to continue its work without the taint of its CIA connection. For reasons that have so far eluded declassification, this does not appear to have happened, but the CIA's illegal spying on the *Ramparts* offices allowed the Asia Foundation to stay a step ahead of investigative reporters.[14]

The day before the *Times* ran its story, the Asia Foundation's trustees issued a misleading statement indicating that the foundation had, "in the past," "knowingly received contributions from private foundations and trusts which have been recently named as having transmitted Central Intelligence Agency funds to private American organizations." The statement claimed that receipt of those funds "in no way" affected the foundation's mission, policies, or programs, and restated the line that the organization was a "non-profit, private philanthropic organization, governed by an independent Board of Trustees" that had "sought and received" funding from US "corporations, foundations, and individuals."[15]

The State Department prepared for a backlash, urging its ambassadors in affected countries to notify local government officials before the news broke. To the administration's surprise, however, the trustees' preemptive statement defused the crisis. The governments of Japan, Malaysia, and South Korea shrugged. Only in India, where the government announced a review of all foreign assistance, did the situation threaten "acute embarrassment." In the CIA's assessment, the revelations "posed no serious threat to TAF [Asia Foundation] operations in Asia." But, ironically, the trustees' statement created a new problem, in that the organization's credibility depended on its ability to continue its programming without the CIA's support. As of May 1967, the Asia Foundation's annual budget of $8.5 million supported 346 Asian and US full-time staff, plus an additional 35 US consultants. Between 1962 and 1967, it issued more than 500 direct grants to Asian voluntary organizations, in amounts from $100 to $300,000. The organization had funds to operate through the end of its fiscal year—July 31, 1967—but then what?[16]

The Katzenbach Committee's report, publicly released a week after the Asia Foundation's brush with scandal, recommended that Congress create a new federal mechanism to provide overt support to deserving US private organizations working abroad. The president appointed a new Committee on Overseas Voluntary Activities, commonly known as the Rusk Commission, after its chair, Secretary of State Dean Rusk, to explore the possibilities. Over the summer of 1967, the Rusk Commission—whose members included Attorney General Ramsey Clark, Budget Director Charles Schultze, several US senators and congressional representatives, university presidents, corporate leaders, and the president-elect of the student organization of the University of North Carolina—met three times to hear about the work of voluntary organizations and consider alternative ways to support their programs. The mix of members posed challenges for a truly open discussion, as the CIA attempted to keep some items, like the extent of US support for Radio Free Europe or the scale of the Asia Foundation's work, secret. Even so, through meetings with more than ninety private individuals and government officials, the Rusk Commission managed to glean a fairly detailed sense of what kind of things the CIA had been sponsoring.[17]

The problem, the Rusk Commission soon learned, went well beyond finding ways to support the work of the so-called CIA orphans. A new, overt funding mechanism that housed *only* CIA orphans would, by definition, out those organizations as recipients of CIA funds. Large private foundations such as Ford and Rockefeller were reluctant to take on these organizations, for the same reason. But more to the point, the commission recognized that the CIA's "flexible" institutional arrangements had allowed the agency to fund politically sensitive operations that would never survive the congressional appropriations process. What was really needed, the Rusk Commission determined, was some sort of public-private group that could support the work of the approximately 100 private voluntary organizations that did "the *kind* of work overseas performed by the CIA orphans, especially in the fields of institutional development." These activities, the commission concluded, were "highly worth-while in terms of U.S. objectives" and deserved support.[18]

The Rusk Commission's final report developed this recommendation into a philosophy of voluntarism. "In developed and developing countries alike," the report announced, "governments cannot and should not do everything." The report noted the historic role of private, nonprofit organizations in building US institutions, calling them a "precious part of the American heritage." In the Rusk Commission's assessment, the problems of poverty, racism, and resource distribution that plagued countries in Africa, Asia, and Latin America—a group of ills shorthanded here as "underdevelopment"—could be traced directly to these nations' lack of private organizations "comparable" to those in the United States. To the Rusk Commission, the solution was obvious. If the United States wanted to improve the lot of developing countries (and, by extension, its standing with those countries), it should and in fact must provide public dollars to support the work of private organizations, and it must do so in a way that maximized flexibility and minimized red tape.[19]

Secretary of State Rusk sent the final version of the commission's report to the White House on June 4, 1968. That night, New York Senator Robert F. Kennedy, Johnson's longtime political rival and the younger brother of the slain president, was shot in the kitchen of the Ambassador Hotel in Los Angeles, after claiming victory in the California primary. He died twenty-six hours later. A note attached to the official version of the Rusk Commission Report in the LBJ Presidential Library says it was never submitted to President Johnson, but it's unclear how seriously the president would have taken its recommendations even under the best of circumstances. Originally asked to, in Nicholas Katzenbach's words, "sweep up the broken china" of the CIA's mess, the Rusk Commission instead issued a sweeping manifesto on the role of voluntary associations in American (and indeed international) life. The only thing missing was a reference to de Tocqueville. Even some of the commission's members thought the language a bit much. Herman B. Wells, the former president of Indiana University, at one point remarked that he "hoped we could get rid of the flavor of the report which sounded like we were going to Americanize the world through luncheon clubs."[20]

The Rusk Commission had misjudged its audience. As originally developed in the late 1940s, the US government's policy of secretly channeling US government funds to private organizations was intended as a tactic in a larger psychological offensive against global Communism. The point of this policy wasn't to replicate US institutions abroad but rather to convince foreign nationals, especially European intellectuals, that they stood to benefit more under liberal democratic systems than under Communism. By the early 1960s, the fight against Communism had moved to Africa, Asia, and Latin America, areas that US policy primarily understood through what they lacked. United States–backed programs such as the Peace Corps, USAID, and the Asia Foundation were designed to bring the so-called Third World into the US orbit by jump-starting economic development, a policy championed by Johnson's national security advisor (and chairman of the 303 Committee), Walt Rostow. But throughout the Cold War, the United States supplemented these psychological and economic strategies with violence, whether by supporting dictators who upheld US policies or through more direct action, including paramilitary force. What had changed in the late 1960s was the willingness of world opinion to tolerate this carrot-and-stick approach.[21]

By 1968, both domestic and foreign critics of US foreign policy were increasingly questioning the possibility of separating "hard" and "soft" power. In Vietnam, the United States funded development projects in the Mekong Delta while simultaneously carrying out devastating bombing campaigns in the North. Vietnam needed many things, but few Americans, and fewer Vietnamese, would have placed voluntary institutions modeled on US civic organizations high on that list. Even Secretary of State Dean Rusk acknowledged this in his unread cover letter to President Johnson, noting that the proposal was likely to receive "a very negative response" in Congress. The CIA scandal had largely blown over; "far better," Rusk suggested, "to let the CIA matter wither away and let a new administration take a fresh look" at the problem. Left unsaid, just two months after 110 US cities had erupted in anger and despair after the assassination of Martin Luther King, Jr., was Americans' diminished faith in their civic institutions to serve as a beacon of light and hope unto the world.[22]

With the Rusk Commission's recommendations shelved, the remaining CIA orphans limped along. In 1967, the 303 Committee agreed to provide so-called surge funding through July 1969 to the CIA's largest threatened assets—the "Radios" and the Asia Foundation—thereby following the letter, if not quite the spirit, of the Katzenbach Committee's requirement that CIA funding cease by December 31. Under pressure from Rostow, USAID and the State Department agreed to pick up approximately $3 million of the Asia Foundation's budget, at least for the short term, leaving foundation officers scrambling to find funding for the rest of its work. "If there were deep sighs for the good old days of straight covert funding," the minute taker at a 303 Committee meeting devoted to the topic observed, "they were not audible over the hum of the air conditioner in the White House Situation Room." Haydn Williams, the Asia Foundation's president, was granted sympathetic meetings with, but no funds from, his counterparts at the Rockefeller Foundation, the Rockefeller Brothers Fund, and the Ford Foundation. The Asia Foundation remained in administrative limbo for several years, but ultimately found an overt funding line through the US Department of State. Federal government sources continue to provide the bulk of its funding (close to 95 percent) at the time of this writing.[23]

Both the Rusk Commission and the scandal that had provoked it soon faded from political memory. Freed from the responsibility of carrying out US covert cultural diplomacy, the CIA recommitted itself to overseas intelligence collection, paramilitary action, and a new (illegal) function: domestic surveillance. The CIA's domestic spying operations had started at Johnson's request, but they expanded dramatically under President Richard Nixon, with Project CHAOS ultimately maintaining files on more than 7,000 US citizens and more than 100 domestic groups. The CIA's covert action arm, meanwhile, stepped up its interference in foreign countries, particularly Latin America, where the United States supported a military coup against Salvador Allende, the democratically elected president of Chile. To the extent that US foreign policymakers retained an interest in soft power, such activities would henceforth be conducted by overt agencies. The CIA was officially out of the hearts-and-minds game.[24]

::

Scientists first encountered reliable information on the CIA's private networks through an earlier wave of reporting that unmasked several academic institutions as well as the CCF. In the last days of April 1966, the New York Times ran a five-part series of extraordinarily detailed articles explaining the origins, purpose, and growth of the intelligence agency, including its involvement in either overthrowing (Iran, Guatemala) or attempting to overthrow (Cuba) foreign governments. The reports noted the existence of the CIA's Division for Science and Technology and quoted the agency's boast that "it could staff any college" from the ranks of its PhDs, who made up 30 percent of agency staff. Column after column of closely packed newsprint explained the CIA's reliance on refugees and its habit of funding, directly or indirectly, newspapers, magazines, and universities, including MIT's Center for International Studies.[25]

For a handful of scholars, however, the response to the allegations, rather than the reporting itself, proved more troubling. The third article in the Times series disclosed that "liberal organizations of intellectuals," including the CCF, had received support from the CIA. In response, four members of the CCF Executive Committee, including J. Robert Oppenheimer, Arthur Schlesinger, the economist John Kenneth Galbraith, and retired diplomat George Kennan, wrote a letter to the Times editor that they intended to serve as a defense of the organization. Based on their sixteen-year involvement with the CCF, they "categorically" defended the group's independence and integrity. "The congress," they wrote, "has had no loyalty except to an unswerving commitment to cultural freedom."[26]

Michael Polanyi received news of this editorial submission in Middletown, Connecticut, where he was spending a year on sabbatical at the Center for Advanced Studies at Wesleyan College. As a member of the CCF Executive Committee, Polanyi was privy to the letter's contents prior to its publication. Like several other members of the CCF's international network, he was deeply concerned—but not about the revelation that the CCF had colluded with the CIA. Instead, Polanyi and his close associates protested to Nicolas Nabokov, who remained the public face of the CCF, that the organization had bungled its defense. The letter, after all, did not deny that the CCF had been associated with the CIA since its inception. "Who is to take responsibility," Polanyi asked, "for the damage done to the cause of all of us in Europe and in the countries intellectually dependent on Europe?"[27]

Polanyi's question, with its explicit condescension toward non-Europeans, correctly predicted who would suffer from the 1966 report. Francisco Sionil José, a prominent Filipino writer associated with the CCF's seminars, wrote from Manila to explain that he would not "be using the CCF name for a long, long time" and asked the secretariat to stop sending him letters on CCF letterhead. Fearing for his safety, he mailed this letter from New York rather than Manila. The United Arab Republic banned *Hiwar*, the CCF's Arab-language literary journal, from the country. For most of the CCF's English-language publishing network, however, the potential of scandal had the surprising effect of producing short-term financial stability. In the fall of 1966, the Ford Foundation, which had been partially supporting the CCF for years, saved the organization's reputation by agreeing to take over its finances. It didn't hurt, of course, that the Ford Foundation's new president, McGeorge Bundy, came directly to his position from the White House, where, as national security advisor, he had chaired the 303 Committee. Bundy, unlike previous administrators at Ford, did not have to be convinced of the value of cultural diplomacy.[28]

All of which is to say that no one involved with the CCF, including its scientists, could plausibly feign surprise when Tom Braden wrote in the *Saturday Evening Post* in 1967 that "we [the CIA] had placed one agent in a Europe-based organization of intellectuals called the Congress for Cultural Freedom." When Mike Josselson, the CIA agent in question, tendered his resignation as executive secretary in an attempt to salvage what was left of the organization's reputation, Polanyi resigned his position as well. But as was the case a year earlier, Polanyi's pique concerned the CCF's response, not the fact of the CIA's involvement. In a letter to Raymond Aron, meant to force the CCF to reconsider its treatment of Josselson, Polanyi pointed out that the CCF leadership included several former Communists "who served Stalin, his murder and lies." "What is this horror of the C.I.A.?" he implored. "Are men like you or me going to declare that in fifteen years we did not notice that we were being manipulated to serve sinister purpose? Are we going to proclaim our awakening, a new version of The God That Failed?" The CCF's new leadership assured Polanyi that Josselson would be provided for until he found a new position, and Polanyi agreed to remain on the board.[29]

Polanyi's enthusiastic defense of Josselson and the CIA may have been an outlier, but his general sentiment—that the CIA's role was tactically unfortunate but not inherently troubling—was not. For scientists who had spent the past twenty years working closely with the US government, whether as intelligence advisors, weapons researchers, or even just domestic policy advisors, news of the CIA's hidden hand seemed more a bump in the road than a downed bridge. Consider, for instance, an exchange between Eugene Rabinowitch and Bernard Feld. Feld wrote to Rabinowitch, asking whether Pugwash had been implicated in the scandal. Rabinowitch wasn't sure; his records indicated that the organization had once, "long ago," received $3,500 from the Kaplan Fund. But more to the point, Rabinowitch didn't necessarily see this as a problem. He explained: "The whole affair reminded me of ONR [Office of Naval Research] support for research after the war, when Congress could not be persuaded to appropriate money for a National Science Foundation, and the only available source of support was the military." Since neither the ONR nor the CIA attempted to steer the direction of the programs they funded, Rabinowitch argued, scientific freedom, and therefore academic integrity, had been maintained.[30]

The leadership of the Biological Sciences Curriculum Study reacted similarly to the (muted) reports of the Asia Foundation's ties to the government. The organization's new executive director, William Mayer, continued to travel on the Asia Foundation's behalf, as did Arnold Grobman in the role of consultant. In 1972, Grobman speculated that science-based programming might be a way to reintroduce the Asia Foundation to India, which had maintained the ban imposed in 1967. In a letter to Edith Coliver, a program manager who had been with the foundation since the Committee for Free Asia days, he wrote, "Science is more nearly neutral politically and might serve, over a period of years, to allay suspicions and pave the way for broad-scaled normal relationships." Grobman also indicated that he might be interested in joining the foundation's staff, "should an appropriate vacancy develop." Grobman clearly not only understood how the US government hoped to use a vision of apolitical science to promote its own agenda but also approved of it.[31]

The Asia Foundation's ties to other institutions, including the NSF, the NAS, and most of the major disciplinary societies (physics, mathematics, chemistry, etc.), potentially implicated a broader community of scientists less inclined to applaud the role of the CIA in international affairs. Since the late 1950s, the Asia Foundation had underwritten foreign memberships in US professional societies, supported international educators' attendance at NSF-sponsored summer teaching institutes, selected and sent Asian scientists on US tours administered by the NAS, and assisted the academy in establishing cooperative scientific programs in Taiwan. The Asia Foundation had a particularly close relationship with the NAS's Pacific Science Board, which coordinated the academy's work throughout Asia. Indeed, the Pacific Science Board and the Asia Foundation shared personnel. Robert Sheeks, associate director of the Pacific Science Board, arrived at his position in 1963 directly from the Asia Foundation's R&D division, where he had spearheaded the foundation's attempt to build science programming. When Sheeks left the NAS in 1968, the Pacific Science Board replaced him with William Eilers, an Asia Foundation program manager.[32]

Given the strength of the ties between the NAS and the Asia Foundation, and the NAS's professed devotion to keeping politics out of science, one might have expected some public discussion of the academy's ties to the CIA. But if anyone at the NAS's 1967 annual meeting protested the academy's relationship with the Asia Foundation, it went unrecorded in the official minutes. The only discernible change at the NAS came three years later, with the retirement of Harold Coolidge, the Pacific Science Board's executive director since 1946. With Coolidge's retirement, most of the Pacific Science Board's activities were folded into the academy's broader international development programs, now supported by USAID. The existing public documentation, alas, does not clarify whether Coolidge, an Office of Strategic Services veteran with extensive ties to the intelligence establishment, was forced out, or whether, at the age of sixty-six, he simply preferred to devote the remaining years of his life to his stated passion, wildlife conservation. (Coolidge was a founding director of the World Wildlife Fund and, at the time of his retirement, president of the International Union for the Conservation of Nature.)[33]

In the late 1960s and early 1970s, the annual meetings of several scientific societies featured antiwar protests and panels on the social responsibilities of scientists. With one exception, however, none of these actions directly confronted professional societies' partnerships with the CIA or the broader concept of using ideas about science to further US interests. Only the American Anthropological Association conducted an investigation into the many and varied ways that the CIA had inserted itself into the day-to-day business of international science. For ethnographers, whose work so closely resembled that of intelligence agents, contact with the CIA was ubiquitous and potentially dangerous. A report prepared for the association's governing council in the fall of 1966 alleged that CIA agents had posed as anthropologists and that some anthropologists worked directly for the CIA. In response, in March 1967, fellows of the society passed a resolution adopting informal ethical standards that promoted financial transparency and condemned secrecy. No other major learned society followed suit.[34]

Why were mainstream US scientists, with the exception of the anthropologists, so unperturbed by the reports that so many of their venerated nongovernmental institutions had ties to the CIA? Human beings justify their actions in many ways, and the historian who speculates on the motives of the dead does so at her peril. Still, Rabinowitch's comparison of support from CIA funds with military research dollars is instructive. In the postwar United States, the vast majority of US scientists, over the course of their careers, had received research funds from one branch or another of the federal government. By 1960, one-third of funding for academic research came directly from the Department of Defense. Having ridden the cresting wave of postwar bureaucracy to positions of power, many senior members of this generation fervently believed in fostering change from within the system. More to the point, having convinced themselves of the possibility of apolitical science, they saw little reason to turn away funds that might help them achieve their own, supposedly enlightened, goals. These assumptions—their belief in technocracy, scientific objectivity, and scientific freedom, and their fundamental faith in the American experiment—left them entirely unprepared for the criticisms of the New Left.[35]

∷

A young activist looking for a cause in the mid- to late 1960s had any number of issues to choose from. The anti–Vietnam War movement, Black Power, Red Power, the Chicano rights movement, the women's movement, renters' strikes, global Marxism, free love, free speech—participants in these and other waves of activism demanded that the United States live up to its stated ideals of liberty, equality, and freedom for all. On college greens, where the protests tended to be whiter than those on the streets, students and faculty alike attempted to theorize what had gone wrong with American democracy. Many of the so-called New Left's critiques focused on "the system," "the establishment," and the role of science and technology in driving the "military-industrial complex."[36]

For the writers and intellectuals of the New Left, the *Ramparts* revelations turned what could have been a self-indulgent exercise into an urgent and public discussion. One of the most damning accounts of how intellectuals had lost their way, written by the cultural critic Christopher Lasch, appeared in a special issue of *The Nation* in September 1967. "The freedom of the American intellectuals as a professional class blinds them to their un-freedom," Lasch wrote, with state subsidies acting as invisible muzzles, worn willingly in exchange for professional autonomy. But with the Vietnam War, intellectuals could no longer ignore that the United States had become a "war machine." It was folly, Lasch argued, for intellectuals to think they could change the system from within, because "the war machine cannot be influenced by the advice of well-meaning intellectuals in the inner councils of government; it can only be resisted."[37]

Lasch's critique—and others like it—shook the foundations of the postwar US consensus on scientific objectivity. Liberals such as Bentley Glass had based their public authority on their ability to draw clear and crisp lines between scientific and political questions. In the late 1950s, Glass himself thought it not only appropriate but necessary to criticize the Atomic Energy Commission's data on fallout even while serving on the AEC's Advisory Committee for Biology and Medicine. This was the mindset that had allowed Rabinowitch to assure Feld that he needn't worry about the possibility that the CIA had funded Pugwash, because the intelligence agency hadn't interfered with the content of the activists' work. The New Left, in contrast, insisted that it was impossible to separate power from process.

The scientists who were receptive to the New Left's challenge responded in one of two ways. Those who were most closely aligned with government advising (defense advisors, nuclear physicists, etc.) interpreted the admonition narrowly: that is, scientists who opposed the war machine could not honorably contribute to defense work. Between 1968 and 1973, several of the nation's most prestigious universities, including Princeton, MIT, Stanford, and the University of Michigan, either banned classified research or limited it to nonlethal forms. For the first time in a generation, large numbers of individual scientists turned away defense contracts. In the fall of 1967, even George Kistiakowsky, former science advisor to President Dwight Eisenhower, decided that he could no longer advise a Department of Defense that seemed hell-bent on escalating the war in Vietnam. For the most part, though, these well-connected scientists limited their criticisms to what they saw as the misuse of objective science rather than the principle of objectivity itself.[38]

In contrast, the scientists who took the critiques of the various 1960s revolutionary movements more deeply to heart sought new forms of scientific organizations that rejected the possibility of a politically neutral science. By 1969, a new organization, Science for the People, was disrupting national scientific meetings, exposing the names of scientists involved in defense projects, and orchestrating so-called citizen science projects. A similar organization, the British Society for Social Responsibility in Science, emerged around the same time in the United Kingdom. Both groups rejected the notion of "objective" science, arguing that the products of science could not be separated from the social and economic conditions under which they were produced. Alleging that scientific (and state) institutions were racist, sexist, classist, and imperialist, Science for the People and its British analog attempted to build partnerships with oppressed peoples. The Chicago chapter of Science for the People, for instance, spearheaded a Science for Vietnam program that surveyed the ecological damage of US defoliants and the broader needs of North Vietnamese scientists. Science for the People groups provided technical assistance to radical community groups, including local chapters of the Black Panther Party, and encouraged scientists to think of themselves as "workers."[39]

The echo of J. D. Bernal's vision of Marxist science is not accidental. By the late 1960s, Marxism and neo-Communism had resurfaced within the academic left, even in the United States. (An elderly Bernal attended the first meeting of the British Society for Social Responsibility in Science in 1969.) With the collapse of the Cold War consensus, US scientists began to reflect more carefully on the potential political uses of the idea of "scientific freedom." Nowhere was this dynamic more on display than in an ongoing dispute involving the NAS, a Nobel laureate, and a racist research foundation.[40]

::

William Shockley's contributions to Cold War computing were legendary. As a coinventor of the junction transistor at Bell Labs, Shockley earned the credit he usually receives for making the semiconductor industry possible. The Nobel Committee's decision to award him one-third of the 1956 Nobel Prize in Physics cemented his reputation. Shockley's interests, however, soon turned elsewhere. Starting in the mid-1960s, he became an outspoken opponent of desegregation. He advocated eugenic policies and argued that black poverty ultimately stemmed from inferior genetics. Despite having no training whatsoever in biology, genetics, or sociology, Shockley became one of the country's leading spokespersons for discredited racial science. His Nobel Prize and the generous financial backing of the Pioneer Fund—a foundation that promoted racist propaganda—amplified his voice.[41]

In 1967, Shockley began a one-man campaign to enlist the NAS in his efforts. He urged the academy to undertake a large-scale study of the genetic basis of socioeconomic inequality in the United States. With the help of several geneticists, the organization's council issued a statement in the fall of 1967 explaining the various reasons why the academy would do no such thing. The bulk of the statement stuck to scientific criteria—especially the impossibility of distinguishing between social and genetic factors in a society where the deck was so obviously stacked—and urged more support for basic research. Behind the scenes, however, academy members acknowledged that Shockley's demands forced the NAS to choose between "its traditional belief in free inquiry and its realization that the formulation of heredity versus environment adds up to a loaded question that might be destructively exploited by racists if the Academy even ratified it as the right question."[42]

The NAS leadership seemed to be inching toward the position that some questions should not be asked, an attitude difficult to reconcile with Cold War notions of scientific freedom. Shockley exploited this discomfort. With the help of the Pioneer Fund's publicity machine, the *Wall Street Journal* ran a column, provocatively entitled "The Lysenko Syndrome," that compared the NAS's canceling of a hearing on the issue to Soviet control of genetics. "It is foolhardy," the columnist wrote, "to suppose that a problem will disappear by refusing to look at it." The NAS was engaged in "wishful thinking science." The academy leadership quite reasonably responded that, unlike in the Soviet Union, the NAS held no monopoly over scientific studies—Shockley was free to find someone else to support his work. The following year, Shockley upped the ante, sending all 800 members of the academy a packet of materials that compared them to German intellectuals who cooperated with the Nazi Party.[43]

A debate about Nazis, Lysenko, and scientific freedom, all happening under the roof of the NAS—here were all the classic ingredients of a Cold War ideological showdown about the nature of science. But this time, the NAS leadership responded by engaging in a several-years-long conversation about the social responsibilities of elite scientists. This discussion took place alongside heated debates about the National Research Council's continued involvement in classified research contracts, including research on nonlethal weapons used in counterinsurgency campaigns. On the right, military advisors like Roger Revelle argued that scientists had an "obligation" to serve their government. On the left, biologist Richard Lewontin—by then closely associated with Science for the People, which took up opposition to racial science as a rallying cry—tendered his resignation from the NAS in protest of its continued sponsorship of classified research. Collectively, these events inspired scientists to reflect on the "social and political" aspects of the academy's work. This was a far cry from Glass's insistence, in 1962, that academy scientists could, through careful attention, separate their political and scientific commitments.[44]

The NAS ultimately stood firm in its refusal to sponsor research on the genetic basis of racial inequality, even as it did little to stop Shockley from promoting damaging claims on his own time. In 1974, the academy allowed Shockley to publish an article in its associated journal, the *Proceedings*, in part on the recommendation of Bentley Glass. Glass, true to form, deemed the "mathematics" of Shockley's article on "gene flow" in "intermixing populations" sound and therefore recommended publication, even though he found Shockley's views on race "scientifically unsound and socially repugnant."[45]

Over the course of the 1970s, the NAS found itself drawn further and further into discussions of scientific values. The organization responded inconsistently, sometimes shying away from "political" problems, sometimes using its institutional sway to push a research agenda on social and environmental issues. If by "Lysenkoism" Shockley and his critics meant sensitivity to the political implications of scientific work, this was a constraint the academy's members were newly ready to accept. Glass's "neutrality" was mostly out; Bernal's "social responsibility" was increasingly in. And with the US government effectively out of the cultural diplomacy business, the academy's own deliberations on the nature of scientific freedom would come to play a surprisingly central role in the last decade of the Cold War.

9 : Scientists' Rights Are Human Rights

The Apollo-Soyuz Test Project (ASTP) was one of the most iconic moments of international scientific cooperation of the entire Cold War. On July 17, 1975, US and Soviet crews successfully docked their respective spacecraft with a special collar that bridged the formerly impassible gap between two nations and their competing views of the world. Once united in orbit, the astronauts and cosmonauts exchanged their famous "handshakes in space," providing an indelible image of détente. Of course, most of the crew members had already met during training exercises, including a joint expedition to Disney World. Nevertheless, the astronauts and cosmonauts embraced the moment as a symbol of the power of scientific cooperation for peace.[1]

While the docked spacecraft orbited overhead, leaders from thirty-five countries were winding down a three-year negotiation that, while less photogenic, would ultimately have a much greater impact on global history. Signed on August 1, 1975, the Helsinki Final Act recognized the legitimacy of Europe's existing postwar borders—long a diplomatic goal of the Soviet Union—and pledged nations to adhere to certain principles of conduct. As foretold by the ASTP, the agreement included extensive measures to increase exchange between the East and the West, including scientific and technical cooperation. But, more importantly, the Helsinki Accords also included a commitment to human rights, with participating states agreeing to "respect human rights and fundamental freedoms, including the freedom of thought, conscience, religion or belief, for all without distinction as to race, sex, language or religion." The Helsinki Accords were not themselves legally binding, but they included provisions for follow-up meetings to track implementation of the agreement.[2]

In one of the most remarkable instances of unintended consequences in political history, the Soviet Union's push for the Helsinki Accords became one of the major drivers of that empire's collapse. The agreement raised the profile of the small but dedicated community of Soviet human rights activists, who increasingly formed partnerships with other activists in Europe, the United States, and Canada. The global response to their work rattled Soviet leadership in the late 1970s and the 1980s. While it would be a stretch to say that the human rights revolution *caused* the collapse of the Soviet Union, Mikhail Gorbachev's series of reforms, including the policy of *glasnost*, or "openness," makes no sense without it.[3]

Many of the Soviet human rights activists were scientists. When physicist Yuri Orlov, mathematician Anatoly Shcharansky, and, later, nuclear scientist Andrei Sakharov were arrested, the US scientific community rallied around their cause. American scientists, after all, had been protesting the plight of science in the Soviet Union since Trofim Lysenko wrested control over genetics in 1948. The arrests of dissident scientists brought together all of the rhetoric that had driven US depictions of Soviet science throughout the Cold War, but with the US government at least partially out of the business of cultural diplomacy, these campaigns were—once again—largely driven by private scientists rather than government officials. But the appropriate response to Soviet human rights violations posed a riddle for US scientists. Their only point of leverage was to pressure the US government to suspend scientific contact with the Soviet Union until the dissident scientists were released. If the problem with Soviet science was that the government wielded political control over its scientists, how could the United States possibly respond by limiting its own scientists' contacts abroad?[4]

Chastened by the critiques of the New Left, US scientists in the 1970s confronted the problem of Soviet scientists' freedoms, and scientific freedom, with a deeper political sophistication than they had two decades earlier. While some scientists, particularly those on the right, relied on the old battle cry of scientific freedom, others embraced the more general language of human rights. In this new, more inclusive framework, Soviet scientists deserved freedom of movement, freedom of speech, and personal autonomy, not because of their special status as scientists, but because of their status as human beings. By the time Sakharov was released from internal exile in 1986, scientists' rights had become human rights, and the Cold War was nearly over.

⁚⁚

According to at least one account, Soviet leader Leonid Brezhnev believed he deserved the Nobel Peace Prize for his work in negotiating the Helsinki Accords. Instead, the 1975 prize went to Andrei Sakharov, the most prominent human rights activist in the Soviet Union, and possibly in the world. Sakharov's wife, the physician Elena Bonner, delivered the Nobel lecture on his behalf in

Oslo that December, as the physicist himself had been denied the proper visa. Sakharov spent the day of the prize ceremony in the company of other dissidents in the lobby of Lithuania's Supreme Court Building while the biologist Sergei Kovalev stood trial inside. Kovalev was being tried for "anti-Soviet agitation and propaganda" for his role in editing a leading dissident publication, the *Chronicle of Current Events*. Sakharov took a break from the trial that evening to listen to Bonner deliver his remarks over shortwave radio. Yuri Orlov and a handful of others stayed behind in the courtroom, only to be dragged out after Kovalev shouted his congratulations to Sakharov from the stand.[5]

For dissidents like Sakharov, Bonner, Kovalev, and Orlov, the Helsinki Accords presented an opportunity for legitimacy. In contrast to more typical rights campaigns, the world over, that urged governments to change unjust laws, the Soviet dissident movement demanded that its government follow the rules, procedures, and protections of *existing* laws. Its leaders were therefore surprised and delighted when the Soviet government distributed more than 20 million copies of the text of the Helsinki Accords, including its human rights provisions, in multiple languages. Now that the Soviet Union had announced its commitment to human rights in *Pravda*, the dissidents planned to hold the government to account.[6]

Soviet dissidents were, by now, experienced in mining the vein of legal rights for maximum publicity. Most participants in the movement dated its origins to a protest in Pushkin Square on December 5, 1965, over the trial of the writers Andrei Sinyavsky and Yuli Daniel. The protesters at this, the first unauthorized public political assembly in Moscow since the end of World War II, weren't calling for the prisoners' freedom; instead, they merely demanded an open trial, in accordance with the Soviet constitution. In the years that followed, dissidents attended and publicized the trials of political prisoners, using petitions, demonstrations, and press conferences to draw attention to legal irregularities. They circulated their findings in texts known as *samizdat* when intended for distribution in the Soviet Union and as *tamizdat* when intended for foreign eyes. The legal approach crested in November 1970, when Sakharov joined several other dissidents in forming a Human Rights Committee dedicated to educating Soviet citizens on the regime's violations of their legal rights.[7]

A separate strand of activism in the Soviet Union protested the treatment of ethnic and religious minorities, especially Jews. As an officially antireligious society, the Soviet Union periodically carried out campaigns against religious institutions and practices. Sometimes this took the form of administrative harassment; at other times, public propaganda. But rather than eradicating a sense of Jewish identity, these campaigns had the effect of pushing observant Jews' religious expression underground and creating a sense of ethnic solidarity among the secular. The climate of anti-Semitism climaxed in January 1953, when Soviet state media accused a group of (mainly Jewish) doctors of plotting to assassinate senior members of the government. Stalin's death two months later relieved fears of a purge, but the incident forced many Soviet Jews to reconsider their place in Soviet society. Some began to dream of moving to Israel.[8]

In 1966, Soviet Premier Alexei Kosygin announced that Jewish citizens with family in Israel could apply for exit visas. Within a year, however, the government had reneged on its promise, issuing only 230 emigration visas for Israel in 1968. The Kremlin's broken promises to its Jewish citizens awakened a sense of purpose in many young activists, including mathematics student and chess prodigy Anatoly (Natan) Shcharansky. Shcharansky recalled in his memoir, "I was a Jew because of anti-Semitism." By the end of 1969, more than 30,000 Soviet Jews had requested emigration visas; less than 10 percent were granted. The growing number of applicants whose requests were declined—the so-called *refuseniks*—built a political movement around their right to leave.[9]

Both activist movements counted publicity, especially Western publicity, as one of their most effective tools against government repression. The Human Rights Committee, for example, announced its launch at a press conference packed with foreign journalists in an organizer's Moscow apartment. (Soviet journalists were invited to these sorts of events but never attended.) Charismatic figures like Sakharov counted on their relationships with the foreign press not only for publicity but also for some measure of protection: the more famous you were, they figured, the less likely you were to be shot. The dissidents similarly welcomed visits from Western dignitaries, including US senators, who carried their stories back to Capitol Hill. With his winning personality and excellent English, Shcharansky often served as translator for both human rights activists and *refuseniks* at these apartment-based summits and press conferences.[10]

American and British publishers, for their part, actively contributed to the Soviet dissidents' celebrity. In July 1968, for example, the *New York Times* devoted three entire pages of newsprint to Sakharov's essay, "Thoughts on Progress, Peaceful Coexistence, and Intellectual Freedom" (most commonly known by the more accurate translation of its Russian title, "Reflections"). The newspaper helpfully explained how to pronounce the nuclear physicist's name and assured readers that "Dr. Sakharov has not been afraid to speak out." Within a year of its original publication in *samizdat*, more than 18 million copies of "Reflections" were in circulation in various languages, including throughout Eastern Europe. Whether the essay found its way behind the Iron Curtain organically or with quiet assistance from the United States is not clear; a decade earlier, the CIA's Soviet Russia Division had been responsible for printing and distributing Russian-language copies of Boris Pasternak's *Doctor Zhivago*, a work originally published in Italian and banned in the Soviet Union.[11]

After some initial high-profile successes, including the release of anti-Lysenkoist biologist Zhores Medvedev from a psychiatric institution in 1970 and a brief spike in the number of exit visas, the dissident movement foundered in the early 1970s. The presence of foreign journalists may have provided safety for the dissidents, but it provoked the Kremlin. By 1974, the number of active dissidents had dwindled, as the government had allowed some *refuseniks* to emigrate and forced other dissidents (including Medvedev) into exile by stripping them of their citizenship during their travels abroad. The *Chronicle of Current Events* briefly suspended publication in the spring of 1973 after several of its editors were arrested. In Sakharov's case, the Soviet authorities chose harassment over arrest, waging a relentless media campaign that decried Sakharov as an Enemy of the People and Bonner as his manipulative puppet master. With most of its members faced with the pressing question of survival, the Human Rights Committee "faded out of existence," in Sakharov's words, in 1974.[12]

The remaining dissidents therefore welcomed the news of the Helsinki Accords as an opportunity to start afresh. In May 1976, eleven of them, including Orlov, Shcharansky, and Bonner, formed the Moscow Helsinki Watch Group to monitor the Soviet Union's compliance with the agreement. Sakharov declined to lend his own name; as he explained to Orlov, "It had been a relief to escape the constraints imposed by the Human Rights Committee, and I had no wish to find myself again saddled with such obligations." Bonner's participation, however, signaled the Nobel laureate's tacit approval and, more importantly, allowed access to his network of journalists. Orlov announced the group's creation on May 12 at a midnight press conference in Sakharov's apartment. Only a single journalist attended. The timing wasn't ideal, but Orlov was trying to stay a step ahead of the KGB. He and his wife returned home and turned off the lights, after which he climbed out of the window to lie low for a few days. He emerged from hiding two days later when he heard a report on Voice of America announcing the group's creation. With Orlov's safety assured, the other members gathered for yet another press conference in Sakharov's apartment.[13]

Under Orlov's leadership, the Helsinki Watch Group brought together the human rights activists, nationalist leaders, and *refuseniks* who had previously operated in largely separate movements. Orlov and Shcharansky were soon besieged with visitors sharing their own stories of oppression. Other members of the group traveled across the various Soviet republics, compiling reports on the conditions of prisons, schools, and religious groups. Inspired by the work of leaders in Moscow, similar Helsinki monitoring groups popped up in other Soviet republics and satellites, including Ukraine, Poland, and Czechoslovakia. But the Moscow group's most consequential act was to build links to independent human rights groups in Europe, the United Kingdom, and the United States. The Commission on Security and Cooperation in Europe (the body that administered the accords) translated the group's reports and distributed them to other interested parties, including Amnesty International and the embassies of countries that had signed the accords. By 1977, the Soviet dissidents had effectively turned a document that was intended to normalize the relationship between the United States and the Soviet Union into a flashpoint for human rights.[14]

: :

The US scientific community's official response to the human rights situation in the Soviet Union had, until this time, been tepid at best. As the organization that coordinated most of the country's scientific exchanges, the National Academy of Sciences was especially invested in maintaining a semi-normal relationship between the United States and the Soviet Union. After some initial stumbling blocks, by the mid-1960s the academy's exchange program had settled into a routine. Twenty to forty US scientists per year traveled to the Soviet Union for short- and long-term visits in a process that slowly built up the Americans' familiarity with Soviet scientific institutions; an equal number of Soviet scientists traveled to the United States. By 1975, more than 350 scientists from the United States had participated in the interacademy exchanges. But in an irony not lost on the NAS leadership, the arrival of détente strained the organization's role in cultural and technical exchange. Between 1972 and 1974, the United States and the Soviet Union signed eleven bilateral agreements for scientific collaboration in areas ranging from environmental protection and artificial heart research to, of course, space. The number of technical exchanges immediately ballooned, from around a hundred persons, total, in 1972 to nearly a thousand in 1974, with the academy exchanges making up a shrinking portion of the total.[15]

That US federal institutions controlled many of the new exchanges highlighted the ideological oddities that had bedeviled the NAS's exchange program since its origins. Promoting scientific freedom in the guise of direct contact between scientists worked as both propaganda and an intelligence strategy. But, from the beginning, the structure of the exchange program undermined the ideological message. The NAS, though not technically a government agency, inserted itself in the middle of these supposedly free interchanges between scientists. The NAS vetted applicants, arranged participants' travel, and approved their itineraries. Over time, this system of centralized administration had become routine. Détente ratified the shift in US foreign policy away from the use of international scientific agreements as messaging opportunities about the virtues of scientific freedom and toward partnerships that could facilitate mutually beneficial scientific and technological progress.[16]

The most important of the new bilateral agreements, the Agreement on Cooperation in Science and Technology, was signed in Moscow on May 24, 1972, as part of an arms reduction summit between Brezhnev and President Richard Nixon. This agreement, and all those that followed (including the agreement on space that served as the basis for the ASTP), explicitly embraced cooperation between the two superpowers on the basis of equality and reciprocity. They therefore emphasized government-to-government initiatives instead of the kinds of facilitated person-to-person exchanges that the United States had advocated since 1958. They also explicitly endorsed political goals over scientific ones.[17]

After all this time, US policymakers had finally acknowledged the politics at the heart of their embrace of scientific internationalism. But there was a hitch. This embrace of scientific *realpolitik* came at the very moment that individual US scientists, particularly those involved in the exchange programs, were becoming more aware of the lived experience of Soviet human rights violations. They began to question whether their participation in US exchange programs tacitly endorsed Soviet behavior. The problems started in 1970, when physicist Wayne J. Holman III was asked to cut short his five-month stay at the Institute of Theoretical Physics in Kiev. While in Kiev, Holman had met Sakharov, who invited him to stop by his apartment in Moscow. A few days after Holman did so, in December, a Soviet official notified the US Embassy that Holman had "violat[ed] the hospitality of the Soviet Union"; his visa was being revoked. On the advice of the embassy, Holman left the country without protest.[18]

Holman's exchange took place under a private agreement, not an interacademy exchange. His experience might not have attracted much attention except that, just three months later, the Soviet Union expelled David Viglierchio, a nematologist, for allegedly "collecting slanderous information" about the USSR from "Zionists." Viglierchio, in other words, had been consorting with *refuseniks*. With six weeks left in his six-month, NAS-sponsored exchange, Viglierchio and his wife headed to the Moscow airport, where the authorities searched both their luggage and their pockets. Then in May, the Soviet newspaper *Literaturnaya Gazeta* accused the US Embassy of using visiting scientists to recruit spies. The report alleged that two scientists participating in the NAS exchanges, mathematician Anthony Tromba and physicist Stephen Salomon, had conspired in an elaborate scheme to introduce a "young scientist" to a cultural attaché at the US embassy.[19]

Of all these episodes, the NAS formally protested only Viglierchio's expulsion. Foreign Secretary Harrison Brown did, however, send a long, private letter to the Soviet Academy warning that such incidents damaged the two countries' "mutual scientific relations." The kinds of problems the scientists from the United States were encountering had nothing to do with science, Brown suggested, and everything to do with politics. As Brown put it, "If [Viglierchio] became aware of any dissatisfaction on the part of Soviet scientists, it was in a passive manner—just as Soviet visitors to Washington on May 1 would have been able to observe the dissatisfaction expressed by many young Americans over the continued involvement of the United States in Viet-Nam." The Soviet Union, in other words, had a lower tolerance for political dissent than did the United States, and that low tolerance was interfering with international scientific cooperation.[20]

Brown's official complaint, with its emphasis on "open exchange of scientific information," echoed what had been the standard US line on scientific freedom for most of the Cold War. But compare Brown's comments, designed to maintain the NAS's existing program, with Viglierchio's own, more personal, letter of protest. In a letter urging the NAS to take stronger action, Viglierchio highlighted the ways that his expulsion violated his *rights* as a scientist. "I feel strongly," he wrote, "that no scientist . . . should be subjected as I was, to a fantasy conjured by some minor official of the security committee." He urged the NAS to respond, to force the Soviet Union to acknowledge that a "frank discussion of professional and personal matters of each others [*sic*] community" was intrinsic to cultural exchange. His expulsion, in other words, presented an opportunity for the United States to draw attention to the hardships and harassment that Soviet scientists, particularly *refuseniks*, faced.[21]

Brown declined this opportunity. The NAS would *not*, he assured the Soviet Academy, be releasing any public statements about Viglierchio's treatment, as any public discussion of the tensions might "poison the atmosphere" for the exchanges. Having spent the past ten years of his life building up the NAS's international programs, Brown was not inclined to let a few restrictions on scientists' freedom undermine a program that he believed contributed so much to the national interest.[22]

The NAS's president, Philip Handler, did make an exception to this policy on behalf of Sakharov, who had been elected a foreign member of the academy in 1969. The NAS could not ignore the harassment of one of its own. In September 1973, Handler sent a telegram to his Soviet counterpart, Mstislav Keldysh, warning that the government's continued harassment of Sakharov would endanger plans for greater scientific and technical cooperation between the two countries. But, as with Brown's protest over Viglierchio, Handler left the threat vague.[23]

The debate resurfaced in the summer of 1975, when the NAS's first formal evaluation of the interacademy exchange program forced the organization to decide what kind of program the exchanges really were. The evaluating panel initially contrasted the exchange program's "scientific" objectives with the more "political" goals of the new bilateral agreements. The real goal of the exchanges, after all, was to "abolish the necessity for formal exchange programs—to conduct exchanges completely freely." A review of the more than 300 reports submitted by US participants, however, convinced the panel that the "social, sociological, and political" benefits of the exchanges outweighed their negligible scientific benefits. Pugwash veteran Paul Doty went further, pointing to the role of the exchanges in gathering scientific intelligence—or, as he more tactfully phrased it, in "monitoring Soviet scientific capabilities or keeping track of scientific progress in the USSR for the general benefit of the US." The panel chair, Carl Kaysen, former deputy national security advisor, reminded the panel that the "very presence of American scientists in the USSR" presented propaganda opportunities. For this reason, the exchanges should continue. Kaysen and Doty, old hands at the political uses of apolitical science, stood by the premise that the NAS, and therefore the United States, could most influence conditions inside the Soviet Union by maintaining the status quo.[24]

The more idealistic members of the panel tried to squirm away from these political calculations. The signing of the Helsinki Accords in August emboldened the skeptics, who wanted their ethical concerns on the record. "It is obvious," the December 1975 interim report observed, "that scientists have been among the most prominent of Soviet dissidents." And, because the Soviet Union allowed only its most loyal scientists to participate in the exchanges, and because visiting Americans so often limited their contact with dissidents to avoid offending their hosts, the NAS's exchange program was actively undermining the work of Soviet human rights groups. The interim report warned that maintaining the interacademy exchange program in its current form risked sacrificing the dissidents' interests to broader political goals, "summarized as détente."[25]

By the mid-1970s, the rank-and-file scientists of the NAS increasingly recognized that their personal politics, and those of their colleagues abroad, did not always neatly overlap with those of the US government. But at least for now, they were willing to distinguish between the two. The panel's report ultimately acknowledged that it would be "inappropriate" for "national scientific bodies like the Academy" to criticize the Soviet Union or its political practices. Instead, the panel endorsed continued gentle pressure—that is, sternly worded letters and telegrams—as the occasion warranted. Nevertheless, the report observed that individual US scientists were free to take stronger action, whether in the form of boycotting the exchanges, protesting the treatment of dissidents, or writing letters or making phone calls on behalf of individuals.[26]

∷

Outside the elite chambers of NAS panel meetings, US scientists were not waiting for permission to participate in the growing global human rights movement. On the left, the human rights movement drew support from those who opposed the US government's relationships with Latin American dictators; on the right, it attracted anti-Communists who opposed détente. In 1966 and 1967, individual members of the NAS urged its leadership to assist students and faculty caught up in political conflict in Argentina and Brazil, respectively. In 1973, 1974, and 1975, US-based advocates for Chilean scientists urged the academy to join private groups in condemning the large-scale human rights violations taking place under dictator Augusto Pinochet. Two independent groups, the Federation of American Scientists and the Committee of Concerned Scientists, made sure to include NAS leadership as part of their own letter-writing campaigns on behalf of Soviet dissidents. By early 1976, academy members had urged Handler to act on behalf of more than two dozen Soviet dissidents, most of whom were *refuseniks*. But with the exception of the letter on behalf of Viglierchio and the public telegram defending Sakharov, the academy consistently chose the path of "quiet diplomacy," with leadership claiming that the NAS, as a scientific body, had no business issuing official statements on political or social issues.[27]

When the American Association for the Advancement of Science passed a human rights resolution at its February 1976 meeting, NAS officers and staff realized that the academy could postpone action no longer. Murray Todd, a staff officer in the academy's Commission on International Relations, offered a damning assessment of the official response to date. "While energetic at first," he wrote, "the reactions of officers and staff . . . yield [to] a detached attitude expressed in ever more routinized responses which have the effect of conveying a lack of serious NAS concern." If the academy received somewhere between three to seven letters, Todd observed, Handler or Brown might take some sort of mild action. But if it received more than that, the leadership—Brown especially—would bristle at what they perceived as outside political pressure. It made no difference whether these inquiries came from academy members or members of the broader public. As a way to avoid what he referred to as "operational bias," Todd recommended that the academy appoint an independent staff member to address the issue.[28]

The NAS Council declined to assign a staff member full-time responsibility for human rights, but the discussion did spur some forward motion. That spring, the council approved new guidelines for the foreign office that, in the mildest possible language, acknowledged that "violations of human rights . . . occur" and "may represent a very high form of the concern of members of the human race for one another." In keeping with the academy's preference for a nonpolitical role, the institution itself would continue to pursue "quiet and informal contact" as its primary strategy; it would attempt to avoid influencing "political polarization." With Todd's urging, however, the council also issued a more strongly worded affirmation of human rights that focused on freedom of scientific inquiry and expression. After asserting that the "search for knowledge" depended on intellectual freedom, freedom to travel, freedom to publish, and freedom from retaliation, the statement ended with a bolder commitment to the more general principle of human rights and "the worth and dignity of each human being." Todd distributed the statement widely, mailing copies to all 1,300 members of the academy and 150 officers of professional societies and securing its publication in the journals *Science* and *Nature*. By the end of 1976, the NAS Foreign Office had received more than 2,300 signed copies of the resolution.[29]

The NAS Council would never have taken even these halting steps without pressure from its members, external organizations, and dedicated staff members like Murray Todd. The charge of inaction stung Handler, who wanted the membership to understand how "remarkable" it was for the stodgy academy to do anything at all. In a letter to biochemist Christian Anfinsen, a Nobel laureate who opposed the NAS Council resolution on account of its toothlessness, Handler defended the statement as a necessary first step. The academy, Handler pointed out, "has never previously stated what it believes with respect to human rights and never previously indicated explicitly that these are a legitimate concern of the institution and therefore a basis for institutional behavior. But now, we have. And that opens the door to the next phase of our behavior."[30]

The academy's new foreign secretary, chemist George Hammond, strode through that door with confidence. Within a month after the council passed the resolution, Hammond had established a Committee on Human Rights, staffed by Todd, to monitor the NAS's compliance with its guidelines and to investigate individual cases brought to the committee's attention. The committee's mandate continued to evolve in its first year, but by March 1977 it had been authorized to issue public statements on individual human rights cases. Following a model pioneered by Amnesty International, the committee would "adopt" individual victims of human rights abuses, with "corresponding members" invited to write letters and make phone calls on their behalf.[31]

Handler approved this plan only grudgingly. He chided the chair of the Committee on Human Rights, geographer Robert Kates, that letters written on behalf of Soviet scientists would "demolish whatever small belief is held in the Soviet Union that this Academy is an autonomous private body." Nevertheless, he assented: "But—so be it." A recent letter of support that newly elected President Jimmy Carter had sent to Sakharov in early February 1977, reiterating US "respect for foreign rights," added to Handler's pique. Carter had used his inaugural address to move human rights to the mainstream of US foreign policy, proclaiming that the United States' "commitment to human rights must be absolute." For Handler, the most pressing question was how the NAS could continue to serve as an open channel of communication between US and Soviet scientists when both the president of the United States and the academy's own members were criticizing the Soviet government. For an outspoken contingent of US scientists, however, the more important question was whether the exchanges should continue at all.[32]

::

On the evening of February 2, 1977, Yuri Orlov once again crept out of his apartment window with the KGB hot on his heels. This time he headed for Tula, an industrial village about 120 miles south of Moscow. He spent several days anxiously drinking tea in the kitchen of a friend's mother's house before his fellow dissidents summoned him back to Moscow. Other members of the Helsinki Watch Group had been arrested; as the group's leader, Orlov had a duty to return home. Disguised in his host's coat and glasses, he made it back to Moscow, where he held a press conference to announce his imminent arrest. He was arrested and taken to Lefortovo prison on February 10.[33]

Just over a month later, Shcharansky, too, was arrested. Shcharansky had been staying at the apartment of another *refusenik* when two journalists, one from the *Baltimore Sun* and the other from the *Financial Times*, arrived to discuss the release of a political prisoner. The journalists asked Shcharansky what it was like to be constantly trailed by the KGB, and Shcharansky offered to show them by taking a walk outside. To the reporters' shock and horror, the KGB shoved Shcharansky into a waiting car. He, too, was taken to Lefortovo.[34]

The two arrests jolted the US scientific community into action. In April, the NAS Committee on Human Rights distributed an information bulletin on Orlov to its corresponding members, asking them to write letters on the physicist's behalf. Press coverage of the group's selection of Orlov, along with seven other victims of human rights violations, noted that the move was a "departure" from the NAS's previous policy of "silent diplomacy." Shcharansky's case, meanwhile, became a cause célèbre for the Committee of Concerned Scientists and the Association of Computing Machinery. Human rights observers were particularly alarmed at the severity of the charges being leveled against the young *refusenik*. While most dissidents, including Orlov and Kovalev, were charged with anti-Soviet agitation and propaganda, Shcharansky was being charged with treason, a capital crime. His endless strings of meetings with US senators and journalists left him exposed to charges of espionage, and the Soviet government claimed that he was working for the CIA.[35]

Knowing what he must have known about the CIA's past use of scientists in intelligence work, Handler at first kept the academy at a distance from Shcharansky's case. After all, as he told the co-chair of the Committee of Concerned

Scientists, the charges might be true. But once President Carter stated unequivocally that the State Department and the CIA had "double-checked" and that Shcharansky "has never had any sort of relationship, to our knowledge, with the CIA," Handler committed the NAS to Shcharansky's fight. In December, Handler cabled Brezhnev to request permission for the NAS to send an observer to Shcharansky's trial. The request went unanswered. Nevertheless, the State Department thanked Handler for his efforts, which "complemented" official diplomatic efforts and signaled the interest of the US scientific community in Shcharansky's fate.[36]

With Orlov and Shcharansky still awaiting trial more than a year after their arrests, US-based human rights activists demanded further action, up to and including a boycott of Soviet science. The campaign for a boycott was led by Valentin Turchin, one of the founders of the Moscow chapter of Amnesty International, now living in exile in New York. Turchin, a computer scientist, began his campaign quietly, recruiting individual high-profile physicists through private, impassioned correspondence. After Handler discouraged his efforts, however, Turchin took his campaign to the pages of *Nature*. A formal boycott, Handler had warned Turchin, "could have the effect of throwing the switch from detente back to the cold war which is but prelude to a hot one." Turchin dismissed this fear as an "exaggeration" and urged resilience: "In the fight for human rights you cannot count on speedy results and lose your nerve when the Soviets refuse to give in." In private, he rebuked Handler in stronger language, asking, "If you are unwilling to use the boycott threat, what is the meaning of adopting Prof. Orlov by the NAS?"[37]

Orlov's conviction in May 1978, after a hasty trial closed to Western observers and to defense witnesses, finally sparked the rebellion that Handler had so diligently been trying to prevent. On the day that the court sentenced Orlov to seven years of hard labor and five years of exile, nineteen US scientists preparing to take part in an NAS delegation to the Seventh Joint US-USSR Symposium on Condensed Matter Theory voted unanimously to cancel their trip. Handler was in the midst of renegotiating the next installment of the interacademy exchange agreement when the boycott was announced. He issued a carefully worded statement clarifying that the *delegation* had canceled the meeting, not the academy. "We have repeatedly warned Soviet authorities that the issue of human rights threatens to erode the willingness of American scientists to cooperate with their Soviet counterparts . . . Our predictions are now being borne out."[38]

The boycott campaign gained momentum over the summer of 1978, with Shcharansky's trial in July sparking more grassroots organizing on the dissidents' behalf. Shcharansky luckily escaped the death penalty, instead receiving three years in prison and ten years of hard labor. A new Berkeley-based organization, Scientists for Shcharansky, brought Shcharansky's wife, Avital, to the United States to promote his case. In a matter of days, the organizers gathered more than 500 signatures from scientists pledging to boycott Soviet science until both Orlov and Shcharansky were released from prison. After changing its name to Scientists for Orlov and Shcharansky in the fall, the new organization started a direct-mail campaign inviting supporters to sign either a "Statement of Conscience" committing to the boycott or, if they preferred, a "Declaration of Principles" that opposed "any enlargement of scientific and technical exchange programs between the Soviet Union and the United States" until the human rights situation improved. Meanwhile, three other delegations scheduled to travel under the bilateral agreements canceled their trips, as did several individuals scheduled under the exchanges.[39]

The question of whether to boycott or merely to not expand the exchange program went beyond tactics. Handler's position—that NAS-sponsored exchanges, symposia, and workshops contributed to international peace—maintained the Cold War fiction that scientists could best contribute to foreign affairs by setting aside politics. This fiction had institutional consequences, in that the National Science Foundation, rather than the State Department, now funded the NAS's international operations. This meant that the exchanges were subject to peer review of proposed projects' scientific merit—not necessarily the right fit for a diplomatic endeavor. As Hammond noted, the types of international exchange programs that "coincide[d] with the optimum vehicle for advancing current government efforts" did not necessarily represent "the best science." Handler nevertheless continued to assert that the academy did not wish to insert itself into politics.[40]

Among human rights activists, however, an emerging ethic demanded that scientists speak truth to power. Jack Minker, one of the vice chairs of the Committee of Concerned Scientists, explored scientists' obligation to speak out in an opinion piece published in a professional journal for computer scientists: "For scientists to say that science should not be involved in politics is to deny reality." Using the universal language of human rights, Minker urged scientists to shoulder their responsibilities: "If we do not speak out for the human rights of our fellow scientists . . . we are bearing silent witness to the destruction of science and humanity."[41]

There was one issue on which nearly everyone involved could agree. The decision of whether or not to boycott—whether or not to commit a political act—had to be made by *individuals*. Almost every press release and direct mail appeal issued by Scientists for Orlov and Shcharansky, for instance, reminded readers that "the actions taken by SOS are a completely spontaneous response to Soviet policy and are not connected to any position or policy of the U.S. government." The NAS Committee on Human Rights echoed this sentiment, suggesting that the boycott demonstrated the power of scientific freedom: "Scientific exchange programs can be negotiated and organized, but individual participation can never be commanded."[42]

Determined to demonstrate the NAS's independence, Handler signed a new interacademy exchange agreement on January 1, 1979, even as more and more US scientists pledged not to participate. When the Soviet authorities exiled Sakharov to Gorky in late January 1980 after the physicist criticized the Soviet invasion of Afghanistan, the NAS Council voted to suspend all bilateral symposia and workshops for six months. The exchanges, however, would continue. To reduce any confusion on the issue, the council explicitly reaffirmed that "interactions between American and Soviet scientists on an individual basis are matters properly left to the consciences of the participating individuals." Even as the US–Soviet relationship dipped to a post-détente nadir and as thousands of US scientists—including 187 members of the academy—pledged to boycott the Soviet Union, the exchanges continued to function as a symbol of open scientific communication. At least as late as 1980, the language of scientific freedom and cooperation still seemed to be compelling as political theater.[43]

∷

With most of the Soviet Union's leading dissidents either in prison or in exile by the early 1980s, the Soviet human rights movement inevitably stalled. United States–based human rights organizations, too, struggled to sustain their members' attention, especially since the new president, Republican Ronald Reagan, lacked Carter's moral fervor for human rights. With limited tools available for publicity, some US scientists returned to the habit of quiet diplomacy. In 1985, the NAS quietly reinstated the bilateral scientific programs, with the hope of reopening the channels of communication.[44]

Elena Bonner thought the US scientists' efforts naive and misguided. Bonner accompanied Sakharov in his exile to Gorky, but until 1984 she was allowed to make frequent trips to Moscow, serving as Sakharov's intermediary with the outside world. On one trip in 1983, she met with Jeremy Stone, president of the Federation of American Scientists. Stone had brought Sakharov a camera, a calculator, and an offer to talk to Soviet leadership on his behalf. Bonner dismissed Stone's efforts. "Tête-à-tête conversations are a game that the Soviet authorities love—they've never done anybody any harm." She worried that US scientists had stopped seeing the dissidents as people and that Sakharov, especially, had become "a symbol, a game, politics, even personal success—a dead concept." On at least one occasion, she had to check herself from asking a concerned scientist, "Do you at least know who Sakharov is?"[45]

But even with the dissidents largely removed from view, the Soviet Union's record on human rights continued to undermine the country's international standing. In 1986, a new generation of Soviet leaders, including General Secretary Mikhail Gorbachev, began to reconsider the calculus of political repression. One by one, the dissidents were released. Shcharansky was first. In February 1986, Soviet authorities—who still claimed that Shcharansky worked for the CIA—included him in a spy exchange. After crossing the famed Glienicke Bridge in Berlin, Shcharansky boarded a plane for Frankfurt and, from there, Israel, where he built a career in politics. Next was Orlov, who in October was stripped of his citizenship and deported to the United States. Orlov immediately fell back into his old role as human rights champion, meeting Reagan at the White House. In 1987, he settled in Ithaca, New York, where he accepted an appointment as a research scientist at Cornell.[46]

Sakharov, the most famous dissident of them all, was released from exile in December 1986. He returned "like a homing pigeon" to his Moscow apartment. His fame as a symbol of reform, of *perestroika*, made a return to his scientific work impossible. Instead, he threw himself into political and economic reform, lending conditional support to Gorbachev's efforts to transform Soviet institutions. Within two years, Sakharov was arguing for democratic elections and a market economy. By the time of his death from a heart attack in December 1989, Sakharov had found an unlikely third act as leader of the democratic opposition in the new parliament, the All-Union Congress of People's Deputies. Bonner rejected Gorbachev's offer of a state funeral.[47]

Less than two years later, the Soviet Union collapsed. This was not, of course, entirely or even mostly the work of the dissidents. Nevertheless, the persistence and courage of the Soviet human rights activists, coupled with their knack for publicity, presented an ongoing problem for Soviet leaders. Silencing the dissidents violated the Helsinki Accords and risked international sanction, but allowing them to speak exposed the rot and repression that gave lie to the Soviet promise. In the fight of and for their lives, the dissidents won.

::

Orlov and Sakharov were theoretical physicists. Turchin had trained as a theoretical physicist before drifting into cybernetics and computer science. Sergei Kovalev and Zhores Medvedev were biologists. Shcharansky started out as a mathematician. So many *refuseniks* were unemployed scientists that they ran scientific seminars out of their apartments, a popular stop on the itinerary for sympathetic Western visitors.[48]

I am hardly the first person to notice that so many of the Soviet dissidents were scientists. In their very ubiquity, the scientists are hard to miss. What are we to make of this? One influential explanation, advanced by historian David Holloway, suggests that the unique privileges granted to Soviet scientists—or at least some Soviet scientists—allowed them to develop the institutions for civil society in the Soviet Union. As elite figures whose work was necessary to the national good, scientists were granted an unusual degree of intellectual autonomy, in which they grew accustomed to wide-ranging, open debate—a necessary condition for democratic reform. At the science city of Chernogolovka, scientists enjoyed screenings of foreign films otherwise unavailable in the Soviet Union; at Akademgorodok, researchers mounted exhibitions of subversive art. Life for scientists in the Soviet Union was decidedly different from that experienced by most of the population.[49]

Given the story told in the rest of this book, I cannot help but notice how much this explanation sounds like the Cold War line that scientists, left to their own devices, naturally turn to democracy. But recent work by historians working in post-Soviet archives has made abundantly clear that the vast majority of Soviet scientists weren't dissidents. In the years following Stalin's death, many Soviet scientists reached an uneasy truce with their government, accepting generous funding and creature comforts in exchange for their acquiescence to the Party's wishes. In this, they were not so very different from their counterparts in the United States.[50]

And yet: It is undeniable that some Soviet dissidents, including Sakharov, marshaled the language of science and freedom to advance their cause. The Soviet Union did not lack for dissidents in the 1970s; Sakharov's peculiar success, both as a thorn in the side of the Kremlin and as an international figure, drew in large part from attitudes about scientific freedom that circulated as part of the US ideological offensive during the Cold War. In this small but important way, cultural diplomacy involving science contributed to the eventual downfall of the Soviet Union.

Sakharov's memoirs give the impression of a man who had internalized an American approach to scientific freedom and internationalism since his days as a graduate student. Sakharov entered the Soviet elite at the tender age of thirty-two, when he was elected to the Soviet Academy of Sciences for his contributions to the development of the Soviet hydrogen bomb. His work was additionally rewarded with the Stalin Prize, which carried a purse of half a million rubles, a dacha outside Moscow, and a medal designating him a Hero of Socialist Labor. By the early 1950s Sakharov was spending most of his time at the isolated Soviet nuclear installation known as Arzamas-16, located about 250 miles east of Moscow. Like the US nuclear installation at Los Alamos during World War II, Arzamas was largely cut off from the world. But since the bombs built there worked, and since the residents had little contact with the rest of the public anyway, the Soviet authorities allowed the country's leading nuclear scientists to express a range of political opinions well beyond what would otherwise have been acceptable. The scientists also had access to certain Western journals, including those that promoted ideals of scientific freedom and objectivity in the West.[51]

Sakharov first crossed Khrushchev at a meeting at the Kremlin in 1961, when the scientist voiced his opposition to restarting nuclear tests. Sakharov would later describe this event as one of the "principal turning points" in his life. His second act of disobedience, however, is perhaps more telling. In 1964, Sakharov opposed the nomination of Nikolai Nuzhdin, one of Lysenko's colleagues, to membership in the Soviet Academy of Sciences. As a physicist, Sakharov had learned the details of Lysenko's abuse of power through Medvedev's *samizdat* manuscript. Speaking from the floor of an assembly of the academy, Sakharov condemned Lysenko for "the dissemination of pseudoscientific views, for adventurism, for the degradation of learning, and for the defamation, firing, arrest, even death, of many genuine scientists." These were strong words for anyone to utter—let alone a Soviet scientist speaking in Lysenko's own presence. Reflecting on his action later, Sakharov speculated that he spoke out because of his respect for the "freedom and integrity of science; after all, science remains a keystone of civilization, and any unwarranted encroachment on its domain is impermissible." Nuzhdin's membership was roundly defeated. Within a year, Lysenko, like Khrushchev, was out of power.[52]

Shortly after this incident, Sakharov began spending more time at his Moscow apartment, where he came in contact with the nascent dissident movement. He began circulating *samizdat*, including, in 1968, the essay that made him internationally famous. In "Reflections," Sakharov called on those in power to use the scientific method to solve social problems. Sakharov saw a "convergence" between capitalism and socialism, driven more by a benign technocracy than by democracy, in which a science freed from the demands of politics could meet all the needs of the people.[53]

Sakharov's message multiplied the US narrative that connected science and freedom in ways that he may not have understood. Consider, for instance, a long article that appeared in the *Bulletin of the Atomic Scientists* in 1971, praising Sakharov as "one of the most valiant and perhaps successful fighters for political freedom and human rights in the Soviet Union." The article made sure to point out that Sakharov's colleagues in physics had recently been rebuked for "showing insufficient hostility to Western ideas" before recounting every incident of political interference in Soviet science then known to the West. We know that the *Bulletin of the Atomic Scientists* was one of a handful of US journals generally available to Soviet nuclear scientists. We also know that the author of the article, Walter Clemens, Jr., worked for the Russian Research Center at Harvard, which, historically, had close ties to the CIA. Is this article an anti-Communist rant, pitched at Americans skeptical of détente? Is it a sincere exposé of scientists' human rights violations in the Soviet Union? Or is it a direct appeal to Soviet scientists, designed to stoke their insecurities about whether Soviet science was, in fact, "real" science? Or all of the above?[54]

By the early 1970s, Sakharov was not just a dissident; he was a world-renowned scientist famous for standing up for scientific freedom within the Soviet Union. With each step that the Soviet Union took to discipline Sakharov, starting with the media campaign in 1973, anti-Communist Western scientists expressed ever-increasing levels of doubt about the wisdom of continuing to cooperate with Soviet scientists. But Soviet scientists, too, caught wind of these conversations. The Soviet Union's continued existence depended on the cooperation of its scientists, who increasingly believed themselves deserving of the kinds of scientific freedoms available to scientists in the West. By the mid-1970s, many individual scientists in the Soviet Union, having interacted with Westerners at Pugwash conferences, scientific symposia, and individual exchanges, would have *agreed* with Detlev Bronk and Philip Handler that science functioned best when it operated free from government interference, even if they kept these attitudes largely to themselves.

The Soviet dissidents' triumph represents more than the global recognition of human rights. Their claims on the public's sympathy depended on a global consensus that scientists were entitled to freedom of movement and freedom of inquiry and that all scientists should be free to base their interpretations of the natural world on empirical research rather than political agendas—in short, that all scientists were entitled to scientific freedom. None of the dissidents were arrested for ideological crimes in science. No one in the Soviet government questioned whether Sakharov's commitment to scientific objectivity was compatible with dialectical materialism. None of the Soviet dissident scientists attempted to distinguish between the different worldviews of "Western science" and "Marxist science." By 1980, there was only one "science," and it looked remarkably like the vision advanced by the West.

Unlike most of the United States' other attempts to destroy Communism through culture, science diplomacy worked. Not in the way that Frank Wisner's Office of Policy Coordination or Michael Josselson's Congress for Cultural Freedom intended—Soviet intellectuals didn't defect en masse or attempt to overthrow the Kremlin because of their faith in scientific internationalism. Nor, as Hal Goodwin might have hoped, did Soviet scientists necessarily throw their support behind US delegates at international scientific meetings because of the Americans' supposed fealty to apolitical scientific discovery. Nevertheless, by the time the Cold War ended, Soviet scientists were more likely to quote H. J. Muller than J. D. Bernal on the question of "party lines" in science.

This particular notion of apolitical science could triumph only through scientists' embrace of politics. Scientists both in the United States and in the Soviet Union participated in international human rights campaigns that looked remarkably like other late-twentieth-century social and political movements. They wrote letters, made phone calls, signed petitions, held press conferences, lobbied their institutions and elected officials, and organized boycotts. Sometimes they held protests, and they raised money for legal defense funds. If asked directly, most of the participants would have sworn that these activities had nothing to do with politics—or, if pressed, they might clarify, nothing to do with "partisan" politics.

Scientists' reluctance to recognize their actions as "political" nearly derailed their ability to participate in one of the great human rights movements of the late twentieth century. Jack Minker's warning bears repeating: "For scientists to say that science should not be involved in politics is to deny reality." As we enter a new, unsettled period of global politics, let us hope that scientists can yoke their faith in freedom to a willingness to fight for freedom and justice for all.[55]

Epilogue

This book started to take shape in 2012, a moment of renewed faith in the potential of science to bridge gaps across politics and culture. Shortly after taking office in 2009, President Barack Obama appointed a series of "science envoys" to strengthen the United States' relationship with Muslim countries. By the end of his second term, these science envoys, acting as private citizens, had collectively visited more than twenty countries, including Bangladesh, Indonesia, Kazakhstan, Morocco, and pre-revolutionary Egypt. Obama's science advisor, John P. Holdren, met regularly with his science policy counterparts from Brazil, China, India, Japan, Korea, and Russia. When, in late 2015, the United States and five other powers reached a nuclear agreement with Iran, the US Department of State largely credited the deal to the shared scientific worldview of US Secretary of Energy Ernest Moniz and Ali Akbar Salehi, the head of Iran's Atomic Energy Organization.[1]

The concept of scientific internationalism had found a second life as "science diplomacy." As defined in an influential 2010 report issued by the American Association for the Advancement of Science and the British Royal Society, science diplomacy encompasses three separate but related roles for science in foreign affairs. "Science in diplomacy" uses scientific and technical knowledge to inform foreign policy decisions—basically, science advising. "Diplomacy for science" uses foreign policy channels to pave the way for international scientific collaboration, such as scientists' participation at international meetings. The third category, "science for diplomacy," positions international scientific cooperation as an explicit form of cultural diplomacy, using scientists' formal and informal contacts to build bridges and strengthen alliances between nations.[2]

I wanted to believe in science diplomacy. Shortly after the Iran nuclear deal, I published an article praising the potential of a politically sophisticated science diplomacy to solve—or at least ameliorate—intractable political conflicts. This piece earned me an invitation to a December 2015 conference in Amman, Jordan, on the potential for regional scientific collaboration to defuse tensions in the Middle East. Senior science advisors from Jordan, Egypt, Lebanon, Sudan, and several Gulf States gathered to discuss possible international solutions to the environmental problems facing their region, from wastewater treatment to climate change. The participants were particularly looking forward to the projected 2017 opening of SESAME, a clever acronym for a particle accelerator officially known as the Synchrotron-light for Experimental Science and Applications in the Middle East. Like its predecessor, the Geneva-based European Organization for Nuclear Research (CERN), SESAME offers researchers from several countries that don't always get along—including Egypt, Israel, Iran, and the Palestinian Authority—shared access to capital-intensive equipment. SESAME's backers explicitly hoped that opportunities for international scientific collaboration would contribute to peace in a region plagued with conflict. This was the dream of science diplomacy, realized.[3]

Then Donald J. Trump won the 2016 US presidential election. The full implications of a Trump presidency have yet to reveal themselves at the time of this writing, but it seems safe to say that the Trump administration does not plan to make science diplomacy—or any kind of diplomacy, really—a high priority. State Department staffing levels plummeted under Trump's first Secretary of State, Rex Tillerson. As of December 2017, 97 of 154 senior State Department positions requiring confirmation by the US Senate had gone unfilled. No one had been nominated for the positions of assistant secretary for democracy, human rights, and labor; assistant secretary for educational and cultural affairs; or assistant secretary for oceans and international, environmental, and scientific affairs. The United States lacked ambassadors to Jordan and South Korea, and the seats on the US Advisory Commission on Public Diplomacy remained unfilled. Tillerson's firing on March 13, 2018, left eight of the top ten positions at the State Department vacant. The administration's repeated attempts to bar citizens of certain Muslim nations from entering the country have been only the most visible manifestation of its larger rejection of a hearts-and-minds approach to foreign policy.[4]

Some of the Trump administration's starkest departures from the post–World War II foreign policy consensus have come in the realm of science. As this book has shown, science-based US cultural diplomacy has consistently promoted certain ideals associated with scientific freedom, including objectivity, research-driven science policy, and scientific (rather than political) control of scientific institutions. But as with the empty desks at the State Department, most senior science positions remained unfilled as of March 2018. Trump had not nominated a presidential science advisor, and the understaffed Office of Science and

Technology Policy was rumored to be operating only with the assistance of a shadow staff of volunteer Obama administration alumni. Those scientists who remain are increasingly unwilling to tolerate what they see as the administration's attack on the values of science. In August 2017, Daniel Kammen, a science envoy holdover from the Obama administration, resigned in a biting letter that called out the president for his failure to condemn white supremacists and for his decision to withdraw the United States from the Paris Climate Accord. The letter, which Kammen published on Twitter, contained a cheeky hidden message: the first letter of each paragraph spelled out I-M-P-E-A-C-H.[5]

But even before Trump took office, the official Twitter account of the House Committee on Science, Space, and Technology was promoting false stories published by Breitbart News, a right-wing online media company whose former CEO, Stephen Bannon, had been installed as the White House chief strategist. Breitbart is not quite state-sponsored or state-endorsed news, but the United States had not seen such close ties between an ideologically motivated media organization and an incoming government in more than a century. The tweets raised the alarming possibility of an official state "line" in science, something that US propagandists had spent much of the Cold War denouncing.[6]

Federal scientists' fears of an upcoming ideological purge made front-page news when climate change researchers announced a "guerilla archiving" event designed to preserve environmental datasets. After the Trump transition team announced a series of cabinet-level appointees who had previously endorsed eliminating the agencies they were now nominated to lead, scientists at the Environmental Protection Agency, the Department of Energy, and the National Oceanic and Atmospheric Association urged their colleagues to download as much data as possible from federal websites, lest the data disappear. As Michael Halpern, speaking on behalf of the scientists' advocacy group Union of Concerned Scientists, told the *Washington Post*, "There is a fine line between being paranoid and being prepared, and scientists are doing their best to be prepared." The Department of Energy, meanwhile, refused a request from the Trump transition team to supply a McCarthy-esque list of names of individual scientists and contractors who had attended scientific meetings related to climate change, as well as a list of certain researchers' professional memberships, publications, and contributions to public websites.[7]

As a historian of science and the Cold War, it's impossible for me not to hear echoes of Trofim Lysenko's power grab in the Trump administration's attitude toward science. And by "Lysenkoism," I mean the actual experience of Soviet genetics post-1948—the shuttered scientific institutions, the blacklisting of scientists, the movement underground of scientific research on banned topics— rather than the sensationalistic version promoted in early Cold War propaganda. When climate change researchers receive death threats, they come from anonymous online sources, not from the secret police. But still, I never thought that this story would end how it began, with a meditation on the limits of state control of science.[8]

This book, in other words, has not turned out to be a prehistory of science diplomacy after all. The United States' recent, short-lived enthusiasm for science diplomacy depended on a belief in the virtues of both scientific freedom and diplomacy, neither of which is much in evidence today. I have found this discouraging, for many reasons, not least of which is that it has forced me to reevaluate my own attitudes toward scientific freedom and scientific internationalism. The Trump administration's "America First" approach to national security strategy has laid bare the usually at least partially hidden power structures that undergird federal science and technology policy. The Obama administration was not cultivating scientific allies simply because it believed in the virtues of scientific internationalism; it also hoped to advance US interests in the world.

Perceptions of science's political neutrality matter, regardless of who occupies the Oval Office. This became clearer to me in the weeks leading up to the first-ever March for Science, held on Earth Day, April 22, 2017. Because the idea for the march emerged shortly after Trump's inauguration, many observers assumed that it would draw attention to the ways that the new administration's policies would harm federal scientific institutions and, by extension, all Americans who rely on evidence-based claims to protect the environment and their health. (That's certainly what I thought when I first signed up.) Others began preparing for an event that would highlight how the administration's attacks on immigrants, minorities, and women tacitly endorsed discrimination and harassment in science and technology—an ongoing problem that continues to belie some scientists' claims that science has the unique ability to escape identity politics.

The organizers quickly distanced themselves from either of these interpretations, even claiming at one point that the march wasn't "political." After an outcry from historians, social scientists, and scientists, many of them nonwhite, the organizers eventually settled on "political but nonpartisan." The critique forced an unusually public discussion of how claims to scientific objectivity have been used to inflict harm. Still, at the time of the march, the movement's ongoing mission statement referred to science "as a pillar of human freedom and prosperity." Among the sea of pun-filled signs on the day of the protest, scientists wearing lab coats announced their faith in scientific objectivity and universalism in signs reading: "Science is Real," "Science Knows No Borders," and even "2 + 2 = 4." The dream of a transcendent science, free from politics, remains alive and well.[9]

Who wins, and who loses, when scientists claim a unique ability to operate free from politics?

∷

In the United States, scientists' appeals to scientific freedom have traditionally depended on their claim to objectivity. As advocates for scientific freedom, men like Vannevar Bush, Michael Polanyi, and Bentley Glass argued that scientists' unique dedication to observed reality—their ability to separate their personal identities and views from their work—earned the scientific community a special dispensation to govern itself. Even as an increasing number of US scientists drew their salaries and research funds from the federal government, the logic went, scientists should still maintain control over policy decisions about funding and research design, because scientists—and only scientists—could be trusted to objectively follow the path of knowledge. That the United States' main postwar enemy, the Soviet Union, so blatantly restricted scientists' freedoms only bolstered US scientists' moral claims.

But make no mistake: the particular version of "scientific freedom" promoted by postwar scientific administrators and US propaganda was a racist, sexist, and antidemocratic vision of science. Scientists (including social scientists) who acknowledged that their research agenda might be influenced by any aspect of their identity, whether the color of their skin, their gender presentation, their political affiliations, or their citizenship, were, by this definition, inherently not objective. Oblivious to the routine discrimination and structural barriers that severely limited women's and racial minorities' access to scientific careers, Vannevar Bush's generation assumed that scientists operated in a meritocracy. Their rhetoric on scientific freedom emphasized that scientists who had risen to the top had earned their position through their research talents, their administrative acumen, and their commitment to fact. It was merely coincidence, they would surely claim, that nearly everyone who promoted scientific freedom on behalf of the United States, whether at home or abroad, happened to be a white man who had either trained at or currently taught at an elite university.[10]

Though this particular critique feels very of the moment, it was already available at the time that the edifice of scientific freedom was being built. Writing in the *American Scholar* in 1939, W. E. B. Du Bois explained, in painstaking detail, the barriers faced by black scientists working in the United States in gaining access to laboratories, libraries, scientific collections, professional recognition, or even paying work. After recounting the biographies of eight men whose careers had been stymied by racism and lack of opportunity, Du Bois observed, "It seems fair to conclude that had these men not been of Negro descent they would have been offered a broader and better chance to carry on scientific work."[11]

Du Bois's conclusion drew on his own experience to explain how white apathy shaped scientific knowledge. Du Bois had to abandon his own, pathbreaking work on the sociology of race, pioneered with his 1899 book, *The Philadelphia Negro*, for lack of funds. Between 1896 and 1910, Du Bois explained, the work of his research group at Atlanta University, "the only institution in the world" to study this concept, was routinely cited by leading anthropologists and sociologists. And yet, Du Bois abandoned the project when he could not raise the $5,000 needed annually to keep the work going. When white institutions took up the problem in the 1930s, however, funding quickly followed. Du Bois concluded: "It is perhaps too much to say that our work failed solely because Negros were doing it, but certainly America was not disposed to help until white folk took it up."[12]

Nor were concerns about the elitist, technocratic bent of "scientific freedom" unvoiced in 1948. President Truman vetoed the original, Vannevar Bush–inspired plans for a National Research Foundation in 1947 on the principle that the proposed federal agency would not be held accountable to the public. Truman had supported a different, more populist bill that included the possibility of prioritizing research based on national needs; input from agriculture, industry, and labor; and a formula to make sure that federal funds found their way to all corners of the country. But as anti-Communist rhetoric heated up, what might once have seemed a reasonable policy of coordinating the nation's research needs became associated with scientific planning and, by association, socialism. In this context, calls for scientific planning or public input on research priorities became toxic, perhaps even evidence of Communist sympathies. The eventual version of the National Science Foundation Act that finally passed in 1950 eschewed coordination and funded projects solely on the basis of scientific merit—not on their relation to national needs, and certainly not the needs of disadvantaged groups.[13]

Both then and now, scientists made a choice when they ignored the conditions that made possible their authority and their professional autonomy in the guise of scientific freedom. During the first two decades of the Cold War, few (white, male) US scientists questioned these choices, confusing a system of privilege with scientific values. But Vietnam changed all that, at least for a time. The rise of radical groups such as Science for the People in the late 1960s and early 1970s amplified the unheralded critiques, like Du Bois's, that had been pushed to the margins but never fully silenced. By the early 1970s, scientific activists and antiwar protesters forced more mainstream scientists to confront how routine scientific practice ignored the needs of marginalized people in favor of defense contracts and access to power. Appeals to objectivity, these activists claimed, were a way to preserve the status quo.

In less than a decade, these critiques of scientific objectivity reached mainstream academic discourse in the form of a new discipline: science studies. One particularly influential strain of this work, the "strong programme" associated with the Science Studies Unit at the University of Edinburgh, argued that scientific theories derived their authority from social consensus rather than objective reality. Feminist historians of science revealed the presence of gender bias in everything from mammalian classification to the very concept of gender difference, while feminist philosophers of science rejected "value-free" science in favor of an "oppositional" science that focused on liberating oppressed peoples. Other scholars, particularly in cultural studies, used the techniques of postmodern literary criticism to "read" scientific texts as one might dissect literature or film. Collectively, these various approaches aimed to decenter science and scientific authority—particularly the kinds of scientific projects that either inadvertently or purposefully advanced racism, sexism, colonialism, and other forms of state power.[14]

Most of these academic critiques, it is safe to say, came from the left. But around this same time, evangelical Christians and movement conservatives also began questioning scientists' role in American life. Opponents of evolution created two new pseudodisciplines—first, creationism; later, intelligent design—and argued that schools should "teach the controversy" instead of accepted science. Small-government conservatives began drawing media attention to what they considered "wasteful" scientific research projects with no perceptible public utility. Democrats got in on the game, too. In 1976, Wisconsin Senator William Proxmire started issuing monthly "Golden Fleece Awards" to ridicule federally funded projects on topics ranging from the nature of love to the search for extraterrestrial life. As federal and state funding for institutions of higher education waned, private corporations stepped into the gap, creating opportunities and pressure for scientists to fund their work by pleasing their sponsors. A corporate industry devoted to debunking established science related to public health and climate change soon followed.[15]

Scientists, feeling attacked, needed a scapegoat, and science studies proved an irresistible target. These events reached a head in 1996, when physicist Alan Sokal published an article titled, "Transgressing the Boundaries: Toward a Transformative Hermeneutics of Quantum Gravity," in a special issue of the journal *Social Text* dedicated to the "science wars." A few weeks later, Sokal revealed that his article had been intended as a hoax, a stunt meant to expose what he considered the intellectual bankruptcy of attacks on scientific objectivity. In justifying his act, Sokal drew heavily on the postwar language of science and freedom. "For most of the past two centuries," he wrote, "the Left has been identified with science and against obscurantism; we have believed that rational thought and the fearless analysis of objective reality (both natural and social) are incisive tools for combating the mystifications promoted by the powerful—not to mention being desirable human ends in their own right." In criticizing objectivity, Sokal suggested, the left was abandoning the surest path to social progress.[16]

Sokal, in other words, claimed to be attacking science studies *from the left*. But the text of his original article belied that claim. His "parody" targeted not only postmodern scholars' penchant for obfuscating jargon but also the desire of feminists and nonwhite scholars for a science based on liberation rather than existing power structures. In language that he fully intended to be read as satire, Sokal parroted the desire of science studies scholars to build a "new and truly progressive science that can serve the needs of such a democratized society-to-be." Sokal, in other words, ridiculed the idea that *anything* about the existing structures of science needed reform. Writing in that same issue of *Social Text*, sociologist of science Dorothy Nelkin observed that scientists were "arguing with extraordinary passion to support their own dispassionate objectivity."[17]

Sokal's critique proved ascendant for the next fifteen years. Under the Republican presidential administration of George W. Bush, US conservatives put ever more pressure on federally funded science. Disturbingly, they increasingly attacked environmental and health protections, using language straight out of science studies, suggesting, for instance, that scientists had not reached "epistemic closure" on human-caused climate change. By 2004, Bruno Latour, one of the founding figures of science studies, was publicly expressing doubts about the real-world consequences of a field with radical roots: "Was I wrong to participate in the invention of this field known as science studies?" At recent annual meetings of the History of Science Society, conversations at the hotel bar frequently turn to the question of historians' and sociologists' culpability in undermining scientific authority. Maudlin feminist historians chide themselves, then order another beer.[18]

The ascent of Trumpism revealed the depth of wishful thinking on both sides of the science wars. Scientists were wrong to blame historians, philosophers, and sociologists of science for their declining fortunes, and science studies scholars much too grandiose in shouldering the blame. The postwar scenario in the United States, in which (white, male) scientists received virtually unlimited research funds to investigate whatever they wanted, so long as their questions didn't upset existing power structures, was a historical anomaly rather than a naturally occurring state of affairs. For a brief twenty-year period, both the public at large and the country's political leadership deferred to elite judgment—including that of scientists—in a tacit agreement that elites would contribute to the national interest. Now that the structures of power no longer value independent thought, a common public good, or global opinion, US scientists are at a loss to explain why the government should support an autonomous scientific community.

The uncomfortable fact of writing histories of the Cold War is that the stories must be told in shades of gray. Both the United States and the Soviet Union engaged in activities that were good, bad, and morally neutral. The belief that science, generously funded and left to investigate questions based on the natural world, could be a force for both international peace and domestic prosperity was one of the brighter lights of the Cold War. Postwar advocates for scientific freedom too rarely asked who lost when US scientists won, but the underlying premise of scientific internationalism continues to offer real potential for scientists to build relationships across national borders that too often create harm.

Even while the institutions of US science routinely reinforced structures of inequality, radical activist groups like Science for the People envisioned alternatives that brought scientific thinking to bear on justice.

When scientific institutions fall short, the solution is not to throw out a belief in "science," with its orientation toward problem-solving and real-world systems, but to work to make the best parts of science better. To paraphrase the science studies scholars whom Alan Sokal so mercilessly mocked, what we need is a liberatory science that acknowledges the existence of power structures even while working to dismantle them through evidence, curiosity, and wonder.

∷

The judgment of history is fickle, but from our contemporary perspective—the only moment from which we can write—it seems clear enough that the United States won the Cold War. It did so both by embracing the idealistic concepts of freedom, democracy, and self-determination and by waging violent military, paramilitary, and economic campaigns. The past is a complicated place. Recognizing the inherent contradictions in historical actors' behavior can be politically liberating, a necessary first step in uncovering a useable past.

Is there a role for a liberatory science in foreign relations? Certainly, understanding "scientific freedom" in the widest possible terms may help us to articulate concepts of the public good that depend less on demonstrating the superiority of the American way of life and more on fostering a truly inclusive, collaborative, international approach to knowledge of the natural world. But by definition, "foreign relations" and "diplomacy" advance the interests of a particular state. This was the fundamental fallacy of arm's-length cultural diplomacy during the Cold War, and it remains the fundamental fallacy of the more recent (and recently thwarted) push for science diplomacy. It is not possible for individuals to participate in "informal diplomacy" or "personal diplomacy" without advancing systems of state power.

The United States, unfortunately, is not the only country whose leadership has embraced racism, misogyny, xenophobia, and crony capitalism in recent years. The rise of extremist antidemocratic movements across the globe has reminded people everywhere of the temptations of power. With the prospect of hate-based states, the moral questions of the Vietnam era have once again bubbled up to the surface. Should an ethical person work within a corrupt system, with the hope of facilitating change, or at least mitigating harm? Or is all cooperation a form of collaboration?

Scientists' professional identity does not grant them a free pass from these ethical dilemmas. Scientists, like economists, doctors, lawyers, and politicians, have choices to make in the ongoing struggle for freedom and equality. The political choices that scientists make today, in their lives and in their work, will affect our global society for years to come. May they choose freedom—but may it be a version of freedom that encompasses freedom and justice for all, not just the few.

Acknowledgments

tarting in 1966 and 1967, reporters at the *New York Times*, the *Washington Post*, and *Ramparts* convinced their editors that some secrets had been kept long enough. Their series of stories on the CIA's vast network of private partners changed how the public understood the work of the US government and how the media reported on it. Their pathbreaking work laid the foundation for subsequent revelations on the nature of the US government's clandestine activities, ranging from the publication of the Pentagon Papers to the hearings of the Church Commission. Muckraking journalists continue this work today, now assisted by declassification specialists operating throughout the US government. This book simply could not have been written without their shared conviction that democracy demands transparency.

Archivists and librarians at institutions across the country patiently fielded my requests and steered me toward useful documents. I am particularly grateful for the assistance of Charles Greifenstein, Valerie-Ann Lutz, and Earle Spamer at the American Philosophical Society; Carol Leadenham at the Hoover Institution Library and Archives; Anna Trammell at the Archives Research Center of the University of Illinois at Urbana-Champaign; Jody Mitchell at the Lilly Library at Indiana University; Jeffrey Flannery at the Library of Congress; John Wilson at the LBJ Presidential Library and Archive; Janice Goldblum at the National Academy of Sciences; Jennifer Dryer, David Fort, Jeff Hartley, and Gene Morris of the National Archives at College Park; Lee Hiltzik at the Rockefeller Archive Center; Mary Markey at the Smithsonian Institution Archives; and Julia Gardner at the Special Collections Research Center at the University of Chicago. Two research assistants, Beth Ann Williams and Zivka Mares Randic, provided essential access to archives from afar. The research portion of this project was made possible by a Scholars Award from the National Science Foundation (award no. 1026715).

My conversations with practitioners of science diplomacy deepened my understanding of why a country might want to use ideas about science and scientists to further its national goals. I am particularly thankful that Cathy Campbell, Bill Colglazier, Nart Dohjoka, Norm Neureiter, Glenn Schweitzer, Gerson Sher, Phill Starling, and Vaughan Turekian shared their knowledge and experiences with me. I owe a special thanks to Linda Staheli, who organized a salon on science diplomacy in her home in Arlington, Virginia, on my behalf. On the topic of science education, my conversations with Joe McInerney and the late Arnold Grobman deepened my understanding of the BSCS's work.

Audiences at several institutions and conferences provided constructive feedback while this project was still taking shape. I am especially grateful for the hospitality and good humor of Richard Immerman, Petra Goedde, and Carly Goodman at Temple University's Center for the Study of Force and Diplomacy; Abram Kaplan, Rachel Rothschild, and Matt Stanley of the New York City History of Science Lecture Series; Bruce Hunt, Abena Osseo-Asare, and Megan Raby, of the University of Texas–Austin; and Matt Brown of the Values in Medicine, Science, and Technology Conference in Dallas, Texas.

My intellectual debts to the community of Cold War historians and historians of science are much larger than can be counted on these pages. The online community of #twitterstorians is unceasingly generous. I am grateful to Alistair Sponsel and Mary Mitchell for sharing research materials from their own collections. Over the years, portions of the manuscript (in various states) have been read by Elena Aronova, William deJong-Lambert, Nikolai Kremenstov, Charlotte Lerg, Patrick Manning, Lisa Onaga, John Rudolph, Mat Savelli, Giles Scott-Smith, Barbara Walker, and three anonymous peer reviewers. I am especially grateful for feedback from Ronald Doel, who read the manuscript in its entirety. Of the countless colleagues who have spent their precious conference time patiently listening to me expound on the uses of science as propaganda, I owe a special thanks to Luis Campos, Angela Creager, Nathan Crowe, Mary Dudziak, Greg Good, Michael Gordin, Jacob Hamblin, Patrick Iber, Gwen Kay, John Krige, Monique Laney, Roger Launius, Joshua Mather, Erika Milam, Teasel Muir-Harmony, Allan Needell, Amy Offner, Michael Robinson, Joy Rohde, Suman Seth, Matt Shindell, Heidi Voskuhl, Zuoyue Wang, and Alex Wellerstein. All errors of fact or interpretation remain my own.

As an editor myself, I am intensely aware that this book would never have seen the light of day without the work of a team of publishing professionals. My agent, Deirdre Mullane, believed in this project from the very beginning. I very nearly abandoned this manuscript at least twice; my writing coach, K. Anne Amienne, helped me remember that I had something to say and taught me to find the time to say it. At Johns Hopkins University Press, Matt McAdam championed the book to his colleagues. Catherine Goldstead and later William Krause, Juliana McCarthy, and Kimberly Johnson kept all the moving parts in order. A special thank-you to copyeditor Linda Strange for her remarkable attention to detail and to indexer Becky Hornyak for uncovering a hidden sense of order underneath all the facts.

I live in Philadelphia, a city graced by the presence of people who believe a better world is possible. I am so grateful for communities associated with Arch Street United Methodist Church, POWER, Tuesdays with T———, and Warrington Community Gardens for drawing me away from this manuscript and back into the city I love. By reminding me that there is more to life than finishing a book, Brett Bonfield, Christina Burris, Michael Donahue, Fae Ehsan, Beth Filla, Sue Gerber, Rabbi Julie Greenberg, Alexandra Gunnison, Reverend Robin Hynicka, David Krueger, J. Robert Lennon, Marisa McClellan, Erin McLeary, Alexandra Moede, Larissa Mogano, Daria Panichas, Ian Petrie, and Amy Salit have, in their wonderfully varied ways, made it possible for me to finish this one.

Finally, words cannot fully express my gratitude to my family members, who have graciously accepted the presence of this book—this uninvited guest—in their lives. My in-laws, including Daniel, Sally, and Ted Chalfen and Jeff and Suzanne Resnick, asked thoughtful questions and shared their own stories. My siblings, Clint and Sarah Wolfe, have been a source of constant comfort. It has been a pleasure to welcome Trevor Murry into our family, and a joy to watch Hattie, Oliver, Sylvia, and Abel grow. The unwavering support of my parents, Tony and Debra Wolfe, makes all things possible. And finally, I thank my husband, Andrew Chalfen, on whose love and companionship I rely everyday.

Note on Sources and Methods

igital cameras have transformed how historians work. In days of yore—the 1990s—historians lingered over documents in the archives. If you were particularly excited about something, you might request a photocopy. Now that an increasing number of archives allow research photography, historians show up armed with cameras and laptops. Some days I used a pocket-sized Nikon. In a pinch, I could use the camera on my iPad. But more often than not, I entered the manuscript reading room with a hefty Canon Rebel DSLR strapped around my neck. The lens was too big for the job, so I spent most days on my feet, hoping to gain sufficient distance from whatever it was I was trying to photograph. I'd leave with a backache and vast stores of data.

On my first day at any new archive, I would ask a staff member about the best way to maximize my time. Some facilities, like the US National Archives in College Park, Maryland, have elaborate rules about pull times and locations and how many carts researchers can have out at once. I learned to game the system. Two carts loaded up with twenty-four boxes and a shelf pull, all from separate record groups? Bliss. Some of these boxes yielded junk, of course. I'd open them up, flip my fingers through the folders, and send them back immediately. I needed to empty that cart. But when I finally found something—anything—remotely promising, I'd be back on my feet, camera in hand. Photograph the box, photograph the folder, photograph the documents. Pound out some notes in my research log, and repeat.

A successful day usually yielded around 800 to 1,000 photographs, typically representing 200 to 500 documents. I continued this process, off and on, for about six years. All told, I collected more than 8,000 documents from institutional and personal manuscript collections at approximately a dozen archives across the country. Once I returned from a trip, I'd reduce the images, convert them into PDFs, and upload them into a sortable bibliographic database. I'd then spend several weeks reviewing these documents from the comfort of my home office. This allowed me to snack and sip coffee while adding in metadata and content tags.

The reviewing process took years. Truth be told, my hard drive (and its multiple backups) still contains unprocessed documents from the last two or three days of various archival trips. I suspect future historians will regard this vacuum-cleaner approach to historical research with deserved suspicion. Hoarders are rarely self-diagnosed. Nevertheless, the maximalist approach to historical research produces certain benefits, provided the writer finds some way to organize and access the data. I used Zotero, a free, open-source reference management tool developed by the Roy Rosenzweig Center for History and New Media. Researchers typically use Zotero for citation management, but if you treat it like a primary research tool, it becomes one. With a click of the mouse, I can view the entire run of Biological Sciences Curriculum Study memos in chronological order, even though I assembled this run from incomplete series in at least four different locations. A name sounds familiar, but I can't place it? Two quick queries later, I realize that not one, but two low-level employees of the National Academy of Sciences came to their positions directly from the Asia Foundation. Over and over again, this massive cache of archival documents allowed me to perceive connections among individuals, institutions, and projects that claimed to have no relationship to one another.

In effect, I was building a personalized database of the United States' attempts to use science as a form of cultural diplomacy during the Cold War. Of course, the value of any database depends on its underlying dataset. The vast majority of items in mine derive from personal and institutional records collected at archival and online repositories located within the United States. Major institutional records include those of the Rockefeller Foundation, the Ford Foundation, the National Academy of Sciences, the Asia Foundation, the Congress for Cultural Freedom, the National Science Foundation, the United States Information Agency, the US State Department, and the National Security Council. Memos, meeting agendas, grant requests and reports, budgets, routine correspondence—all of this is the stuff of institutional archives. Fairly frequently, researchers can fill in missing bits of the institutional archives from the personal papers of individuals involved with the organization in question. The personal papers of Wallace Brode, Detlev Bronk, Bentley Glass, Arnold Grobman, H. J. Muller, Michael Polanyi, and Eugene Rabinowitch provided particularly rich leads into the overlapping worlds of cultural diplomacy, scientific internationalism, and anti-Communism during the Cold War.

Presidential libraries offer researchers and members of the public alike rich documentation on the most powerful political office in the United States. The archival collections at presidential libraries typically include not only materials gathered from the West Wing but also those gathered from key individuals associated with the administration. The Lyndon Baines Johnson Library and Museum in Austin, Texas, for instance, includes essential materials on the work of the Katzenbach Commission as well as the personal papers of Donald Hornig, President Johnson's science advisor. The websites of most presidential libraries provide audio and video recordings of key historical moments. The American Presidency Project, hosted by the University of California, Santa Barbara, offers an online, text-searchable database of presidential papers, party platforms, and other resources related to presidential politics. For a more behind-the-scenes look at presidential history, the University of Virginia's Miller Center provides online access to the secret White House tapes recorded by presidents Roosevelt, Truman, Eisenhower, Kennedy, Johnson, and Nixon.

When I explain this project to new acquaintances, they don't ask me about my data management practices or which archive has the best cafeteria (that would be the National Academy of Sciences in Washington, DC). They all want to know one thing: how do you deal with the problem of classified documents? I feel like I'm letting them down when I share the dirty little secret of writing about cover organizations: the most successful cover organizations operated in plain sight. The majority of the operational files of the Congress for Cultural Freedom and the Asia Foundation have never been classified; several hundred boxes' worth of their records are now open to researchers at the University of Chicago's Joseph Regenstein Library and the Hoover Institution Library and Archives, respectively. Unwitting and witting individuals who served as consultants to these organizations, or as members of their executive committees, often kept copies of key documents in their own files. The Michael Josselson Papers at the Harry Ransom Center of the University of Texas–Austin, for instance, include several years' worth of the CCF's audits. Organizations that received funding from the CIA's instruments for covert cultural diplomacy listed the grants in their annual reports.

The trick to writing a history of covert cultural diplomacy is in understanding that the organization you are writing about is, in fact, a cover. Fortunately for researchers, this information has been released to the public many times, starting with reports published in *Ramparts* and the *New York Times* in the late 1960s, as discussed in chapter 8. The published reports of the US Senate's Select Committee to Study Governmental Operations with Respect to Intelligence Activities (known as the Church Committee, after Chairman Frank Church, D-Idaho) provide crucial background information on the CIA's wide range of covert activities, from assassinations and domestic wiretapping to the funding of cultural programming. You may download public materials associated with the Church Report, including text-searchable PDFs of the published reports, through the "Resources" pages of the US Senate Select Committee on Intelligence (intelligence.senate.gov).

There are, of course, limits to the contents of these publicly available archives. Each archive was presumably sanitized, and researchers will search these files in vain for top-level budgetary information or any straightforward discussion of funding. After years of stonewalling, the CIA has finally declassified significant caches of materials related to its covert cultural diplomacy programs. Members of the public may access select declassified materials through the CIA's online electronic reading room (cia.gov/library/readingroom). This online collection includes access to articles published in *Studies in Intelligence* as well as select documents declassified through the Freedom of Information Act process. If you're seeking information on either the Congress for Cultural Freedom or the Asia Foundation, you will find your search more efficient if you use the assigned operation codenames: QKOPERA for the Congress for Cultural Freedom, and DTPILLAR for the Asia Foundation.

As required by law, the CIA also declassifies historically valuable records on a rolling twenty-five-year basis. Upon declassification, the documents are entered into a full-text searchable electronic system known as CREST (CIA Records Search Tool). During the time that I was researching this book, CREST could be accessed only via four locked computers in a separate room at the US National Archives in College Park, Maryland. While the CIA generously provided unlimited free toner and paper (no downloading allowed), all parties involved recognized the inconveniences of this system. In January 2017, the CIA made the contents of CREST available remotely, via the CIA's online electronic reading room. Be warned that the CREST search system is notoriously clunky; the migration online does not seem to have improved the situation.

The Office of the Historian at the US Department of State systematically reviews, declassifies, and publishes important documents related to US foreign policy. The almost 500 volumes in the *Foreign Records of the United States* series collectively comprise a documentary record of major and minor decisions relating to foreign affairs, national defense, intelligence, and diplomacy. Following conventions in diplomatic history, I have included the volume information and document number in my citations of documents published in *FRUS*. That being said, the Office of the Historian's website has recently undergone a major upgrade that reduces the bibliographic importance of volume information. All *FRUS* content is now available online in full-text searchable form (history.state.gov); a feature that allows researchers to sort search results chronologically is in alpha-stage testing at the time of this writing.

Most of the individuals involved in the events discussed in this book died well before I began this work. A few, like Grobman, died shortly thereafter. While this circumstance unfortunately closed off one investigative channel, it inadvertently opened another: anyone may request the FBI and CIA files of deceased persons. Alas, these files must be treated with interpretive caution. Besides the obvious ethical problems associated with using materials too often gathered through illegal surveillance, intimidation, and hearsay, the documents are simply not that interesting. Harried FBI agents used stock language to capture their informants' description of their subject, for instance, so that one learns over and over again that Lloyd Berkner's former coworkers described his "character, reputation, loyalty, and associates" as "beyond reproach," or that Bentley Glass's neighbors considered him "a staunch and loyal American citizen with no use whatsoever for Communism—or any other foreign ideology." Did real people talk like this? Even in 1953? Nevertheless, FBI files do provide a useful index to an individual's career trajectory in public service, as every request for a security clearance requires an FBI check. The CIA is less forthcoming. Most often, the agency answers FOIA requests for records on an individual with a letter that neither confirms nor denies the existence of records. That being said, the CIA *will* confirm employment records for certain classes of employees, a technique that I found useful in identifying at least one of the CIA's most likely contacts at the National Academy of Sciences for most of the 1950s.

While I came to this story too late to interview most of the key characters, I made use of oral histories conducted by prior historians with better timing. I am particularly grateful for a set of interviews with early science attachés conducted by historians Ronald Doel and Allan Needell in the 1990s, available at the Niels Bohr Library and Archives of the American Institute of Physics in College Park, Maryland. Transcripts of these interviews are available through the AIP's website (aip.org). Most presidential libraries also include at least a small oral history collection; the interviews at the Harry S. Truman Presidential Library and Museum are particularly valuable for understanding the reasoning behind early psychological warfare practices.

Postwar US science administrators, retired or disgruntled CIA officers, and Soviet dissidents all shared a passion for writing memoirs. Given the various interpretive issues associated with all three sets of actors, I treat these documents cautiously and have cited them sparingly. As windows into these actors' worlds, however, they can't be beat. Vannevar Bush, James Conant, James Killian, Jerome Wiesner, Cord Meyer, Philip Agee, Elena Bonner, Andrei Sakharov—all wrote memoirs, as did many of their friends and associates. As literary objects, two of the most interesting are Yuri Orlov's *Dangerous Thoughts* and Arthur Koestler's two-volume set, *Arrow in the Blue* and *The Invisible Writing*. Orlov's account overtly plays with Russian literary traditions, including changes in time and narrative perspective. Given Koestler's career as a propagandist, it's hard to take anything in his memoirs at face value, but the man was a talented prose stylist. Perhaps he never should have left science writing.

Aside from the usual newspapers and magazines, I found two periodicals particularly helpful in understanding the relationship between scientific freedom and scientific internationalism. First published in 1945 as a newsletter for the atomic scientists' movement, the *Bulletin of the Atomic Scientists* became a major forum in which US scientists and their overseas allies worked out notions of scientific freedom, scientific internationalism, and apolitical science. At the time of this writing, full-text searchable content from all issues of *BAS* published between 1945 and 1998 is available through Google Books (books.google.com). The second periodical is perhaps more surprising: *Chemical and Engineering News*. In part because of the personal networks of Roger Adams at the University of Illinois, chemists played a disproportionate role in postwar science diplomacy and intelligence. *Chemical and Engineering News* closely followed the ups and downs of the Office of the Science Advisor at State and, to a lesser extent, the CIA's Office of Scientific Intelligence, as news of the profession.

A full list of archives used in the research for this book appears at the beginning of the Selected Bibliography. Full citations to all manuscript sources used appear in the notes. All documents cited in this volume either were never classified or have undergone declassification review; prior classification level (where applicable) is identified in brackets. Readers should be advised that some file locations have changed since the time of my research. The American Philosophical Society has reprocessed several manuscript collections in the very recent past, including the Bentley Glass Papers and the Leslie C. Dunn Papers. The documents cited represent only a small fraction of the materials consulted during the research for this book. My hope is that this note on sources will inspire future researchers to explore the role of science in US foreign policy in more detail.

Notes

Introduction

1. Memoranda attached to Arnold B. Grobman to Bentley Glass, 18 April 1963, box 158, BSCS, International Activities, H. Bentley Glass Papers, American Philosophical Society (hereafter Glass Papers). Grobman recounted this story to me unprompted. Interview with Arnold B. Grobman, 3 October 2011, Gainesville, FL.

2. The literature on science and the Cold War is vast. For entry points, see Audra J. Wolfe, "For Further Reading," in *Competing with the Soviets: Science, Technology, and the State in Cold War America* (Baltimore: Johns Hopkins University Press, 2013), 143–59; and Naomi Oreskes, "Introduction," in *Science and Technology in the Global Cold War*, ed. Naomi Oreskes and John Krige (Cambridge, MA: MIT Press, 2014), 1–9. For the particular career challenges of science under McCarthyism, see Jessica Wang, *American Science in an Age of Anxiety: Scientists, Anticommunism, and the Cold War* (Chapel Hill: University of North Carolina Press, 1999); and Lawrence Badash, "Science and McCarthyism," *Minerva* 38, no. 1 (2000): 53–80, https://doi.org/10.1023/A:1026516820931. For a broader discussion of identity and Cold War career troubles, see Ellen W. Schrecker, *No Ivory Tower: McCarthyism and the Universities* (New York: Oxford University Press, 1986); David K. Johnson, *The Lavender Scare: The Cold War Persecution of Gays and Lesbians in the Federal Government* (Chicago: University of Chicago Press, 2004); and Margaret W. Rossiter, *Women Scientists in America: Before Affirmative Action, 1940–1972* (Baltimore: Johns Hopkins University Press, 1995). There is some evidence that federal employment presented opportunities as well as challenges for minority scientists. See, for example, Margot Lee Shetterly, *Hidden Figures: The American Dream and the Untold Story of the Black Women Mathematicians Who Helped Win the Space Race* (New York: William Morrow, 2016); and Richard Paul, *We Could Not Fail: The First African Americans in the Space Program* (Austin: University of Texas Press, 2015).

3. By far the best single-volume history of the Cold War is Odd Arne Westad, *The Cold War: A World History* (New York: Basic Books, 2017).

4. Westad's prior book, *The Global Cold War* (New York: Cambridge University Press, 2005), pushed historians to grapple with the consequences of the Cold War in the Global South. For the turn toward psychological warfare and propaganda, see Walter L. Hixson, *Parting the Curtain: Propaganda, Culture, and the Cold War, 1945–1961* (New York: St. Martin's Press, 1997); Scott Lucas, *Freedom's War: The American Crusade against the Soviet Union* (New York: New York University Press, 1999); Sarah-Jane Corke, *US Covert Operations and Cold War Strategy: Truman, Secret Warfare, and the CIA, 1945–1953* (New York: Routledge, 2008);

Nicholas J. Cull, *The Cold War and the United States Information Agency: American Propaganda and Public Diplomacy, 1945–1989* (New York: Cambridge University Press, 2008); Kenneth Osgood, *Total Cold War: Eisenhower's Secret Propaganda Battle at Home and Abroad* (Lawrence: University Press of Kansas, 2006); and Laura A. Belmonte, *Selling the American Way: U.S. Propaganda and the Cold War* (Philadelphia: University of Pennsylvania Press, 2008).

5. In addition to the titles listed in the previous note, my understanding of "cultural diplomacy" is indebted to discussions in Frank A. Ninkovich, *The Diplomacy of Ideas: U.S. Foreign Policy and Cultural Relations, 1938–1950* (New York: Cambridge University Press, 1981); and Emily S. Rosenberg, *Financial Missionaries to the World: The Politics and Culture of Dollar Diplomacy, 1900–1930* (Cambridge, MA: Harvard University Press, 1999). Other historians prefer to describe these activities as "public diplomacy." See, for example, Justin Hart, *Empire of Ideas: The Origins of Public Diplomacy and the Transformation of U.S. Foreign Policy* (New York: Oxford University Press, 2013). For the classic account of the relationship between race and foreign policy, see Mary L. Dudziak, *Cold War Civil Rights: Race and the Image of American Democracy* (Princeton, NJ: Princeton University Press, 2000).

6. Badash, "Science and McCarthyism"; J. Wang, *American Science*, 52–54. The actual situation of Soviet scientists will be discussed more fully in this book, but for an accessible introduction to the topic, see Simon Ings, *Stalin and the Scientists: A History of Triumph and Tragedy, 1905–1953* (New York: Atlantic Monthly Press, 2016).

7. The most comprehensive history of the CIA's role in covert cultural diplomacy is Hugh Wilford, *The Mighty Wurlitzer: How the CIA Played America* (Cambridge, MA: Harvard University Press, 2008). The literature on the CCF is voluminous; I return to it in chapter 4. In contrast, only a handful of scholarly accounts even mention the Asia Foundation, let alone consider its broader role in cultural diplomacy. One thoughtful exception is Becky Shelley, *Democratic Development in East Asia* (New York: RoutledgeCurzon, 2005), 115–35. In part, this lack of scholarship on the Asia Foundation reflects lingering secrecy. The organization's archives, now located at the Hoover Institution Library and Archives, opened as recently as 2011. I return to the Asia Foundation's work in more detail in chapters 7 and 8.

8. For a concise overview of key figures' understanding of "mass culture," see Volker R. Berghahn, *America and the Intellectual Cold Wars in Europe: Shepard Stone between Philanthropy, Academy, and Diplomacy* (Princeton, NJ: Princeton University Press, 2001), 92–107.

9. Peter Mandler, *Return from the Natives: How Margaret Mead Won the Second World War and Lost the Cold War* (New Haven, CT: Yale University Press, 2013), 56–70; David C. Engerman, *Know Your Enemy: The Rise and Fall of America's Soviet Experts* (New York: Oxford University Press, 2009), 43–70. For an in-depth accounting of anthropologists' relationship with the US state in World War II and the Cold War, see David H. Price, *Anthropological Intelligence: The Deployment and Neglect of American Anthropology in the Second World War* (Durham, NC: Duke University Press, 2008); and David H. Price, *Cold War Anthropology: The CIA, the Pentagon, and the Growth of Dual Use Anthropology* (Durham, NC: Duke University Press, 2016).

10. "Document 85: Note by the Executive Secretary to the National Security Council on United States Objectives and Programs for National Security (NSC 68) [TOP SECRET/declas-

sified]," in *Foreign Relations of the United States 1950*, vol. 1 (individual volumes hereafter *FRUS*); Wolfe, *Competing with the Soviets*, 60–61.

11. Fa-ti Fan, "The Global Turn in the History of Science," *East Asian Science, Technology, and Society* 6, no. 2 (2012): 252, https://doi.org/10.1215/18752160-1626191. For a sharp critique of accounts of transnationalism that neglect the scale and scope of US power, see Paul A. Kramer, "Power and Connection: Imperial Histories of the United States in the World," *American Historical Review* 116, no. 5 (2011): 1348–91, https://doi.org/10.1086/ahr.116.5.1348.

12. On the establishment of scientific intelligence in the United States, see Ronald E. Doel, "Scientists as Policymakers, Advisors, and Intelligence Agents: Linking Contemporary Diplomatic History with the History of Contemporary Science," in *The Historiography of Contemporary Science and Technology*, ed. Thomas Söderqvist (Amsterdam: Harwood Academic Press, 1997), 215–44; Ronald E. Doel and Allan A. Needell, "Science, Scientists, and the CIA: Balancing International Ideals, National Needs, and Professional Opportunities," in *Eternal Vigilance? 50 Years of the CIA*, ed. Rhodri Jeffreys-Jones and Christopher Andrew (Portland, OR: Frank Cass, 1997), 59–81.

13. Two essential discussions of the meanings of "scientific internationalism" in the postwar United States are Joseph Manzione, " 'Amusing and Amazing and Practical and Military': The Legacy of Scientific Internationalism in American Foreign Policy, 1945–1963," *Diplomatic History* 24, no. 1 (2000): 21–55, https://doi.org/10.1111/1467-7709.00197; and John Krige, "Atoms for Peace, Scientific Internationalism, and Scientific Intelligence," *Osiris* 21, no. 1 (2006): 161–81, https://doi.org/10.1086/507140.

14. Rexmond C. Cochrane, *The National Academy of Sciences: The First Hundred Years, 1863–1963* (Washington, DC: National Academy of Sciences, 1978); A. Hunter Dupree, *Science in the Federal Government* (Baltimore: Johns Hopkins University Press, 1986), 305–25.

15. Schrecker, *No Ivory Tower*; Peter J. Kuznick, *Beyond the Laboratory: Scientists as Political Activists in 1930s America* (Chicago: University of Chicago Press, 1987); Badash, "Science and McCarthyism"; J. Wang, *American Science*; Kelly Moore, *Disrupting Science: Social Movements, American Scientists, and the Politics of the Military, 1945–1975* (Princeton, NJ: Princeton University Press, 2008).

16. For background on Glass, see Sarah Brady Siff, "Atomic Roaches and Test-Tube Babies: Bentley Glass and Science Communication," *Journalism & Communication Monographs* 17, no. 2 (2015): 88–144, https://doi.org/10.1177/1522637915577107. Details on Glass's career are taken from his CV, located in box 235, Glass Papers, and from additional materials scattered throughout the collection. See also Frank C. Erk, "Remembering Bentley Glass (1906–2005)," *Quarterly Review of Biology* 80 (2005): 165–73, https://doi.org/10.1086/433053.

17. Audra J. Wolfe, "The Organization Man and the Archive: A Look at the Bentley Glass Papers," *Journal of the History of Biology* 44, no. 1 (March 2011): 147–51, https://doi.org/10.1007/s10739-011-9276-6.

18. Federal Bureau of Investigation, "Subject File: H. (Hiram) Bentley Glass (116-256352, 133-233, and 121-27487)," 28 September 2001 (in the author's possession).

19. Hiden T. Cox to Sigmund Suskind, 12 September 1957, box 145, AIBS, Glass Papers. For more on Coolidge and the Pacific Science Board, see Gary Kroll, "The Pacific Science

Board in Micronesia: Science, Government, and Conservation on the Postwar Pacific Frontier," *Minerva* 41, no. 1 (2003): 25–46, https://doi.org/10.1023/A:1022205821483.

20. Bentley Glass to Howard A. Meyerhoff, n.d. 1950, box 12, Science, 1950–1951, Glass Papers; "Memorandum of Conversation: Meeting of U.S. Participants at Pugwash Conference in Moscow, November 27, 1960," 14 November 1960, box 381, 14G 43a, A1 3008, RG 59, US National Archives, College Park, MD (hereafter NARA II); Harrison Brown, Bernard Feld, and Bentley Glass to A. V. Topchiev, 24 October 1962, box 201, Pugwash, 1962–1963, Glass Papers.

21. I discuss these issues at length in chapters 7 and 8.

22. I return to Polanyi, Bronk, and Rabinowitch in chapters 4, 5, and 6, respectively. For the specific articles mentioned here, see Michael Polanyi, "The Republic of Science: Its Political and Economic Theory," *Minerva* 1, no. 1 (1962): 54–73, https://doi.org/10.1007/BF01101453; and Eugene Rabinowitch, "Editorial: International Cooperation of Scientists," *Bulletin of the Atomic Scientists* (hereafter *BAS*) 2 (September 1946): 1.

23. The story of the BSCS's international scientific programming is told in chapter 7.

24. "Funds in abundance": James G. Dickson to Hiden T. Cox, 15 May 1961, box 171, BSCS Committee on Foreign Utilization, Glass Papers.

25. "Conway: Press Secretary Gave 'Alternative Facts,'" *Meet the Press* (NBC News, 22 January 2017), http://www.nbcnews.com/meet-the-press/video/conway-press-secretary-gave-alternative-facts-860142147643.

1 : Western Science vs. Marxist Science

1. J. D. Bernal, *The Social Function of Science* (1939; repr., Cambridge, MA: MIT Press, 1967).

2. Diane B. Paul, "A War on Two Fronts: J. B. S. Haldane and the Response to Lysenkoism in Britain," *Journal of the History of Biology* 16, no. 1 (1983): 1–37, https://doi.org/10.1007/BF00186674.

3. The literature on the Lysenko affair is vast, but the following English-language works give a taste of how historians' understanding of the episode changed from the Cold War era to the opening of the Soviet archives: Z. A. Medvedev, *The Rise and Fall of T. D. Lysenko*, trans. I. Michael Lerner (New York: Columbia University Press, 1969); David Joravsky, *The Lysenko Affair* (Cambridge, MA: Harvard University Press, 1970); Nikolai Krementsov, *Stalinist Science* (Princeton, NJ: Princeton University Press, 1997); Nils Roll-Hansen, "Wishful Science: The Persistence of T. D. Lysenko's Agrobiology in the Politics of Science," *Osiris*, 23 (2008): 166–88, https://doi.org/10.1086/591873; and William deJong-Lambert and Nikolai Krementsov, "On Labels and Issues: The Lysenko Controversy and the Cold War," *Journal of the History of Biology* 45, no. 3 (2012): 373–88, https://doi.org/10.1007/s10739-011-9292-6.

4. Robert E. Kohler, *Lords of the Fly: Drosophila Genetics and the Experimental Life* (Chicago: University of Chicago Press, 1994), 138–41, 160–61; Nikolai Krementsov, *International Science between the World Wars: The Case of Genetics* (New York: Routledge, 2005).

5. Elof Axel Carlson, *Genes, Radiation, and Society: The Life and Work of H. J. Muller* (Ithaca, NY: Cornell University Press, 1981), 4, 10, 27, 32, 51–69.

6. Ibid., 120–30; Nikolai Krementsov, "Big Revolution, Little Revolution: Science and Politics in Bolshevik Russia," *Social Research* 73, no. 4 (2006): 1173–1204.

7. Michael David-Fox, *Showcasing the Great Experiment: Cultural Diplomacy and Western Visitors to the Soviet Union, 1921–1941* (New York: Oxford University Press, 2012), 1; H. J. Muller, "Observations of Biological Science in Russia," *Scientific Monthly* 16, no. 5 (1923): 539, 541, 543; Loren R. Graham, *Science in Russia and the Soviet Union: A Short History* (New York: Cambridge University Press, 1993), 242.

8. Carlson, *Genes, Radiation, and Society*, 151–83; Graham, *Science in Russia*, 242.

9. Carlson, *Genes, Radiation, and Society*, 184–92, 194; Diane B. Paul and Costas B. Krimbas, "Nikolai V. Timoféef-Ressovsky," *Scientific American* 266 (February 1992): 86–92, https://doi.org/10.1038/scientificamerican0292-86; Krementsov, *Stalinist Science*, 320–21, n. 15.

10. Carlson, *Genes, Radiation, and Society*, 194, 202; Kuznick, *Beyond the Laboratory*, 121–22; Susan Gross Solomon and Nikolai Krementsov, "Giving and Taking across Borders: The Rockefeller Foundation and Russia, 1918–1928," *Minerva* 39, no. 3 (2001): 265–98, https://doi.org/10.1023/A:1017909113410.

11. Krementsov, *International Science*, 40–43, 162 n. 43.

12. Graham, *Science in Russia*, 99–103.

13. Ibid., 121–23. The Soviet papers presented at this conference, all of which discuss issues in the history and philosophy of science from the perspective of dialectical materialism, are available as N. I. Bukharin et al., eds., *Science at the Cross Roads: Papers from the Second International Congress of the History of Science and Technology 1931* (New York: Routledge, 2013).

14. David Joravsky, "Soviet Marxism and Biology before Lysenko," *Journal of the History of Ideas* 20, no. 1 (1959): 85–104, https://doi.org/10.2307/2707968; Medvedev, *Rise and Fall*, 5–17, 78–85. On Nazi eugenics, see Sheila Faith Weiss, *The Nazi Symbiosis: Human Genetics and Politics in the Third Reich* (Chicago: University of Chicago Press, 2010).

15. Hermann J. Muller, "Lenin's Doctrines in Relation to Genetics," in *Science and Philosophy in the Soviet Union*, in Loren R. Graham (New York: Knopf, 1972), 453–69; Krementsov, *Stalinist Science*, 55–60.

16. H. J. Muller, *Out of the Night: A Biologist's View of the Future* (New York: Vanguard Press, 1935); John Glad, "Hermann J. Muller's 1936 Letter to Stalin," *Mankind Quarterly* 43, no. 3 (2003): 305–19; Carlson, *Genes, Radiation, and Society*, 233. For a discussion of "socialist eugenics," see Diane B. Paul, "Eugenics and the Left," *Journal of the History of Ideas* 45, no. 4 (1984): 567–90, https://doi.org/10.2307/2709374.

17. Carlson, *Genes, Radiation, and Society*, 242–43; Krementsov, *Stalinist Science*, 61; Diane B. Paul, "H. J. Muller, Communism, and the Cold War," *Genetics* 119, no. 2 (1988): 223–25. For more background on the Rockefeller Foundation's support for genetics in the United States, see Robert E. Kohler, *Partners in Science: Foundations and Natural Scientists, 1900–1945* (Chicago: University of Chicago Press, 1991).

18. Paul, "War on Two Fronts"; Krementsov, *International Science*, 108–9, 112–14, 138–40.

19. Joravsky, *Lysenko Affair*, 320–28; Medvedev, *Rise and Fall*, 123–31.

20. Krementsov, *Stalinist Science*, 179–83; Medvedev, *Rise and Fall*, 128–29; Ethan Pollock, "From Partiinost' to Nauchnost' and Not Quite Back Again: Revisiting the Lessons of

the Lysenko Affair," *Slavic Review* 68, no. 1 (2009): 95–115, https://doi.org/10.1017/S003767 7900000103; Mark B. Adams, "Networks in Action: The Khrushchev Era, the Cold War, and the Transformation of Soviet Science," in *Science, History, and Social Activism: A Tribute to Everett Mendelsohn*, ed. Garland E. Allen and Roy M. MacLeod (Dordrecht, Netherlands: Springer-Science+Business Media, 2001), 255–76.

21. Krementsov, *Stalinist Science*, 81; Joravsky, "Soviet Marxism." For histories of the so-called evolutionary synthesis, see Ernst Mayr and William B. Provine, eds., *The Evolutionary Synthesis: Perspectives on the Unification of Biology* (Cambridge, MA: Harvard University Press, 1980); Vassiliki Betty Smocovitis, *Unifying Biology: The Evolutionary Synthesis and Evolutionary Biology* (Princeton, NJ: Princeton University Press, 1996).

22. For the "1 million" number, I have relied on Michael Ellman, "Soviet Repression Statistics: Some Comments," *Europe-Asia Studies* 54, no. 7 (2002): 1151–72, https://doi.org/10.1080 /0966813022000017177. See also Robert Conquest, *The Great Terror: A Reassessment* (New York: Oxford University Press, 1990). For the effects of the purges in specific areas of science, see Paul R. Josephson, *Physics and Politics in Revolutionary Russia* (Berkeley: University of California Press, 1991); Robert A. McCutcheon, "The 1936–1937 Purge of Soviet Astronomers," *Slavic Review* 50, no. 1 (1991): 100, https://doi.org/10.2307/2500602; and Asif A. Siddiqi, "The Rockets' Red Glare: Technology, Conflict, and Terror in the Soviet Union," *Technology and Culture* 44, no. 3 (2003): 470–501, doi.org/10.1353/tech.2003.0133.

23. For examples of British Marxists' contortions on this topic, see F. Le Gros Clark, "On Soviet Genetics," *Modern Quarterly* 3, no. 1 (1947): 93; J. B. S. Haldane, "Biology and Marxism," *Modern Quarterly* 3, no. 4 (1948): 2–11; J. D. Bernal, "The Biological Controversy in the Soviet Union and Its Implications," *Modern Quarterly* 4, no. 3 (1949): 203–17.

24. For an in-depth discussion of both Zhebrak's visit and this first US campaign against Lysenko's rise, see Nikolai Krementsov, "A 'Second Front' in Soviet Genetics: The International Dimension of the Lysenko Controversy, 1944–1947," *Journal of the History of Biology* 29, no. 2 (1996): 229–50, https://doi.org/10.1007/BF00571083. Lerner sent similar letters to several geneticists; the quotes here are from I. Michael Lerner to Barbara McClintock, 27 June 1945, USSR Correspondence with Geneticists, 1945–1946, Leslie C. Dunn Papers, American Philosophical Society (hereafter Dunn Papers).

25. Sophia Dobzhansky Coe, "Theodosius Dobzhansky: A Family Story," in *The Evolution of Theodosius Dobzhansky: Essays on His Life and Thought in Russia and America*, ed. Mark B. Adams (Princeton, NJ: Princeton University Press, 1994), 13–28; Theodosius Dobzhansky to L. C. Dunn, 4 July 1945, Dunn Papers.

26. Dobzhansky to Dunn, 4 July 1945. For Dobzhansky as a *nevozvrashchenets*, see Mark B. Adams, "Introduction: Theodosius Dobzhansky in Russia and America," in Adams, *Evolution of Theodosius Dobzhansky*, 5. Background on Dunn can be found in Melinda Gormley, "Scientific Discrimination and the Activist Scientist: L. C. Dunn and the Professionalization of Genetics and Human Genetics in the United States," *Journal of the History of Biology* 42, no. 1 (December 2009): 33–72, https://doi.org/10.1007/s10739-008-9170-z. For more detailed discussions of the events that followed, see Audra J. Wolfe, "What Does It Mean to Go Public? The American Response to Lysenkoism, Reconsidered," *Historical Studies in the Natural Sci-

ences 40, no. 1 (2010): 48–78, https://doi.org/10.1525/hsns.2010.40.1.48; and Rena Selya, "Defending Scientific Freedom and Democracy: The Genetics Society of America's Response to Lysenko," *Journal of the History of Biology* 45, no. 3 (2012): 415–42, https://doi.org/10.1007/s10739-011-9288-2.

27. Theodosius Dobzhansky to L. C. Dunn, 31 July 1945 and 18 August 1945, both in Dobzhansky 1943–1945, Dunn Papers; Dunn to Henry Silver, 11 November 1945, Lysenko Controversy in the US (2), Dunn Papers.

28. Theodosius Dobzhansky to L. C. Dunn, 4 September 1945, Dobzhansky 1943–1945, Dunn Papers; Henry Wallace to Dunn, 12 August 1945, Lysenko Controversy in the United States (2), Dunn Papers. The book was published as T. D. Lysenko, *Heredity and Its Variability*, trans. Theodosius Dobzhansky (New York: King's Crown Press, 1946).

29. L. J. Stadler to L. C. Dunn, 28 December 1945, Lysenko Controversy in the US (2), Dunn Papers; J. B. S. Haldane to H. J. Muller, 15 May 1946, Correspondence Series, Hermann J. Muller Papers, Lilly Library, Indiana University (hereafter Muller Papers); L. C. Dunn, "Review of Heredity and Its Variability, by T. D. Lysenko," *Science* 103, no. 2667 (1946): 180–91, https://doi.org/10.1126/science.103.2667.180; Theodosius Dobzhansky, "Lysenko's 'Genetics': A Review," *Journal of Heredity* 37, no. 1 (1946): 5–9, https://doi.org/10.1093/oxfordjournals.jhered.a105536. In Britain, where a disproportionate number of geneticists were either socialists or communists, the response to Lysenkoism was extremely muted until 1948. See Paul, "War on Two Fronts."

30. L. C. Dunn to Waldemar Kaempffert, 31 January 1946, Lysenko Controversy in the US (2), Dunn Papers; Waldemar Kaempffert, "Man and His Milieu," *New York Times* (hereafter *NYT*), 3 March 1946; William deJong-Lambert, *The Cold War Politics of Genetics Research: An Introduction to the Lysenko Affair* (New York: Springer, 2012), 63–70. Kaempffert returned to this theme frequently; see, for instance, Waldemar Kaempffert, "Science—and Ideology—in Soviet Russia; Russian Scientists Are Servants of the State and in Their Work Must Follow Marxian Laws," *NYT*, 15 September 1946.

31. See, for instance, "Soviet Biologists Ousted from Jobs," *NYT*, 28 August 1948.

32. Carlson, *Genes, Radiation, and Society*, 308; "Gloomy Nobelman," *Time*, 11 November 1946; "Genetic Death," *Time*, 22 September 1947.

33. William L. Laurence, "U.S. Scientist Quits Moscow Academy," *NYT*, 30 September 1948; H. J. Muller, "The Destruction of Science in the U.S.S.R.," *Saturday Review of Literature*, 4 December 1948; H. J. Muller, "Back to Barbarism—Scientifically," *Saturday Review of Literature*, 11 December 1948.

34. John D. Hickerson to H. J. Muller, 6 October 1948, Subject files, Lysenko, 1939–1950, Muller Papers.

35. The "stones shall speak" reference appears in both Theodosius Dobzhansky to L. C. Dunn, 9 September 1948, Dobzhansky 1948–1949, Dunn Papers, and in Dobzhansky to H. J. Muller, 9 September 1948, Correspondence Series, Dobzhansky 1948–1950, Muller Papers. These letters are nearly identical except that the one sent to Muller tellingly omits a reference to red-baiting. For Dobzhansky's draft statement, see Dobzhansky to Muller, 7 October 1948, Correspondence Series, Dobzhansky 1948–1950, Muller Papers.

36. Selya, "Defending Scientific Freedom and Democracy"; Wolfe, "What Does It Mean to Go Public?"; Ralph Cleland to M. R. Irwin, 6 November 1948, box 5, 1948, AIBS #2, Records of the Genetics Society of America, American Philosophical Society (hereafter GSA Records); E. G. Butler et al., "A Statement of the Governing Board of the AIBS," *Science* 110, no. 2848 (1949): 124–25, https://doi.org/10.1126/science.110.2848.124-a.

37. "Dear Teacher," *Time*, 9 September 1948; "Scientists' Choice," *Time*, 27 September 1948; "Watch Your Quantum Theory," *Time*, 6 December 1948; "Why Russia Purges Scientists," *U.S. News and World Report*, 14 January 1949; Muller, "Destruction of Science"; Muller, "Back to Barbarism."

38. Though now forty years old, the best introduction to physical scientists' role in World War II remains Daniel J. Kevles, *The Physicists: The History of a Scientific Community in Modern America* (New York: Knopf, 1977), 287–323. My thoughts on this group's collective attitudes toward scientific freedom are indebted to David A. Hollinger, "Free Enterprise and Free Inquiry: The Emergence of Laissez-Faire Communitarianism in the Ideology of Science in the United States," *New Literary History* 21, no. 4 (1990): 897–919, https://doi.org/10.2307/469191.

39. Samuel A. Goudsmit, *ALSOS* (New York: Henry Schuman, 1947), xi–xiii. The actual situation of physics in Germany was, of course, considerably more complicated than this. See Mark Walker, *Nazi Science: Myth, Truth, and the German Atomic Bomb* (New York: Plenum Press, 1995), esp. 243–47.

40. G. Pascal Zachary, *Endless Frontier: Vannevar Bush, Engineer of the American Century* (New York: Free Press, 1997), 213; Vannevar Bush, *Science, the Endless Frontier: A Report to the President* (Washington, DC: Government Printing Office, 1945), https://www.nsf.gov/od/lpa/nsf50/vbush1945.htm; Wolfe, *Competing with the Soviets*, 24–27.

41. Vannevar Bush, *Modern Arms and Free Men: A Discussion of the Role of Science in Preserving Democracy* (New York: Simon and Schuster, 1949), 201; Vannevar Bush, "Our Moral Armor," *Life*, 5 December 1949.

42. James B. Conant, *On Understanding Science: An Historical Approach* (New Haven, CT: Yale University Press, 1947), 14–16; James Hershberg, *James B. Conant: Harvard to Hiroshima and the Making of the Nuclear Age* (New York: Knopf, 1993), 407–11; Christopher Hamlin, "The Pedagogical Roots of the History of Science: Revisiting the Vision of James Bryant Conant," *Isis* 107, no. 2 (2016): 282–308, https://doi.org/10.1086/687217.

43. For scientists as public intellectuals during this time period, see Paul Boyer, *By the Bomb's Early Light: American Thought and Culture at the Dawn of the Atomic Age* (New York: Pantheon, 1984), 49–64; and Jessica Wang, "Scientists and the Problem of the Public in Cold War America, 1945–1960," *Osiris* 17 (2002): 323–47, https://doi.org/10.1086/649368.

44. Bernal, "Biological Controversy in the Soviet Union," 203.

2 : Ambassadors for Science

1. Leonard Eyges and Jeffries Wyman, diary notes, 1954, entry 430, P83, RG 59, NARA II. For more on Oppenheimer's relationship with Chevalier, the source of many of Oppenheimer's later troubles, see Kai Bird and Martin J. Sherwin, *American Prometheus: The Triumph and Tragedy of J. Robert Oppenheimer* (New York: Knopf, 2005), 195–201, 476–77, 498–537.

2. Interview of Joseph B. Koepfli by Elizabeth Hodes, transcript, 8 November 1983, 85–86, California Institute of Technology Archives.

3. For a survey of secrecy in postwar science, see Peter Galison, "Removing Knowledge," *Critical Inquiry* 31, no. 1 (2004): 229–43, https://doi.org/10.1086/427309.

4. Karl H. Weber, *The Office of Scientific Intelligence, 1948–1968* [TOP SECRET/declassified], vol. 1 (Washington, DC: Central Intelligence Agency, 1972), 1–2; Vincent C. Jones, *Manhattan: The Army and the Atomic Bomb* (Washington, DC: US Army Center of Military History, 1985), 280–91.

5. Jones, *Manhattan*, 290; Goudsmit, *ALSOS*, 93, 101, 112.

6. John Gimbel, *Science, Technology, and Reparations: Exploitation and Plunder in Postwar Germany* (Stanford, CA: Stanford University Press, 1990), 17, 63; Annie Jacobsen, *Operation Paperclip: The Secret Intelligence Program That Brought Nazi Scientists to America* (Boston: Little, Brown, 2014).

7. R. W. Home and Morris F. Low, "Postwar Scientific Intelligence Missions to Japan," *Isis* 84, no. 3 (1993): 527–37, https://doi.org/10.1086/356550.

8. Ray S. Cline, *Secrets, Spies, and Scholars: Blueprint of the Essential CIA* (Washington, DC: Acropolis Books, 1976), 91; Weber, *Office of Scientific Intelligence*, 1: annex II; Arthur B. Darling, *The Central Intelligence Agency: An Instrument of Government, to 1950* (University Park: Pennsylvania State University Press, 1990), 228–32.

9. Doel and Needell, "Science, Scientists, and the CIA," 62; Jeffrey T. Richelson, *The Wizards of Langley: Inside the CIA's Directorate of Science and Technology* (Boulder, CO: Westview Press, 2002), 2–4; Darling, *Central Intelligence Agency*, 161–65.

10. Michael J. Hogan, *A Cross of Iron: Harry S. Truman and the Origins of the National Security State, 1945–1954* (New York: Cambridge University Press, 1998), 23–69; Doel and Needell, "Science, Scientists, and the CIA," 62–63.

11. Donald S. McClure, *Wallace Reed Brode*, Biographical Memoirs 82 (Washington, DC: National Academy Press, 2002), http://www.nasonline.org/publications/biographical-memoirs/memoir-pdfs/brode-wallace-r.pdf; William B. Fretter, *Robert Bigham Brode*, Biographical Memoirs (Washington, DC: National Academy of Sciences, 1992), http://www.nasonline.org/publications/biographical-memoirs/memoir-pdfs/brode-robert-b.pdf; Wallace R. Brode to Mr. Richard, 25 September 1947, box 9, folder 1, Wallace R. Brode Papers, Library of Congress (hereafter Brode Papers).

12. Wallace R. Brode, "CIA Scientific Intelligence Units [RESTRICTED/declassified]," 1947, box 9, folder 1, Brode Papers.

13. Darling, *Central Intelligence Agency*, 231; Howard Simons, "Scientist of Statecraft: Symbol of New Power," *Saturday Review*, 1 March 1958.

14. For Condon's troubles, see J. Wang, *American Science*, 130–45; "Rockefeller Gift Helps Scientists Open Door to Russian Knowledge," *NYT*, 8 February 1947; and "Secret Hearings Held over Condon," *NYT*, 25 May 1948.

15. J. Wang, *American Science*, 132; George E. Sokolsky, "The Case of Dr. Dunn," *New York Sun*, 12 March 1948; L. C. Dunn to Walter Landauer, 14 March 1948, series 1, box 16, folder 25, Dunn Papers.

16. J. Wang, *American Science*, 139; National Academy of Sciences, "A Statement by Mem-

bers of the National Academy of Sciences Concerning a National Danger," 31 March 1948, and Alfred N. Richards to L. C. Dunn, 6 April 1948, both in series 1, box 5, Condon, Edward U., 1946–1952, Dunn Papers.

17. Wallace R. Brode to Admiral Hillenkoetter [RESTRICTED/declassified], 4 August 1948, box 9, folder 1, and Wallace R. Brode, [Draft Letter to Karl Compton], September 1948, box 9, folder 2, both in Brode Papers; Doel and Needell, "Science, Scientists, and the CIA," 65. For background on the Bureau of Standards, see Elio Passaglia, *A Unique Institution: The National Bureau of Standards, 1950–1969* (Washington, DC: US Department of Commerce, 1999), https://nvlpubs.nist.gov/nistpubs/Legacy/SP/nistspecialpublication925.pdf; and Thomas C. Lassman, "Government Science in Postwar America: Henry A. Wallace, Edward U. Condon, and the Transformation of the National Bureau of Standards, 1945–1951," *Isis* 96, no. 1 (2005): 25–51, https://doi.org/10.1086/430676.

18. Wallace R. Brode to Karl Compton, November 1948, Wallace R. Brode Papers, box 9, folder 1, Brode Papers; "Document 428: National Security Council Intelligence Directive No. 8 (May 25, 1948) [SECRET/declassified]," in *FRUS 1945–1950, Intel*. For more on the Research and Development Board, see Dan Kevles, "Cold War and Hot Physics: Science, Security, and the American State, 1945–56," *Historical Studies in the Physical and Biological Sciences* 20, no. 2 (1990): 239–64, https://doi.org/10.2307/27757644.

19. Weber, *Office of Scientific Intelligence*, 1:5–8.

20. Doel and Needell, "Science, Scientists, and the CIA," 65–67; Weber, *Office of Scientific Intelligence*, 1:10; "Dr. Willard Machle," *NYT*, 9 May 1976; "Document 399: Memorandum from the Assistant Director for Scientific Intelligence (Machle) to Director of Central Intelligence Hillenkoetter (September 29, 1949) [TOP SECRET/declassified]," in *FRUS 1945–1950, Intel*. For a fascinating account of the US intelligence failures associated with the Soviet bomb, see Michael D. Gordin, *Red Cloud at Dawn: Truman, Stalin, and the End of the Atomic Monopoly* (New York: Farrar, Straus, and Giroux, 2009).

21. Weber, *Office of Scientific Intelligence*, 1:12–17.

22. National Security Council, "National Security Council Intelligence Directive No. 10: Collection of Foreign Scientific and Technological Data [SECRET/declassified]," 18 January 1949, http://www.foia.cia.gov/sites/default/files/document_conversions/50/NSCID_No_10_Collection_of_Foreign_S%26T_Data_18_Jan_1949.PDF; Weber, *Office of Scientific Intelligence*, 2: 1–5; 1:17.

23. National Security Council, "NSCID No. 10."

24. W. Park Armstrong to James E. Webb [SECRET/declassified], 7 February 1949, box 64, Science and Foreign Relations, A1 3008A, RG 59, NARA II.

25. John E. Peurifoy to James E. Webb [SECRET/declassified], 27 May 1949, box 64, Science and Foreign Relations, A1 3008A, RG 59, NARA II; Allan A. Needell, *Science, Cold War, and the American State: Lloyd V. Berkner and the Balance of Professional Ideals* (Amsterdam: Harwood Academic Press, 2000), 109–22. For background on Webb, see W. Henry Lambright, *Powering Apollo: James E. Webb of NASA* (Baltimore: Johns Hopkins University Press, 1995).

26. The complete (open) portion of the Berkner Report was published as Lloyd V. Berkner, Douglas Merritt Whitaker, and National Research Council (US), *Science and Foreign Relations: International Flow of Scientific and Technological Information*, General Foreign Policy Series 30

(Washington, DC: US Department of State, 1950). For a more concise (and accessible) version, see Lloyd Berkner, "Science and Foreign Relations: Berkner Report to the U.S. Department of State," *BAS*, October 1950, 293–98. For the NRC's contribution to the report, including the names of scientists surveyed, see Douglas Whitaker, "National Research Council Report on Studies for the International Science Policy Survey of the Department of State," 7 January 1950, box 5, NRC Division of International Relations, A1 1549, Lot 58D606, RG 59, NARA II.

27. F. H. Russell to Edward W. Barrett and James E. Webb, 12 May 1950, box 10, A1 1950, RG 59, NARA II.

28. Cover note stapled to Russell to Barrett and Webb; Joseph Koepfli to R. Borden Reams, "On the Function of Science Attachés," 24 July 1951, box 1, Bern Science Attaché, P83, RG 59, NARA II; R. Gordon Arneson to Lloyd Berkner, 5 May 1950, box 64, Science and Foreign Relations, A1 3008A, RG 59, NARA II.

29. Lloyd Berkner, "Appendix: Scientific Intelligence [SECRET/declassified]," 18 April 1950, box 64, Science and Foreign Relations, A1 3008A, RG 59, NARA II.

30. Interview of Joseph B. Koepfli by Ronald E. Doel, transcript, 3 August 1995, Niels Bohr Library and Archives, American Institute of Physics, https://www.aip.org/history-programs /niels-bohr-library/oral-histories/31375.

31. Doel and Needell, "Science, Scientists, and the CIA"; National Security Council, "NSCID 10." The first published discussion of the Berkner Appendix by a non-intelligence official who had actually seen it is Needell, *Science, Cold War, and the American State*.

32. "Joseph Koepfli, 100, Museum Benefactor, Dies," *NYT*, 21 November 2004 (oddly, Koepfli's obituary does not mention his tenure at the State Department); Ronald E. Doel, email message to author, 20 December 2015.

33. Berkner, "Science and Foreign Relations"; Berkner, "Appendix: Scientific Intelligence."

34. Koepfli interview (Hodes), 63, 71.

35. Berkner, "Science and Foreign Relations."

36. Berkner, "Appendix: Scientific Intelligence," 10.

37. Walter M. Rudolph to C. S. Piggot, 21 May 1951, box 2, folder 2, ASTA, A1 1549, Lot 58D606, RG 59, NARA II.

38. Walter M. Rudolph to Wallace W. Atwood [SECRET/declassified], 6 March 1951; Joseph Koepfli to William C. Johnstone and Warren Kelchner [RESTRICTED/declassified], 18 May 1951; "American Scientists Traveling Abroad, June 13, 1951," 13 June 1951, all in box 2, ASTA, A1 1549, Lot 58D606, RG 59, NARA II.

39. William C. Johnstone to Joseph Koepfli, 25 June 1951; Warren Kelchner to Koepfli, 6 June 1951; Curt Stern to Koepfli, 28 May 1951; and Edmund Rowan to Robert L. Loftness, 25 September 1951, all in box 2, ASTA, A1 1549, Lot 58D606, RG 59, NARA II.

40. As an example of a debriefing report conducted by a science attaché, see J. Wallace Joyce, "Report of Dr. William R. Amberson," 3 December 1952, box 2, ASTA, A1 1549, Lot 58D606, RG 59, NARA II. For the CIA's request for a list of scientists traveling abroad, see Wallace W. Atwood to Joseph Koepfli [CONFIDENTIAL/declassified], 21 August 1951, in the same folder (declassified through my Freedom of Information Act [FOIA] request). Note that all identifying information about the originating agency and the identity of Atwood's correspondent has been redacted, but the FOIA code used, "50 U.S.C. 403g–CIA," indicates a de-

nial on the basis of protecting a given agency's function. It is therefore reasonable to assume that this is correspondence with the CIA.

41. Wallace W. Atwood to Edmund Rowan, 31 May 1951, box 2, ASTA, A1 1549, Lot 58D606, RG 59, NARA II. For Bronk's role in the Berkner Report, see Needell, *Science, Cold War, and the American State*, 144–45.

42. Atwood to Rowan, 31 May 1951; Roger Adams, "The Academy–Research Council Program in International Relations," 27 April 1954, and NAS–NRC, "Minutes of the 3rd Meeting of the Policy Committee of the Office of International Relations, NAS–NRC," 11 April 1953, both in box 13, NRC, International Relations Committee Materials, Subject files of Alan T. Waterman (hereafter NSFW), RG 307, NARA II.

43. For background on atomic espionage, see Gordin, *Red Cloud at Dawn*.

44. Koepfli interview (Hodes); Koepfli interview (Doel); Ronald E. Doel, "Roger Adams: Linking University Science with Policy on the World Stage," in *No Boundaries: University of Illinois Vignettes*, ed. Lillian Hoddeson (Urbana: University of Illinois Press, 2004), 124–44.

45. Interview of Richard T. Arnold by Ronald E. Doel, transcript, 10 August 1994, Niels Bohr Library and Archives, American Institute of Physics, https://www.aip.org/history-programs /niels-bohr-library/oral-histories/31208.

46. Koepfli to Reams, "On the Functions of Science Attachés"; Koepfli interview (Hodes), 71; Arnold interview.

47. Eyges and Wyman, "Diary Notes," entry 110; Koepfli interview (Doel).

48. Arnold interview.

49. Joseph Koepfli, memo to file [CONFIDENTIAL/declassified], 7 October 1952, box 2, ASTA, A1 1549, Lot 58D606, RG 59, NARA II; Eyges and Wyman, diary notes, entry 430.

50. J. Wang, *American Science*, 274, 278–79; John Krige, *American Hegemony and the Postwar Reconstruction of Science in Europe* (Cambridge, MA: MIT Press, 2006), 139–40.

51. Koepfli interview (Hodes), 80. For a discussion of the importance of Proviso 9 exceptions for psychological warfare, see "Document 142: Memorandum of Conversation by Mr. Robert P. Joyce of the Policy Planning Staff (April 6, 1950) [SECRET/declassified]," in *FRUS 1950*, vol. 4.

52. Koepfli interview (Hodes), 92–93; "Turmoil Inside State Department," *U.S. News and World Report*, 18 December 1953, 23. The magazine's campaign was part of a broader smear on the State Department's public affairs programming spearheaded by anti-Communist Congressional Republicans. See Hart, *Empire of Ideas*, 178–197.

3 : A War of Ideas

1. "Document 85: Note by the Executive Secretary to the National Security Council on United States Objectives and Programs for National Security (NSC 68) (April 14, 1950) [TOP SECRET/declassified]," in *FRUS 1950*, vol. 1.

2. Michael J. Hogan, *The Marshall Plan: America, Britain, and the Reconstruction of Western Europe, 1947–1952* (New York: Cambridge University Press, 1987); Hixson, *Parting the Curtain*; Lucas, *Freedom's War*, 75.

3. For scientists as technical advisors in psychological warfare, see Allan A. Needell,

"'Truth Is Our Weapon': Project Troy, Political Warfare, and Government-Academic Relations in the National Security State," *Diplomatic History* 17, no. 3 (1993): 399–420, https:// doi.org/10.1111/j.1467-7709.1993.tb00588.x; and Engerman, *Know Your Enemy*, 43–70.

4. For concise histories of US information campaigns prior to the Cold War, see Cull, *Cold War and the USIA*, 1–22; and Hart, *Empire of Ideas*, 15–70.

5. Harry S. Truman, "Special Message to the Congress on Greece and Turkey: The Truman Doctrine, March 12, 1947," American Presidency Project, http://www.presidency.ucsb.edu/ws /index.php?pid=12846; Krige, *American Hegemony*, 19–21.

6. On the Cominform, see Wilford, *Mighty Wurlitzer*, 23, 70; and Hixson, *Parting the Curtain*, 5, 7–10. On Lysenko and anti-America campaigns, see Krementsov, "'Second Front.'" For communication restrictions, see "Document 530: The Ambassador to the Soviet Union (Smith) to the Secretary of State (January 29, 1948)," in *FRUS 1948*, vol. 4.

7. "Document 244: Memorandum from DCI Hillenkoetter to the State-Army-Navy Force Coordinating Committee (October 22, 1947) [SECRET/declassified]"; "Document 250: Memorandum of Discussion at the Second National Security Council (November 14, 1947) [TOP SECRET/declassified]"; "Document 252: Memorandum from the Executive Secretary (Souers) to the Members of the National Security Council (December 9, 1947) [CONFIDEN-TIAL/declassified]"; "Document 257: Memorandum from the Executive Secretary (Souers) to DCI Hillenkoetter (December 17, 1947) [TOP SECRET/declassified]," all in *FRUS 1945–1950, Intel*. On passage of the Smith-Mundt Act, see Cull, *Cold War and the USIA*, 36–41; and Belmonte, *Selling the American Way*, 26–33. On the legal basis of covert action, see Rhodri Jeffreys-Jones, *The CIA and American Democracy* (New Haven, CT: Yale University Press, 1989), 41, 50.

8. Cull, *Cold War and the USIA*, 43–45; Lucas, *Freedom's War*, 43–45; Trevor Barnes, "The Secret Cold War: The CIA and American Foreign Policy in Europe, 1946–1956. Part I," *Historical Journal* 24, no. 2 (1981): 399–415, https://doi.org/10.1017/S0018246X00005537; Wendy Wall, "America's 'Best Propagandists': Italian Americans and the 'Letters to Italy' Campaign," in *Cold War Constructions: The Political Culture of United States Imperialism, 1945–1966*, ed. Christian G. Appy (Amherst, MA: University of Massachusetts Press, 2000), 89–109; Tim Weiner, "F. Mark Wyatt, 86, C.I.A. Officer, Is Dead," *NYT*, 6 July 2006.

9. NSC 10/2 is published as "Document 292: National Security Council Directive on Office of Special Projects (June 18, 1948) [TOP SECRET/declassified]," in *FRUS 1945–1950, Intel*. For a sampling of the differences of opinion among State, the CIA, and the various defense agencies on this issue, see Documents 274–76 in this same volume of *FRUS*.

10. "Document 292: National Security Council Directive."

11. "Document 298: Memorandum of Conversation and Understanding (August 6, 1948) [TOP SECRET/declassified]," and "Document 299: Memorandum for the President for Discussion at the Eighteenth Meeting of the National Security Council (August 19, 1948) [TOP SE-CRET/declassified]," both in *FRUS 1945–1950, Intel*.

12. Thomas W. Braden, "I'm Glad the CIA Is 'Immoral,'" *Saturday Evening Post*, 20 May 1967, 11; Wilford, *Mighty Wurlitzer*, 19; Michael Warner, "The CIA's Office of Policy Coordination: From NSC 10/2 to NSC 68," *International Journal of Intelligence and CounterIntelligence* 11, no. 2 (June 1998): 211–20, https://doi.org/10.1080/08850609808435373; Hixson,

Parting the Curtain, 58–63; Tim Weiner, *Legacy of Ashes: The History of the CIA* (New York: Doubleday, 2007), 49–51; Corke, *US Covert Operations*, 82–100.

13. On refugees, see Ben Shephard, *The Long Road Home: The Aftermath of the Second World War* (New York: Knopf, 2011). Much of the scholarship on intellectuals' role in the cultural Cold War tries to work out who was using whom. The essential texts in this argument are Frances Stonor Saunders, *The Cultural Cold War: The CIA and the World of Arts and Letters* (New York: New Press, 1999); and Wilford, *Mighty Wurlitzer*. For a particularly satisfying discussion of mutual benefit, see Patrick Iber, *Neither Peace Nor Freedom: The Cultural Cold War in Latin America* (Cambridge, MA: Harvard University Press, 2015).

14. Sean McMeekin, *The Red Millionaire: A Political Biography of Willi Münzenberg, Moscow's Secret Propaganda Tsar in the West* (New Haven, CT: Yale University Press, 2003); Stephen Koch, *Double Lives: Stalin, Willi Münzenberg, and the Seduction of the Intellectuals*, revised and updated edition (New York: Enigma Books, 2004), esp. 38–39.

15. Michael Scammell, *Koestler: The Literary and Political Odyssey of a Twentieth-Century Skeptic* (New York: Random House, 2009), 13–30, 68–74; Arthur Koestler, *Arrow in the Blue: The First Volume of an Autobiography, 1905–1953* (New York: Macmillan, 1952), 355, 371, 380. For a history of science writing, see Dorothy Nelkin, *Selling Science: How the Press Covers Science and Technology* (New York: W. H. Freeman, 1987).

16. Biographical details in this and the next paragraph are taken from Arthur Koestler, *The Invisible Writing: The Second Volume of an Autobiography: 1932–1940* (New York: Macmillan, 1954); and Scammell, *Koestler*.

17. Mary Jo Nye, *Michael Polanyi and His Generation: Origins of the Social Construction of Science* (Chicago: University of Chicago Press, 2011), 198–200; Saunders, *Cultural Cold War*, 60–61; Arthur Koestler, *Darkness at Noon* (New York: Macmillan, 1941).

18. Richard Crossman, ed., *The God That Failed* (New York: Harper and Brothers, 1949); Saunders, *Cultural Cold War*, 63–66.

19. Hugh Wilford, *The CIA, the British Left, and the Cold War: Calling the Tune?* (2003; repr., New York: Routledge, 2013), 27–34.

20. Arthur Schlesinger, Jr., "The U.S. Communist Party," *Life*, 29 July 1946; Arthur Schlesinger, Jr., *The Letters of Arthur Schlesinger, Jr.*, ed. Andrew Schlesinger and Stephen Schlesinger (New York: Random House, 2013), 17–19, 20–21; Arthur Schlesinger, Jr., *A Life in the Twentieth Century: Innocent Beginnings, 1917–1950* (Boston: Houghton Mifflin, 2000), 399–401, 411–13; Saunders, *Cultural Cold War*, 91.

21. Wilford, *Mighty Wurlitzer*, 70–71; Sidney Hook, *Out of Step: An Unquiet Life in the 20th Century* (New York: Harper & Row, 1987), 382–96; "Hook to Algernon D. Black, March 7, 1949," and "Hook to Thomas Mann, March 14, 1949," in *Letters of Sidney Hook: Democracy, Communism, and the Cold War*, ed. Edward Shapiro (Armonk, NY: M. E. Sharpe, 1995), 121–23.

22. Hook, *Out of Step*, 392–94; Sidney Hook to Thomas Mann, 14 March 1949, and Hook to H. J. Muller, telegram, 18 March 1949; both in Correspondence Series, Sidney Hook, 1949–1958, Muller Papers. For lively secondary accounts of the conference and counter-conference, see Saunders, *Cultural Cold War*, 45–56; and Vincent Giroud, *Nicolas Nabokov: A Life in Freedom and Music* (New York: Oxford University Press, 2015), 216–21.

23. H. J. Muller, "Destruction of Biological Science in the U.S.S.R.," 27 March 1949, Correspondence Series, Sidney Hook, 1949–1958, Muller Papers. The speech was based on Muller's article by the same name in the *Saturday Review of Literature*, 4 December 1948.

24. Hook, *Out of Step*, 385–96 ("mimeograph" quote, p. 387; Shapley story, pp. 391–92); Saunders, *Cultural Cold War*, 45–53.

25. All accounts of the Waldorf meeting and counter-conference mention Dubinsky's presence in the Americans for Intellectual Freedom's suite, but only Saunders explicitly claims the funds came from the OPC. Saunders cites an interview with Melvin Lasky, who would later play a key role in the CCF's early operations, for this claim. See Saunders, *Cultural Cold War*, 54–55. A more recent history of the Congress for Cultural Freedom suggests that the OPC admired the event but did not actively support it: Sarah Miller Harris, *The CIA and the Congress for Cultural Freedom in the Early Cold War: The Limits of Making Common Cause* (New York: Routledge, 2016), 63. On Burnham's relationship with the CIA, see Wilford, *Mighty Wurlitzer*, 74–77. The "official" history of the origins of the CIA's support of the CCF, an article written by CIA historian Michael Warner, suggests that the CIA took inspiration from but did not actually support Hook's counter-conference: Michael Warner, "Origins of the Congress for Cultural Freedom [SECRET/declassified]," *Studies on Intelligence* 38, no. 5 (1995): 89–98, https://www.cia.gov/library/center-for-the-study-of-intelligence/kent-csi/vol38 no5/pdf/v38i5a10p.pdf. That being said, most researchers who work on this topic agree that the Warner article contains important omissions. Since 2008, the CIA has repeatedly denied historian Patrick Iber's requests for the agency to declassify the larger report on which Warner's short article is based (Iber, email message to the author, 21 July 2016). Given the presence of Dubinsky and the involvement of James Burnham, who was already working for the CIA, the OPC's involvement seems more likely than not.

26. For State's interest in and monitoring of the conference, see Lucas, *Freedom's War*, 94–95. Josselson's presence is recounted in Saunders, *Cultural Cold War*, 54, quoting Nicolas Nabokov's memoir, *Bagázh: Memoirs of a Russian Cosmopolitan* (New York: Atheneum, 1975). The exact timing of Josselson's joining the OPC has long remained a mystery, but a recent biography suggests autumn of 1949: Harris, *CIA and the Congress for Cultural Freedom*, 60.

27. Harry S. Truman, "Address on Foreign Policy at a Luncheon of the American Society of Newspaper Editors, April 20, 1950," American Presidency Project, http://www.presidency.ucsb.edu/ws/index.php?pid=13768.

28. "Document 85: Note by the Executive Secretary to the National Security Council." The relationship of NSC 68 to national security objectives has been a major research question for historians of the Cold War since the document was first declassified in 1975. Much of the discussion focuses on the extent to which NSC 68 repudiated or extended George Kennan's earlier strategies and whether the new strategy should be classified as "containment," "liberation/rollback," "Titoism," or something else. The classic account of containment is John Lewis Gaddis, *Strategies of Containment: A Critical Appraisal of American National Security Policy during the Cold War*, revised and expanded edition (New York: Oxford University Press, 2005). For important works on liberation and Titoism, respectively, see Walter L. Hixson, *George F. Kennan: Cold War Iconoclast* (New York: Columbia University Press, 1989); and

Lorainne Lees, *Keeping Tito Afloat: The United States, Yugoslavia, and the Cold War* (University Park: Pennsylvania State University Press, 1997). More recent works question the existence of a coherent strategic framework during this time. See especially Sarah-Jane Corke, "History, Historians, and the Naming of Foreign Policy: A Postmodern Reflection on American Strategic Thinking during the Truman Administration," *Intelligence and National Security* 16, no. 3 (September 2001): 146–65, https://doi.org/10.1080/02684520412331306250; and Scott Lucas and Kaeten Mistry, "Illusions of Coherence: George F. Kennan, US Strategy and Political Warfare in the Early Cold War, 1946–1950," *Diplomatic History* 33, no. 1 (2009): 39–66, https://doi.org/10.1111/j.1467-7709.2008.00746.x.

29. Needell, "Truth Is Our Weapon," 404, n. 25; "Document 8: Memorandum by the Assistant Director for Policy Coordination of the Central Intelligence Agency (Wisner) (May 8, 1950) [TOP SECRET/declassified]," in *FRUS 1950–1955, Intel.*

30. Needell, "Truth Is Our Weapon."

31. Ibid.; Engerman, *Know Your Enemy*, 48–70. For a moderately redacted version of the Project Troy report, see "Document 59: Memorandum from Robert J. Hooker of the Policy Planning Staff to the Director of the Policy Planning Staff (Nitze) (March 26, 1951) [TOP SECRET/declassified]," in *FRUS 1950–1955, Intel.* For the broader story of social scientists' relationship with military and intelligence agencies, see Joy Rohde, *Armed with Expertise: The Militarization of American Social Research during the Cold War* (Ithaca, NY: Cornell University Press, 2013); and Price, *Cold War Anthropology.*

32. "Document 49: Memorandum from the Assistant Secretary of State for Public Affairs (Barrett) to Secretary of State Acheson (February 13, 1951) [SECRET/declassified]," in *FRUS 1950–1955, Intel.*

33. "Document 52: Memorandum from the Deputy Assistant Secretary of State for Administration (Scott) to the Under Secretary of State (Webb) [n.d.] [CONFIDENTIAL/declassified]," in *FRUS 1950–1955, Intel*; "Document 16: Directive by the President to the Secretary of State (Acheson), the Secretary of Defense (Marshall), and the Director of Central Intelligence (Smith) (April 4, 1951) [SECRET/declassified]," in *FRUS 1951*, vol. 1; "Document 74: Department of State Press Release (June 20, 1951)," in *FRUS 1950–1955, Intel*; Harry S. Truman, "Directive Establishing the Psychological Strategy Board (June 20, 1951)," American Presidency Project, http://www.presidency.ucsb.edu/ws/index.php?pid=13808.

34. "Document 78: Minutes of a Meeting of the Psychological Strategy Board (July 2, 1951) [SECRET/declassified]," in *FRUS 1950–1955, Intel*; Oral History Interview of Gordon Gray by Richard D. McKenzie, transcript, 18 June 1973, Harry S. Truman Presidential Library, http://www.trumanlibrary.org/oralhist/gray.htm. For more on the institutional history of the PSB, see Shawn J. Parry-Giles, *The Rhetorical Presidency, Propaganda, and the Cold War, 1945–1955* (Westport, CT: Praeger, 2002), 49–58; and Scott Lucas, "Campaigns of Truth: The Psychological Strategy Board and American Ideology, 1951–1953," *International History Review* 18, no. 2 (1996): 279–302, https://doi.org/10.1080/07075332.1996.9640744.

35. "Document 68: Memorandum from the Executive Secretary of the National Security Council (Lay) to the Director of the Policy Planning Staff (Nitze) et al. (May 14, 1951) [TOP SECRET EYES ONLY/declassified]"; "Document 90: Note from the Executive Secretary of the National Security Council (Lay) to the National Security Council [NSC 10/5] (October 23,

1951) [TOP SECRET/declassified]"; and "Document 104: Memorandum from the Director of the Psychological Strategy Board (Allen) to Board Members (February 20, 1952) [TOP SECRET/declassified]," all in *FRUS 1950–1955, Intel*; Ludwell Lee Montague, *General Walter Bedell Smith as Director of Central Intelligence, October 1950–February 1953* (University Park: Pennsylvania State University Press, 1992), 204–15.

36. "Document 333: Foreign Service Information and Educational Exchange Circular (November 1, 1951) [SECRET/declassified]," and "Document 334: Memorandum by the Assistant Secretary of State for Public Affairs (Barrett) (November 13, 1951) [SECRET/declassified]," both in *FRUS 1951*, vol. 1. On Barrett's philosophy, see Cull, *Cold War and the USIA*, 51–67; and Edward W. Barrett, *Truth Is Our Weapon* (New York: Funk & Wagnalls, 1953).

37. John D. Hickerson to H. J. Muller, 6 October 1948, Subject Series, Lysenko, 1939–1950, Muller Papers. On the Office of International Information, see Cull, *Cold War and the USIA*, 41.

38. Cull, *Cold War and the USIA*; Hixson, *Parting the Curtain*, 31–37; "Document 534: The Ambassador in the Soviet Union (Smith) to the Secretary of State (February 5, 1948) [SECRET/declassified]," in *FRUS 1948*, vol. 4.

39. Karl Compton to H. J. Muller, 1 July 1949; Muller to Compton, 6 July 1949; Evelyn Eisenstadt to J. R. Cominsky, 20 December 1948; and Muller to R. C. Cook et al., 9 December 1949, all in Chronological Series, Muller Papers; H. J. Muller, "A Message to Scientists beyond the Iron Curtain" (United States Information Service, US Department of State, 5 May 1952), Writing Series, Muller Papers; Milislav Demerec to Colleagues, 10 January 1949, Lysenko Controversy, Milislav Demerec Papers, American Philosophical Society; Theodosius Dobzhansky to Alexander Frankley, 10 October 1949, Dobzhansky (7), Dunn Papers.

40. For a more detailed discussion of the Mendel celebration, see Audra J. Wolfe, "The Cold War Context of the Golden Jubilee; Or, Why We Think of Mendel as the Father of Genetics," *Journal of the History of Biology* 45, no. 3 (2012): 389–414, https://doi.org/10.1007/s10739-011-9291-7. Whether or not Mendel's work was "rediscovered" is another question altogether. See Jan Sapp, "The Nine Lives of Gregor Mendel," in *Experimental Inquiries: Historical, Philosophical and Social Studies of Experimentation in Science*, ed. H. E. LeGrand (Dordrecht, Netherlands: Kluwer Academic Publishers, 1990), 137–66.

41. Details of the ceremony are recounted in various press releases, as well as in Paul C. Mangelsdorf to L. C. Dunn, 8 August 1950, box 7, 1950—Golden Jubilee Correspondence #7, and Paul Mangelsdorf to M. R. Irwin, 11 August 1950, box 7, 1950—Golden Jubilee Irwin Correspondence #3, both in GSA Records. Mangelsdorf's conference presentation was published as Paul C. Mangelsdorf, "Hybrid Corn: Its Genetic Basis and Its Significance in Human Affairs," in *Genetics in the 20th Century: Essays on the Progress of Genetics during Its First 50 Years*, ed. L. C. Dunn (New York: Macmillan, 1951), 555–72. For more on Mangelsdorf's international corn-breeding work, see Karin Matchett, "At Odds over Inbreeding: An Abandoned Attempt at Mexico / United States Collaboration to 'Improve' Mexican Corn, 1940–1950," *Journal of the History of Biology* 39, no. 2 (2006): 345–72, https://doi.org/10.1007/s10739-006-0007-3. It is unclear how much of the event was broadcast, but a thank-you letter from Dunn confirms that VOA recorded the proceedings in their entirety: Dunn to Roger Lyons, 29 December 1950, box 7, 1950—Golden Jubilee L. C. Dunn Correspondence #3, GSA Records.

42. The personal connections are particularly suggestive. In 1960, Ralph Cleland, a key member of the Golden Jubilee's organizing committee, performed consulting work for the Asia Foundation, a CIA asset (see chapter 7). Cleland's correspondence with Muller moreover suggests an unusual level of familiarity with the US government's ability to spirit useful German scientists out of occupied territories through the Top Secret Project Paperclip program. David C. Rife, the local organizer, later served as a US agricultural advisor to the Thai government. Even L. C. Dunn had connections to the State Department, having been recruited as a potential science attaché in 1947 before his links to the American-Soviet Science Society disqualified him for the position. Ralph Cleland to H. J. Muller, 6 December 1948, Subject Series, Lysenko, Oct. 26–Dec. 1950, Muller Papers; Cleland to Robert Blum, 30 January 1960, box P-128, Social and Economic: Science Advisor, Cleland, R.E., 1960–1961, Asia Foundation Records, Hoover Institution Library and Archives (hereafter TAF); "Biographical Data on Mr. David C. Rife," 1959, box 30, Office of Special International Programs, NSFW, RG 307, NARA II; Dunn to Walter Landauer, 12 November 1947, series 1, box 16, folder 24, Dunn Papers; Melinda B. Gormley, "Geneticist L. C. Dunn: Politics, Activism, and Community" (PhD dissertation, Oregon State University, Corvallis, 2007), 386.

43. "Document 85: Note by the Executive Secretary to the National Security Council," 15. See also Paul Erickson et al., *How Reason Almost Lost Its Mind: The Strange Career of Cold War Rationality* (Chicago: University of Chicago Press, 2013).

44. Krige, *American Hegemony*, 15–73; Andrew L. Yarrow, "Selling a New Vision of America to the World: Changing Messages in Early U.S. Cold War Print Propaganda," *Journal of Cold War Studies* 11, no. 4 (2009): 22, https://doi.org/10.1162/jcws.2009.11.4.3; Sam Lebovic, "From War Junk to Educational Exchange: The World War II Origins of the Fulbright Program and the Foundations of American Cultural Globalism, 1945–1950," *Diplomatic History* 37, no. 2 (2013): 303, https://doi.org/10.1093/dh/dht002; Gerhard J. Drechsler, "The U.S. State Department and World Science," *BAS*, April 1951. President Truman's so-called Point Four program, which provided foreign aid for technical development to countries in the Global South, might also be considered an attempt to use science and technology to sway hearts and minds. For Point Four as a failed propaganda program, see Jason Parker, *Hearts, Minds, Voices: US Cold War Public Diplomacy and the Formation of the Third World* (New York: Oxford University Press, 2016), 29–41; for the broader agenda of the program, see Stephen C. Macekura, "The Point Four Program and U.S. International Development Policy," *Political Science Quarterly* 128, no. 1 (2013): 127–160, doi.org/10.1002/polq.12000.

4 : *Science and Freedom*

Portions of this chapter are adapted from Audra J. Wolfe, "*Science and Freedom*: The Forgotten Bulletin," in *Campaigning Culture and the Global Cold War: The Journals of the Congress for Cultural Freedom*, edited by Giles Scott-Smith and Charlotte Lerg (New York: Palgrave Macmillan, 2017), 27–44.

1. Paul Koenig to George Polanyi, 2 October 1956, box 5, folder 13, Committee on Science and Freedom Records, Joseph Regenstein Library, Univeristy of Chicago (hereafter CSF Records).

2. Rockefeller Foundation, "Grant File GA 55061," 12 January 1955, series 2.1, RG 6.1 (Paris Field Office), box 65, folder 607, Rockefeller Foundation Archives, Rockefeller Archive Center, Sleepy Hollow, NY (hereafter RF); Price Waterhouse & Co. for the Congress for Cultural Freedom, "Statement of Receipts and Disbursements for the Year Ended December 31, 1960," box 19, Audit Files, 1960–1962, 1966, Michael Josselson Papers, Harry Ransom Center of the University of Texas, Austin.

3. Essential histories of the CCF include Peter Coleman, *The Liberal Conspiracy: The Congress for Cultural Freedom and the Struggle for the Mind of Postwar Europe* (New York: Free Press, 1989); Saunders, *Cultural Cold War*; Giles Scott-Smith, *The Politics of Apolitical Culture: The Congress for Cultural Freedom, the CIA, and Post-war American Hegemony* (New York: Routledge, 2002); Wilford, *Mighty Wurlitzer*; and Harris, *CIA and the Congress for Cultural Freedom*. With the exception of a handful of paragraphs, these accounts do not discuss science. See, however, Elena Aronova, "The Congress for Cultural Freedom, *Minerva*, and the Quest for Instituting 'Science Studies' in the Age of Cold War," *Minerva* 50, no. 3 (September 2012): 307–37, https://doi.org/10.1007/s11024-012-9206-6.

4. Warren D. Manshel to Nicolas Nabokov, n.d. 1955 [July], box 274, folder 1, International Association for Cultural Freedom Records, Joseph Regenstein Library, University of Chicago (hereafter IACF).

5. Warner, "Origins of the CCF," 92.

6. Daniel F. Harrington, *Berlin on the Brink: The Blockade, the Airlift, and the Early Cold War* (Lexington: University Press of Kentucky, 2012).

7. Warner, "Origins of the CCF"; Coleman, *Liberal Conspiracy*, 15–19; Wilford, *Mighty Wurlitzer*, 78–79. On Lasky and *Der Monat*'s complicated relationship with the CCF, see Michael Hochgeschwender, "*Der Monat* and the Congress for Cultural Freedom: The High Tide of the Intellectual Cold War, 1948–1971," in Scott-Smith and Lerg, *Campaigning Culture*, 71–89.

8. Brendan Nash, "The Titania-Palast," *Cabaret Berlin* (blog), 10 December 2012, http:// www.cabaret-berlin.com/?p=850; Coleman, *Liberal Conspiracy*, 27–28; Hook, *Out of Step*, 440–42.

9. Warner, "Origins of the CCF"; Coleman, *Liberal Conspiracy*, 32. The entire text of the Manifesto for Freedom is reprinted on pp. 249–51 of Coleman's book.

10. "Arthur Schlesinger to W. Averell Harriman, July 19, 1950," in Schlesinger, *Letters*, 31–33; Warner, "Origins of the Congress," 96, 97. On Schlesinger's "wittingness," see Saunders, *Cultural Cold War*, 91.

11. Saunders, *Cultural Cold War*, 85–94; Coleman, *Liberal Conspiracy*, 33–46.

12. On the publications network, see the essays in Scott-Smith and Lerg, *Campaigning Culture*.

13. Coleman, *Liberal Conspiracy*, 9. Most accounts of Koestler's break with the CCF are based on rumor and hearsay. A welcome exception, based on archival sources, is Scammell, *Koestler*, 362–86. The claims about Burnham's matchmaking with the CIA are on pp. 372 and 383; the quotation from Burnham's report to the CIA is on p. 384.

14. Coleman, *Liberal Conspiracy*, 28. Muller's speech is reprinted in full as H. J. Muller, "Science in Bondage," *Science* 113, no. 2924 (12 January 1951): 25–29, https://doi.org/10.1126 /science.113.2924.25. Muller's postwar speeches frequently conflated communism and fascism; for more on this phenomenon, see Les K. Adler and Thomas G. Paterson, "Red Fascism: The

Merger of Nazi Germany and Soviet Russia in the American Image of Totalitarianism, 1930's–1950's," *American Historical Review*, 75, no. 4 (1970): 1046–64, https://doi.org/10.1086/ahr/75.4.1046.

15. "Atomic Physicist Scraps Defense of Reds at Cultural Talk as a Result of Korea Attack," *NYT*, 28 June 1950. Nachtsheim's remarkably brazen paper is reprinted in full as Hans Nachtsheim, "For a New Academy," *Science* 113, no. 2924 (12 January 1951): 30–31, https://doi.org/10.1126/science.113.2924.30. For Nachtsheim's wartime work, see Weiss, *Nazi Symbiosis*, 115–20.

16. See the cover letter and attachments for Congress for Cultural Freedom, "Manifesto Unanimously Adopted in Berlin, Germany, on June 30, 1950," 9 November 1950, Organizations/Committees Series, Congress for Cultural Freedom, Muller Papers.

17. H. J. Muller to Sidney Hook, 17 October 1950, Correspondence Series, Sidney Hook, 1949–1958, Muller Papers; Muller to Pearl Kluger, 15 January 1951, 1 March 1951, and 6 March 1951, all in Organizations/Committee Series, Congress for Cultural Freedom, Jan.–March 20, 1951, Muller Papers; Muller to Pearl Kluger, 8 April 1952, Organizations/Committee Series, Congress for Cultural Freedom, 1952–1955, Muller Papers. Muller's papers contain many more gossipy letters like these, mostly addressed to Kluger, executive secretary of the American Committee for Cultural Freedom. As late as 1960, Muller continued to send tidbits on the status of science in the Soviet Union to Josselson. See, for instance, items discussed in Michael Josselson to Edward Shils, 28 April 1961, box 202, folder 2, IACF. For the CCF's campaign against Lysenko in Japan, conducted with Muller's help, see Kaori Iida, "A Controversial Idea as a Cultural Resource: The Lysenko Controversy and Discussions of Genetics as a 'Democratic' Science in Postwar Japan," *Social Studies of Science* 45, no. 4 (2015): 546–69, https://doi.org/10.1177/0306312715596460.

18. Rockefeller Foundation, "Materials Related to the Science and Freedom Conference (1953)," 1953, Series 100.D, RG 1.2, box 25, folder 179, RF; Henry Margenau to James Franck, 8 January 1953, box 124, Congress for Cultural Freedom, Correspondence, 1952–1954, Sidney Hook Papers, Hoover Institution Library and Archives. For Weissberg's own telling of his imprisonment, see Alex Weissberg, *The Accused*, trans. Edward Fitzgerald (New York: Simon and Schuster, 1951).

19. Rockefeller Foundation, "Materials Related to the Science and Freedom Conference (1953)."

20. "Hamburg Congress of Science and Freedom," 23 July 1953, and MMCongren, Hamburg, Germany to Department of State, Washington, "RE: Hamburg Congress of Science and Freedom," 11 August 1953, both in box 9, Science and Freedom, A1 1549, Lot 58D606, RG 59, NARA II; Michael Polanyi, "Protests and Problems," *BAS*, November 1953; Edward Shils, "The Scientific Community: Thoughts after Hamburg," *BAS*, May 1954.

21. For background on Polanyi, see Nye, *Michael Polanyi*. For the history of the Society for Freedom in Science, see William McGucken, "On Freedom and Planning in Science: The Society for Freedom in Science, 1940–46," *Minerva* 16, no. 1 (1978): 42–72, https://doi.org/10.1007/BF01102181.

22. Hollinger, "Free Enterprise and Free Inquiry"; Nye, *Michael Polanyi*, 200, 209–12; Arthur Koestler, *The Yogi and the Commissar and Other Essays* (New York: Macmillan, 1945); Rockefeller Foundation, "Grant File GA 55061."

23. Nicolas Nabokov to Michael Polanyi, 19 October 1953, box 273, folder 8; Nabokov to M. Polanyi, 21 May 1954, box 273, folder 11; and Warren Manshel to Michael Josselson, 7 June 1954, box 272, folder 7, all in IACF.

24. George Polanyi to Warren Manshel, 6 June 1954, and Manshel to Michael Josselson, 7 June 1954, both in box 272, folder 7, IACF.

25. Warren Manshel to George Polanyi, 22 June 1954, and Manshel to G. Polanyi, 14 July 1954, both in box 272, folder 7, IACF; "W. D. Manshel, 66, Magazine Publisher Was an Ambassador," *NYT*, 27 February 1990.

26. According to Saunders, the CIA sent both Manshel (in 1954) and Hunt (in 1956) to keep a tighter grip over the CCF's operations. Saunders, *Cultural Cold War*, 242–43. Warren Manshel to George Polanyi, 6 December 1954, box 272, folder 7, IACF; Manshel to G. Polanyi, 29 March 1955 and 4 April 1955, both in box 272, folder 9, IACF.

27. Michael Josselson to Priscilla Polanyi, 16 September 1954 and 3 November 1954, both in box 272, folder 7, IACF. See also Nicolas Nabokov to George Polanyi, 5 July 1956, box 273, folder 1, IACF; Michael Josselson to G. Polanyi, 21 August 1957, and Marion Bieber to G. Polanyi, 13 May 1957, both in box 273, folder 3, IACF.

28. Committee on Science and Freedom, "Report on the First Year's Activities: July 1954–August 1955," box 9, folder 12, CSF Records.

29. *Science and Freedom* 3 (1956).

30. For example, on Soviet presentations at a recent philosophy of science meeting, see W. Mays, Marcello Boldrini, and A. Buzzati-Traverso, "Encounters with Soviet Thought," *Science and Freedom* 2 (1955). For the University of Tasmania, see *Science and Freedom* 4 (1956); for British universities, *Science and Freedom* 7 (1957); for Spain, *Science and Freedom* 5 (1956); for Hungary, *Science and Freedom* 8 (1957). The telegram campaign on behalf of Hungarian scientists is especially notable in that it represents one of the Committee on Science and Freedom's only concrete accomplishments beyond publication of the bulletin.

31. Committee on Science and Freedom, "Study Group: Science and Freedom," 4 March 1956, box 273, folder 2, IACF; Michael Josselson to George Polanyi, 8 March 1956, box 273, folder 2, IACF; Sidney Hook to Michael Polanyi, 16 March 1956, box 118, Committee on Science and Freedom (118.2), Sidney Hook Papers, Hoover Institution Library and Archives; Josselson to G. Polanyi, 17 March 1956, box 273, folder 3, IACF; Rockefeller Foundation, "Grant File GA G 5604," 4 May 1956, series 100, RG 1.2, box 6, folder 40, RF. The papers were reproduced as *Science and Freedom* 11 (June 1958).

32. Hunt and Josselson had been urging Polanyi to consider the problem of Chinese universities for at least three years. For a typical example, see Michael Josselson to George Polanyi, 21 August 1957, box 273, folder 3, IACF. The published article was William H. Newell, "Universities in Modern China," *Science and Freedom* 14 (1960): 17–25. On Kristol's recommendation, see G. Polanyi to Josselson, 3 May 1960, box 273, folder 6, IACF; Hunt's response, John Hunt to G. Polanyi, 8 March 1960, is in the same folder.

33. "Notes on the Discussion of the Bulletin of the Committee on 'Science and Freedom,' Paris, 23 January 1955," 23 January 1955, box 273, folder 9, IACF; Committee on Science and Freedom, "Report on the First Year's Activities: July 1954–August 1955."

34. Committee on Science and Freedom, "Report on Activities in the Period September

1956–April 1957," April 1957, box 9, folder 15, CSF Records; Committee on Science and Freedom, "Report on Activities in the Period September–November 1955," November 1955, series 100.D, RG 1.2, box 25, folder 180, RF. Committee on Science and Freedom, "Report on the Work Carried out by the Secretariat of the Committee on Science and Freedom during January and February 1960," 1960, box 9, folder 19, CSF Records.

35. George Polanyi to Professor Duyvendak, 19 April 1955, box 5, folder 4; M. E. McKinnon to G. Polanyi, 20 July 1955, box 5, folder 5; Alexander Szalai to G. Polanyi, 20 October 1956, box 5, folder 10; and M. G. J. Minnaert to G. Polanyi, 20 April 1957, box 6, folder 7, all in CSF Records.

36. See Coleman, *Liberal Conspiracy*, 139–57, 199–209, and essays by Eric Pullin, Elizabeth Holt, and Asha Rogers in Scott-Smith and Lerg, *Campaigning Culture*. P. Gisbert to George Polanyi, 9 April 1956, box 5, folder 14, CSF Records.

37. George Polanyi to Michael Josselson, 6 May 1958, and Josselson to G. Polanyi, 9 May 1958, both in box 273, folder 4, IACF; G. Polanyi to Josselson, 25 May 1959; Priscilla Polanyi to John Hunt, 26 June 1959; and Hunt to P. Polanyi, 3 July 1959, all in box 273, folder 5, IACF; Hunt to P. Polanyi, 19 February 1960, box 273, folder 6, IACF.

38. Priscilla Polanyi to John Hunt, 23 May 1961, box 273, folder 7, IACF. It's unclear whether Hunt delivered this news in person or through correspondence. The archival record includes only the draft of such a letter: John Hunt to Michael Polanyi, "Draft Letter," n.d. [May 1961], box 273, folder 11, IACF. A later letter from Michael Polanyi to Hunt, asking how George and Priscilla took the news, makes clear that Michael agreed with the decision: M. Polanyi to Hunt, 17 June 1961, box 274, folder 11, IACF. On *Minerva*, see Aronova, "Congress for Cultural Freedom."

39. Nicolas Nabokov to Michael Polanyi, 19 October 1953. Among other things, Josselson repeatedly urged Polanyi to hire a secretary at the CCF's expense. Michael Josselson to Michael Polanyi, 27 March 1962, box 202, folder 9, IACF. Saunders, in a footnote, suggests that Josselson might have shut down the committee because "[Michael] Polanyi himself was showing all the signs of mental illness" (Saunders, *Cultural Cold War*, 449, n. 1). Given that Michael Polanyi and the secretariat discussed the need to straighten out George Polanyi's operation, this seems unlikely. Saunders's passing description of *Science and Freedom* identifies Michael, rather than George, as the editor (p. 214), suggesting that her account of the Committee on Science and Freedom should be treated with caution.

40. George Polanyi to Michael Josselson, 5 February 1956, box 273, folder 2, and G. Polanyi to John Hunt, 26 May 1959, box 273, folder 5, both in IACF. The CCF's audit for 1960 does not include a budget line for salary expenses associated with the Committee on Science and Freedom. Price Waterhouse & Co, "Statement of Receipts and Disbursements, 1960."

5 : Science for Peace

1. "Soviet Embassy Guests Hear of Satellite from an American as Russians Beam; Feat Called 'Remarkable,'" *NYT*, 5 October 1957; Victor K. McElheny, "Anatoly A. Blagonravov Dies; Soviet Space Spokesman, 80," *NYT*, 6 February 1975; Needell, *Science, Cold War, and the American State*, 346.

2. For the post-*Sputnik* surge in US science funding, see Wolfe, "Big Science," chapter 3 in *Competing with the Soviets*.

3. Dwight D. Eisenhower, "Dwight D. Eisenhower: Annual Message to the Congress on the State of the Union," American Presidency Project, 9 January 1958, http://www.presidency. ucsb.edu/ws/?pid=11162; Cull, *Cold War and the USIA*, 152.

4. Federal Council for Science and Technology, "Strengthening the Free World Position in Science and Technology," 22 November 1960, box 12, Strengthening the Free World Position in Science and Technology, A1 1568F, RG 59, NARA II.

5. Quoted in Osgood, *Total Cold War*, 53; for a broader discussion of Eisenhower's passion for psychological warfare, see pp. 48–53.

6. Ibid., 55, 82; Cull, *Cold War and the USIA*, 81–82, 90–91; "Document 370: Report to the President by the President's Committee on International Informational Activities (June 30, 1953) [TOP SECRET/declassified]," in *FRUS 1952–1954, NatSec*; Lampton Berry to the Acting Secretary [TOP SECRET/declassified]," 16 September 1953, box 8, Misc. 1953–1956, A1 1568A, RG 59, NARA II.

7. "Document 370: Report to the President by the President's Committee on International Informational Activities"; James S. Lay, "Progress Report on Implementation of the Recommendations of the Jackson Committee [TOP SECRET/declassified]," 1 October 1953, box 8, Jackson Committee, A1 1568F, RG 59, NARA II. While the version of the Jackson Report reproduced in *FRUS* is widely accessible, readers should be aware that the original document contained at least twice as many recommendations as those published. The "Progress Report" document, while still sanitized, includes updates from the State Department and the USIA on their efforts to implement forty-two recommendations for the overt information programs. The portion of this document discussing the CIA's and State Department's progress on covert recommendations remains classified, with the single exception of a plan to increase NATO allies' involvement in covert political and propaganda action.

8. Elmer B. Staats, "Terms of Reference for Working Group on the U.S. Doctrinal Program," 26 May 1954, and USIA, "United States Doctrinal Program [SECRET/declassified]," 15 January 1954, both in box 37, Doctrinal Program, A1 1568C, RG 59, NARA II.

9. Elmer B. Staats, "New Name for the 'Doctrinal Program,'" 25 August 1954, box 37, Doctrinal Program, A1 1568C, RG 59, NARA II; Staats, "Outline Plan of Operations for the U.S. Ideological Program (D-33) [CONFIDENTIAL/declassified]," 16 February 1955, box 39, Ideological Programs, A1 1568C, RG 59, NARA II; "Document 187: Circular Airgram from the Department of State to Certain Diplomatic Missions (April 1, 1955)," in *FRUS 1955–1957, Foreign Economic Policy*.

10. Elmer B. Staats, "Progress Report on the U.S. Ideological Program [CONFIDENTIAL/declassified]," 24 August 1955, box 39, Ideological Programs, A1 1568C, RG 59, NARA II. For a survey of what the Ideological Program did accomplish, see Osgood, *Total Cold War*, 288–322; and Greg Barnhisel, "Cold Warriors of the Book: American Book Programs in the 1950s," *Book History* 13, no. 1 (2010): 185–217, https://doi.org/10.1353/bh.2010.0010.

11. C. E. Sunderlin, "Support of U.S. Participation in International Organizations," 16 December 1954, box 10, International Science, NSFW, RG 307, NARA II; Neil Carothers III, "Interim Criteria on International Scientific Activities," 13 July 1955, box 21, Neil Carothers

III—Chron Files, 1955, NSF Historian's Files (hereafter NSFH), RG 307, NARA II; Alan T. Waterman to William J. Hoff, 9 August 1955, box 2, ATW Notes 1955, NSFH, RG 307, NARA II. The NSF, founded in 1950, remained a small federal agency until after *Sputnik*. See Wolfe, *Competing with the Soviets*, 49.

12. Wallace W. Atwood, "Preliminary Notes on an Expanded Program in International Relations for the NAS," 3 February 1954, and "Materials for the 5th Meeting of the Policy Committee of the Office of International Relations, NAS," 11 February 1956, both in box 13, NRC, International Relations Committee Materials, NSFW, RG 307, NARA II; "What's Happened to Science in State?," *C&EN*, 9 January 1956.

13. Osgood, *Total Cold War*, 156; OCB, "Status Report," 14 December 1954, Series O, box 62, folder 518, Nelson A. Rockefeller Papers, Rockefeller Archive Center, Sleepy Hollow, NY (hereafter NAR Papers); Elmer B. Staats, "Progress Report on Nuclear Energy Projects and Related Information Programs [SECRET/declassified]," 1 December 1954, box 35, Atomic and Nuclear Energy, 1954, A1 1568C, RG 59, NARA II; Staats, "Progress Report on Nuclear Energy Projects and Related Information Programs [SECRET/declassified]," 20 June 1955 and 13 February 1956, box 35, Atomic and Nuclear Energy, 1955, A1 1568C, RG 59, NARA II.

14. John Krige, "Atoms for Peace." Radioisotopes *were* useful, just not quite as useful as US propaganda suggested. See especially Angela N. H. Creager, *Life Atomic: A History of Radioisotopes in Science and Medicine* (Chicago: University of Chicago Press, 2013); and Helen Ann Curry, *Evolution Made to Order: Plant Breeding and Technological Innovation in Twentieth-Century America* (Chicago: University of Chicago Press, 2016).

15. Donald Irwin to Nelson Rockefeller, "Geneva Atomic Conference," 25 August 1955, Series O, box 74, folder 578, NAR Papers; H. J. Muller to Michael Polanyi, 7 December 1955, Organizations/Committee Series, Congress for Cultural Freedom, 1952–1955, Muller Papers. I return to the issue of the genetic dangers of fallout in chapter 6. The story of Muller's exclusion from the official program of the Geneva Conference is usually told as an example of McCarthyism. That story was convenient for supporters of Muller's legacy, in that it allowed him to be painted as a victim rather than a perpetrator of anti-Communist fervor; see, for instance, Carlson, *Genes, Radiation, and Society*, 355–67. Archival evidence makes clear, however, that the AEC attempted to stifle *any* discussion of the dangers of fallout prior to 1955, regardless of the scientist's personal politics. Nor did anyone at the AEC apparently grasp the extent of Muller's role in the United States' ideological campaigns or of the potential blowback from censoring a critic of Lysenkoism. H. Bentley Glass, "Handwritten Notes from the 53rd Meeting of the Advisory Committee for Biology and Medicine," 30 November 1955, box 117, ACBM, 53rd Meeting Packet, Glass Papers.

16. Staats, "Progress Report," 13 February 1956; Staats, "Progress Report on Nuclear Energy Projects and Related Information Programs (Including NSC 5507/2) [CONFIDENTIAL/declassified]," 15 August 1956, box 35, Atomic and Nuclear Energy, 1956, A1 1568C, RG 59, NARA II; David Z. Beckler, "Documents for Next Meeting of the Science Advisory Committee," 30 June 1954, box 15, folder 10, Detlev Bronk Papers, RU 303-U, Rockefeller Archive Center, Sleepy Hollow, NY (hereafter Bronk Papers). On partner nations' enthusiasm for nuclear power, see John Krige, *Sharing Knowledge, Shaping Europe: US Technological Collaboration and Nonproliferation* (Cambridge, MA: MIT Press, 2016), 25–26.

17. Needell, *Science, Cold War, and the American State*, 297–323.

18. "Document 340: National Security Report (May 20, 1955) [SECRET/declassified]," in *FRUS 1955–1957, UN*, vol. 11; Asif A. Siddiqi, "Korolev, Sputnik, and the International Geophysical Year," in *Reconsidering Sputnik: Forty Years since the Soviet Satellite*, ed. Roger D. Launius, John M. Logsdon, and Robert W. Smith (Amsterdam: Harwood Academic Press, 2000), 43–72; Elmer B. Staats, "Memo for the Record [TOP SECRET/declassified]," 9 June 1955, box 1, OCB III, A1 1568A, RG 59, NARA II.

19. Elmer B. Staats, "Minutes of the Operation Coordinating Board Meeting of June 19, 1957 [SECRET/declassified]," 24 June 1957, box 2, OCB V, A1 1568A, RG 59, NARA II; Robert A. Divine, *The Sputnik Challenge* (New York: Oxford University Press, 1993).

20. USIA Office of Research and Intelligence, "World Opinion and the Soviet Satellite: A Preliminary Evaluation," 17 October 1957, box 9, Satellites: Sputnik, P243, RG 306, NARA II.

21. Zuoyue Wang, "The Politics of Big Science in the Cold War: PSAC and the Funding of SLAC," *Historical Studies in the Physical and Biological Sciences* 25, no. 2 (1995): 329–356, https://doi.org/10.2307/27757748; James R. Killian, Jr., *Sputnik, Scientists, and Eisenhower: A Memoir of the First Special Assistant to the President for Science and Technology* (Cambridge, MA: MIT Press, 1977), 26; Alan T. Waterman, "Position Paper on Government Organization for Science," 13 March 1958, box 3, ATW Notes 1958, NSFH, RG 307, NARA II.

22. Wallace R. Brode to Christian A. Herter, 6 January 1958, box 24, folder 3, Brode Papers; "Science Reborn at State," *C&EN*, 20 January 1958; Howard Simons, "Scientist of Statecraft: Symbol of New Power," *Saturday Review*, 1 March 1958; Joseph Koepfli to Brode, 15 January 1958, box 23, folder 8, Brode Papers.

23. Brode to Herter, 6 January 1958; Wilton Lexow, "The Science Attaché Program [SECRET/declassified]," *Studies in Intelligence* 10 (Spring 1966): 21–27, https://www.cia.gov/library/center-for-the-study-of-intelligence/kent-csi/vol10no2/html/v10i2a02p_0001.htm; W. J. Murphy, "Brode to Head Scientific Liaison," *C&EN*, 20 January 1958; "U.S. Program to Post Scientists Abroad Lags," *NYT*, 8 May 1958; "Status of Staffing, Science Program, FY 59," 9 September 1958, box 24, folder 4, Brode Papers; "Embassies Get Science Attachés," *C&EN*, 27 December 1958.

24. J. Y. Smith, "Science Author Harold Goodwin Dies at Age 75," *Washington Post* (hereafter *WaPo*), 21 February 1990; Harold Goodwin to Mr. Harkness, "'Project Proposals' Included in the Paper on Development of a Science Program," 28 February 1958, box 3, IOP/A Basic Paper 1958, P 243, RG 306, NARA II. For Goodwin's subsequent work at NASA, see Teasel Muir-Harmony, "Project Apollo, Cold War Diplomacy and the American Framing of Global Interdependence" (PhD dissertation, MIT, 2014).

25. Harold Goodwin to the Director, "Memorandum to the Director on 'Scientific Breakthrough Projects,'" 20 November 1958, box 4, ICA, P 243, RG 306, NARA II; Henry Loomis to Mr. Bradford, "Basic Science Paper," 26 September 1958, and George V. Allen to All USIS Posts, "Science and Technology: A Basic Guidance Paper," 18 November 1958, both in box 3, IOP/A Basic Paper 1958, P 243, RG 306, NARA II.

26. Allen to All USIS Posts, "Science and Technology."

27. Ibid.

28. Schrecker, *No Ivory Tower*, 296–97.

29. James S. Lay, "Admission to the U.S. of Certain European Non-Official Temporary Visitors Excludable under Existing Law [CONFIDENTIAL/declassified]," 8 February 1955, series O-9, box 4, folder 94, NAR Papers.

30. US Department of State, "Press Release: Regulation of Travel by Soviet Citizens in the United States," 3 January 1955, box 25, folder 3, Brode Papers.

31. Lauren Soth, "If the Russians Want More Meat," *Des Moines Register*, 10 February 1955.

32. US Department of State, Office of Intelligence and Research, "Intelligence Report No. 6884: U.S. and Soviet Gains from Agricultural Exchange," 17 October 1955, series O-7, box 89, folder 669, NAR Papers.

33. Hixson, *Parting the Curtain*, 105–10; "Document 104: Statement of Policy on East–West Exchanges (NSC 5607, June 29, 1956) [CONFIDENTIAL/declassified]," in *FRUS 1955–1957, Soviet Union*, vol. 24; "US–USSR Exchange Policy [CONFIDENTIAL/declassified]," 1 July 1957, box 37, East–West Exchange, Box 37, A1 1586C, RG 59, NARA II; US Department of State, "Press Release: Regulation of Travel of Soviet Citizens in the United States," 15 November 1957, box 25, folder 3, Brode Papers.

34. National Academy of Sciences, "Press Release: Six American Women Doctors to Visit USSR," March 1958, and Helen Taussig, "Report on Trip to Russia," 7 June 1958, both in ADM: IR: US–USSR Exchange of Scientists: Women Physicians: Americans to USSR: 1958, Archives of the National Academy of Sciences, Washington, DC (hereafter NAS Archives).

35. Helen Taussig, "Impression Gained from Four Weeks Inside the U.S.S.R.," July 1958, ADM: IR: US–USSR Exchange of Scientists: Women Physicians: Americans to USSR: 1958, NAS Archives.

36. Dr. John M. Weir to S. D. Cornell, 20 January 1958; Dorothy St. John to Rita Brandt, 7 January 1958; and R. Keith Cannan, "Telephone Call from Mr. Merrill, State Department, Concerning Dr. Thelma Dunn," 25 March 1958, all in ADM: IR: US–USSR Exchange of Scientists: Women Physicians: Americans to USSR: 1958, NAS Archives.

37. Hixson, *Parting the Curtain*, 152–54; US Department of State, "Travel Regulations in the Soviet Union and the United States," 24 May 1958, box 25, folder 4, Brode Papers. For broader context, see Yale Richmond, *Cultural Exchange and the Cold War: Raising the Iron Curtain* (University Park: Pennsylvania State University Press, 2003). I am grateful to historian Barbara Walker for sharing portions of her book-length study of the transformative effects of these exchanges on both US and Soviet culture with me. Barbara Walker, "A War of Experts: Soviet and American Knowledge Networks in Cold War Competition and Collaboration" (unpublished manuscript shared with the author, 2017).

38. Wallace W. Atwood to S. D. Cornell, "Moscow Conversations on Exchanges," 26 September 1958, ADM: IR: US–USSR Exchange of Scientists Interacademy Agreement: Beginning of Program: 1958, NAS Archives; "Protocol of Negotiations between Delegations from the National Academy of Sciences of the USA and the Academy of Sciences of the USSR . . . ," 9 July 1959, ADM: IR: US–USSR Exchange of Scientists Interacademy Agreement: Agreement: 1959, NAS Archives; Atwood, "Agreement on the Exchange of Scientists between the National Academy of Sciences of the USA and the Academy of Sciences of the USSR," 10 August 1959, box 4, USSR Scientific Exchange Program, 1959, Eugene Rabinowitch Papers, University of Illinois Archives (hereafter Rabinowitch Papers [UI]).

39. Detlev Bronk, "To the Members of the National Academy of Sciences," 22 July 1959, ADM: IR: US–USSR Exchange of Scientists Interacademy Agreement: Agreement: 1959, NAS Archives; National Science Board, "Approved Minutes of the Sixtieth Meeting of the National Science Board," 24 May 1959, box 120, folder 7, Bronk Papers; Hugh Dryden, "Minutes of the Business Session of the National Academy of Sciences Annual Meeting, April 26, 1960," 26 April 1960, box 216, NAS, 1961–1965, Glass Papers; PSAC Panel on Science and Foreign Affairs, "Reciprocity Policy in Scientific Exchanges with the USSR: Summary," 3 November 1959, box 16, folder 15, Bronk Papers; "Status of Implementation of Bronk–Nesmeyanov Exchange Agreement," 13 April 1960, box 16, folder 17, Bronk Papers.

40. For the broader context of the United States' rivalry with the People's Republic of China, see Gregg A. Brazinsky, *Winning the Third World: Sino-American Rivalry during the Cold War* (Chapel Hill: University of North Carolina Press, 2017).

41. Alan T. Waterman to Robert Murphy, August 24, 1959, and Waterman to Senior Staff, 20 July 1959, both in box 3, ATW Notes 1959, NSFW, RG 307, NARA II.

42. "Summary of Meeting, November 12, 1957," 12 November 1957; H. D. Smyth to Robert Murphy, "Membership in International Scientific Unions, as Discussed on Tuesday, November 12, 1957," 19 November 1957; and Albert Noyes to Detlev Bronk, November 1957, all in box 332, Scientific Research and Development 4, US Participation in Scientific Meetings Involving Unrecognized Regimes, A1 3008A, RG 59, NARA II.

43. Wallace R. Brode, "Reply to Dr. Killian's Letter of February 14," 10 March 1958, and Brode to Gerard Smith [SECRET/declassified], 8 February 1960, both in box 332, Scientific Research and Development 4, US Participation in Scientific Meetings Involving Unrecognized Regimes, A1 3008A, RG 59, NARA II.

44. Francis O. Wilcox to Mr. Murphy, "Accreditation of U.S. Delegations to Meetings of International Organizations to Which Unrecognized Regimes Adhere or Are Seeking Adherence," 21 November 1957, and Mr. Henderson to Miss Bacon, Mr. Nunley, and Mr. Gibson, 9 December 1959, both in box 332, Scientific Research and Development 4, US Participation in Scientific Meetings Involving Unrecognized Regimes, A1 3008A, RG 59, NARA II. The issue remained unresolved at the point of handover to the Kennedy administration. PSAC Panel on Science and Foreign Affairs, "Science and Foreign Affairs Panel Statement," 5 January 1961, box 16, folder 19, Bronk Papers.

45. Arthur Roe, "Annual Report to the Director on Activities of the Office of Special International Programs during FY61," 14 July 1961, box 41, International Science Annual Reports, NSFH, RG 307, NARA II; Federal Council for Science and Technology, "Strengthening the Free World Position"; George Kistiakowsky, "International Scientific Activities: Presentation to the National Security Council on December 1, 1960 [SECRET/declassified]," 1 December 1960, box 8, International Scientific Activities, 1960, A1 1568F, RG 59, NARA II.

46. Federal Council for Science and Technology, "Strengthening the Free World Position."

47. Kistiakowsky, "International Scientific Activities."

48. "Department of State Comment on Proposed NSC Discussion Paper on International Scientific Activities [SECRET/declassified]," 4 May 1960, box 332, Scientific Research and Development 4, US Participation in Scientific Meetings Involving Unrecognized Regimes, A1 3008A, RG 59, NARA II; Wallace R. Brode, "Development of a Science Policy," *Science* 131,

no. 3392 (1960): 9–15, https://doi.org/10.1126/science.131.3392.9; Wallace R. Brode, "Handwritten Notes," 21 June 1960, box 24, folder 2, Brode Papers.

49. NSC Planning Board, "Discussion Paper on International Scientific Cooperation [SECRET/declassified]," 14 March 1960, box 332, Scientific Research and Development 4, US Participation in Scientific Meetings Involving Unrecognized Regimes, A1 3008A, RG 59, NARA II; Frank Brink, Jr., *Detlev Wulf Bronk, 1897–1975,* Biographical Memoirs (Washington, DC: National Academy of Sciences, 1978), 52; National Science Board, "Approved Minutes of the Sixtieth Meeting of the National Science Board"; Harrison Brown, "Program Status Report, Office of the Foreign Secretary, National Academy of Sciences," October 1962, box 66, NAS Correspondence, 1950–1962, Bronk Papers.

50. Harry Kelly, "Suggested Possibilities for Consideration of Increased Cooperation with the Soviet Union," 2 March 1961, box 383, Whitman Files, 14G, US–USSR Scientific Cooperation, 1961, A1 3008A, RG 59, NARA II; Angelina Long Callahan, "Sustaining Soviet-American Collaboration, 1957–1989," in *NASA in the World: Fifty Years of International Collaboration in Space,* in John Krige, Angelina Long Callahan, and Ashok Maharaj (New York: Palgrave Macmillan, 2013), 127–30; Walter Whitman to L. T. Merchant, "OCB Working Group on Science and Technology," 11 January 1961, box 10, OCB Luncheon Items, 7/60–1/61, A1 1568A, RG 59, NARA II; International Committee of the Federal Council for Science and Technology, "International Scientific and Technological Activities," 20 June 1961, box 4, Image Panel, P 243, RG 306, NARA II.

51. Melvin Kranzberg, "Documents on the Embassy Science Bulletins," 16 May 1961, box 2, Embassy Scientific Bulletin, P 243, RG 306, NARA II; "Engineers and Scientists Committee: The People-to-People Program," 1965, box 21, folder 9, Brode Papers.

6 : Science for Diplomacy

1. The best overall biography of Pauling is Thomas Hager, *Force of Nature: The Life of Linus Pauling* (New York: Simon and Schuster, 1995).

2. "Loyal Defiance" (editorial), *WaPo*, 1 July 1960.

3. Bentley Glass, "Scientists in Politics," *BAS*, May 1962.

4. For a summary of the exchange between Glass and Pauling, see Hager, *Force of Nature,* 525–26. More background and details can be pieced together from the two sides of the correspondence in the Glass and Pauling papers: box 35, Correspondence 1962–1965, and box 36, Pauling, both in Glass Papers; and box 3.058, folders 58.4 and 58.8, Ava Helen and Linus Pauling Papers, Oregon State University Special Collections and Archives (hereafter Pauling Papers). (All documents from the Pauling Papers are courtesy of Mary X. Mitchell.) For histories of Pugwash, see Matthew Evangelista, *Unarmed Forces: The Transnational Movement to End the Cold War* (Ithaca, NY: Cornell University Press, 1999); Lawrence S. Wittner, *Resisting the Bomb: A History of the World Disarmament Movement, 1954–1970,* vol. 2, *The Struggle against the Bomb* (Stanford, CA: Stanford University Press, 1997); Andrew Brown, *Keeper of the Nuclear Conscience: The Life and Work of Joseph Rotblat* (Oxford: Oxford University Press, 2012); and Joseph Rotblat, *Scientists in the Quest for Peace: A History of the Pugwash Conferences* (Cambridge, MA: MIT Press, 1972).

5. On the atomic scientists' movement, see Alice Kimball Smith, *A Peril and a Hope: The Scientists' Movement in America, 1945–47* (Chicago: University of Chicago Press, 1965); and Boyer, *By the Bomb's Early Light*, 49–106. On Szilard and the idea of the bomb, see Ray Monk, *Robert Oppenheimer: A Life Inside the Center* (New York: Doubleday, 2012), 448.

6. J. Wang, *American Science*, 10–25; Shane J. Maddock, *Nuclear Apartheid: The Quest for American Atomic Supremacy from World War II to the Present* (Chapel Hill: University of North Carolina Press, 2010), 48–68.

7. Peter Galison and Barton Bernstein, "In Any Light: Scientists and the Decision to Build the Superbomb, 1952–1954," *Historical Studies in the Physical and Biological Sciences* 19, no. 2 (1989): 267–347, https://doi.org/10.2307/27757627.

8. Casualty estimates calculated using historian Alex Wellerstein's Nukemap, http://nu clearsecrecy.com/nukemap. On the politics of testing locations, see Gabrielle Hecht and Paul N. Edwards, *The Technopolitics of Cold War: Toward a Transregional Perspective* (Washington, DC: American Historical Association, 2007), 32–34.

9. "The H-Bomb and World Opinion: Chairman Strauss's Statement on Pacific Tests," *BAS*, May 1954; Robert A. Divine, *Blowing on the Wind: The Nuclear Test Ban Debate, 1954–1960* (New York: Oxford University Press, 1978), 3–13, 27–31. The Marshall Islanders' fight for justice continues. See Mary X. Mitchell, "Test Cases: Reconfiguring American Law, Techno-science, and Democracy in the Nuclear Pacific" (PhD dissertation, University of Pennsylvania, Philadelphia, 2016).

10. "Statement by the Indian Prime Minister (Nehru) to Parliament Regarding Nuclear Tests [Extracts] (April 2, 1954)," in *Documents on Disarmament, 1945–1959*, vol. 1 (Washington, DC: US Department of State, 1960), 408–11; Divine, *Blowing on the Wind*, 18–19; Evangelista, *Unarmed Forces*, 46–49. For other lay movements against the bomb, see Michael Egan, *Barry Commoner and the Science of Survival: The Remaking of American Environmentalism* (Cambridge, MA: MIT Press, 2007), 47–78; Moore, *Disrupting Science*, 96–129; and Vincent J. Intondi, *African Americans against the Bomb: Nuclear Weapons, Colonialism, and the Black Freedom Movement* (Stanford, CA: Stanford University Press, 2015).

11. The Department of Energy finally declassified the hearings in full in October 2014. "J. Robert Oppenheimer Personnel Hearing Transcripts," https://www.osti.gov/opennet/hear ing.jsp. For more on Oppenheimer and his long history of conflicts with J. Edgar Hoover and Lewis Strauss, see Kai Bird and Martin J. Sherwin, *American Prometheus: The Triumph and Tragedy of J. Robert Oppenheimer* (New York: Knopf, 2005); and Monk, *Oppenheimer*.

12. *BAS*, May 1954.

13. Edward Shils, "The Scientific Community: Thoughts after Hamburg," *BAS*, May 1954; "International Conference on Science and Public Affairs," July 1954, box 90, folder 6, IACF; David Hill to Eugene Rabinowitch, 5 November 1954, box 1, folder 2, Eugene Rabinowitch Papers, Joseph Regenstein Library, University of Chicago (hereafter Rabinowith Papers [UC]).

14. Brown, *Keeper of the Nuclear Conscience*, 102–5, 110–12; Jacob Darwin Hamblin, " 'A Dispassionate and Objective Effort': Negotiating the First Study on the Biological Effects of Atomic Radiation," *Journal of the History of Biology* 40, no. 1 (July 2007): 147–77, https://doi .org/10.1007/s10739-005-6531-8. Rabinowitch had trouble with a security clearance in 1946. Historian Jessica Wang has discussed Rabinowitch's subsequent defense of scientific interna-

tionalism and scientific liberties as an implicit critique of US security policies. Rabinowitch did oppose McCarthyism, but he does not seem to have objected to the war against Communism on a global scale. See J. Wang, *American Science*, 89–94.

15. Joseph Rotblat to Eugene Rabinowich, 29 October 1954, box 1, folder 2, Rabinowitch Papers (UC); Brown, *Keeper of the Nuclear Conscience*, 123. The CIA kept tabs on the leadership and activities of the World Federation of Scientific Workers as a Communist front. CIA, "Communism: Exploitation of the International Communist Movement by the Soviet Intelligence Services," July 1954, 58–59, CIA-RDP78-00915R000300090002-7, and CIA, "World Federation of Scientific Workers: A Compilation of Available Basic Reference Data," 1 October 1956, CIA-RDP78-00915R000600140011-8, both in CREST.

16. Brown, *Keeper of the Nuclear Conscience*, 128. Audio of Russell's speech is available online as Bertrand Russell, "Man's Peril," https://www.youtube.com/watch?v=oZzm6x_IMFE.

17. "The Russell-Einstein Manifesto," in Rotblat, *Scientists in the Quest for Peace*, 137–40.

18. The confluence of these events has created confusion in the secondary literature on the history of the Pugwash movement, in that most accounts of Pugwash's early days list one or two, but not all three, of these meetings. Wittner, *Resisting the Bomb* (p. 34), mentions the London meeting; Evangelista, *Unarmed Forces* (p. 37) and Rotblat, *Scientists in the Quest for Peace* (p. 1), mention London and Geneva. Brown, *Keeper of the Nuclear Conscience* (p. 127), mentions all three but offers no further details or sources for the Paris meeting. I am grateful to Sandra Butcher, Executive Director of the Pugwash Conferences, for sharing an untitled, unpublished manuscript on Pugwash's origins that clarified the relationship between the three events. Butcher also pointed me toward Rabinowitch's account linking them: Eugene Rabinowitch, "Pugwash—History and Outlook," *BAS*, September 1957. For more on the Geneva meeting, see Krige, "Atoms for Peace."

19. For a history of the world federalism movement that focuses on the United States, see Fritz Bartel, "Surviving the Years of Grace: The Atomic Bomb and the Specter of World Government, 1945–1950," *Diplomatic History* 39, no. 2 (2015): 275–302, https://doi.org/10.1093/dh/dhu005. For the British version and its relationship to Russell, see Lord Boyd Orr, "The British Parliamentary Group for World Government," *BAS*, January 1956.

20. Rabinowitch, "Pugwash—History and Outlook."

21. Eugene Rabinowitch, "International Cooperation of Atomic Scientists," *BAS*, February 1956, 37; and Rabinowitch, "Pugwash—History and Outlook," 243. On what the Soviets made of *BAS*'s reporting on how science should operate, see Evangelista, *Unarmed Forces*, 39–40. Evangelista makes extensive use of post-Soviet Russian-language archive sources and memoirs in making this point.

22. H. J. Muller to Cyrus Eaton, 18 June 1957, and 19 July 1957, both in Organizations/Committee Series, Pugwash Correspondence, 1957, Muller Papers; Rotblat, *Scientists in the Quest for Peace*, 4–6. A memoir by Fay Ajzenberg-Selove contains a memorable description of her outdrinking Topchiev at this meeting. Fay Ajzenberg-Selove, *A Matter of Choices: Memoirs of a Female Physicist* (New Brunswick, NJ: Rutgers University Press, 1994), 136.

23. CIA, "The International Communist Fronts in 1958," September 1958, 164–66, CIA-RDP78-00915R000900320080-9, CREST; Victor Weisskopf, *The Joy of Insight: Passions of a Physicist* (New York: Basic Books, 1991), 206, as quoted in Evangelista, *Unarmed Forces*, 33.

See also Evangelista's broader discussion of how the Soviet leadership saw the role of Pugwash on pp. 32–40.

24. Halvor Ekern to John J. Hanley, 18 August 1958, box 381, 14G 43a, Pugwash Conference, A1 3008A, RG 59, NARA II.

25. Pugwash's relationship with both the CIA and State is further discussed below, but around this time information gathered at Pugwash also starts appearing in routine CIA intelligence estimates. See, for example, CIA, "Scientific Intelligence Digest No. 247 [SECRET/declassified]," 8 March 1960, CIA Electronic Reading Room, https://www.cia.gov/library/read ingroom/docs/DOC_0001303417.pdf. The most thorough discussion of the global antinuclear movement during this time is Wittner, *Resisting the Bomb*.

26. Maddock, *Nuclear Apartheid*; Evangelista, *Unarmed Forces*, 53–57.

27. Useful exceptions to a literature that usually takes scientists' participation in test ban debates as a natural consequence of their technical expertise are Paul Rubinson, "'Crucified on a Cross of Atoms': Scientists, Politics, and the Test Ban Treaty," *Diplomatic History* 35, no. 2 (2011): 283–319, https://doi.org/10.1111/j.1467-7709.2010.00950.x; Zuoyue Wang, *In Sputnik's Shadow: The President's Science Advisory Committee and Cold War America* (New Brunswick, NJ: Rutgers University Press, 2008); and Benjamin P. Greene, *Eisenhower, Science Advice, and the Nuclear Test Ban Debate, 1945–1963* (Stanford, CA: Stanford University Press, 2006). See also Rubinson's discussion of how Teller, especially, positioned US nuclear policy as apolitical: Paul Rubinson, *Redefining Science: Scientists, the National Security State, and Nuclear Weapons in Cold War America* (Amherst: University of Massachusetts Press, 2016), 73–76.

28. On the American delegation to the detection conference, see Divine, *Blowing on the Wind*, 215–17. For Scoville's analysis, see CIA Study Group, "Intelligence Analysis of the Geneva Conference to Study the Methods of Detecting Violations of a Possible Agreement on the Suspension of Nuclear Tests [CIA/SI 205-58] [SECRET/declassified]," 28 October 1958, CIA Electronic Reading Room, https://www.cia.gov/library/readingroom/docs/DOC_0000818707.pdf.

29. Bentley Glass to Eugene Rabinowich, 30 June 1958, box 199, Kitzbühel Conference, Glass Papers.

30. Eugene Rabinowich to Jerome Wiesner, 23 February 1960, box 203, Pugwash 1959–1960, Glass Papers; Rabinowich to Walter G. Whitman, 23 February 1960, box 7, folder 6, Rabinowitch Papers (UC); Rabinowich to Col. Richard Leghorn, 24 February 1960, box 203, Pugwash 1959–1960, Glass Papers; Rabinowich to J. Robert Oppenheimer, 11 July 1960, box 199, Pugwash 1958–1960, Glass Papers.

31. Saunders, *Cultural Cold War*, 333, 357; "Reports at Annual Meeting 11 May 1960," *Records of the Academy (American Academy of Arts and Sciences)* 1959/1960 (1960): 92–120.

32. Lloyd K. Belt, "Staff Paper for Assistant Director, Scientific Intelligence: 6th Pugwash Conference, Moscow [SECRET/declassified]," 30 January 1961, box 382, 14G 43d, Pugwash Conference, A1 3008A, RG 59, NARA II. This is the State Department's copy of a document prepared for the CIA's Office of Scientific Intelligence.

33. Divine, *Blowing on the Wind*, 310–14; Maddock, *Nuclear Apartheid*, 137–39.

34. Eugene Rabinowich to Advisors and US Section of the International Continuing Committee for Pugwash Conferences, 27 May 1960; Rabinowich, Bentley Glass, and Harri-

son Brown to A. V. Topchiev, 3 August 1960; and Joseph Rotblat to Rabinowich, 8 August 1960, all in box 199, Pugwash 1958–1960, Glass Papers.

35. Charles Homans, "Track II Diplomacy: A Short History," *Foreign Policy*, 20 June 2011.

36. "Memorandum of Conversation: Meeting of U.S. Participants at Pugwash Conference in Moscow, November 27, 1960," 14 November 1960, box 381, 14G 43a, Pugwash Conferences, A1 3008A, RG 59, NARA II; Eugene Rabinowich to Walter Whitman, 19 October 1960, box 199, Pugwash 1958–1960, Glass Papers.

37. Department of State to American Embassy Moscow, 23 November 1960; American Embassy Moscow to Secretary of State, 29 November, 2 December, and 2 December [b] 1960; Llewellyn Thompson to Department of State, "Soviet Restricted Telephone Exchange," 5 December 1960; Thompson to Secretary of State, 6 December 1960, all in box 381, 14G 43a, Pugwash Conferences, A1 3008A, RG 59, NARA II. For the meeting itself, see Rotblat, *Scientists in the Quest for Peace*, 44–46.

38. Department of State, "Report on Pugwash Conference by Professor W. W. Rostow," 6 December 1960, and Department of State, "Pugwash Meeting in Moscow," 6 December 1960, both in box 381, 14G 43a, Pugwash Conferences, A1 3008A, RG 59, NARA II; Belt, "Staff Paper." Rubinson suggests that the British Foreign Office maintained a similarly close-but-not-quite-captive relationship with British Pugwash attendees (presumably not including Russell). Rubinson, " 'Crucified on a Cross of Atoms,' " 293, n. 36.

39. Eugene Rabinowich to Bertrand Russell, 29 September 1959, and Bentley Glass to Russell, 2 October 1959, both in box 204, Pugwash Continuing Committee, Glass Papers.

40. Hager, *Force of Nature*, 490–506; Toshihiro Higuchi, "Tipping the Scale of Justice: The Fallout Suit of 1958 and the Environmental Legal Dimension of Nuclear Pacifism," *Peace & Change* 38, no. 1 (2013): 33–55, https://doi.org/10.1111/pech.12002; Mitchell, "Test Cases."

41. Hager, *Force of Nature*, 488–89, 512–26. The June subpoena is available online in the Pauling Papers, http://scarc.library.oregonstate.edu/coll/pauling/peace/papers/bio2.021.3.html.

42. Committee on the Judiciary, US Senate, *The Pugwash Conferences: A Staff Analysis Prepared for the Subcommittee to Investigate the Administration of the Internal Security Act and Other Internal Security Laws of the Committee on the Judiciary* (Washington, DC: Government Printing Office, 1961).

43. Linus Pauling to Premier Khrushchev, telegram, 2 September 1961, and Linus Pauling, "Statement by Linus Pauling on the Russian Bomb Tests, Issued to the Press on August 31, 1961," 2 September 1961, both in box 204, Pugwash 1961 (5), Glass Papers; Wayland Young, "Note on the Western Press Coverage of COWSA 7 and 8," November 1961, box 203, Pugwash 1961 (3), Glass Papers.

44. Linus Pauling to Members of the US Planning Committee for the COWSA, 16 October 1961, and Harrison Brown to Pauling, 9 November 1961, both in box 204, Pugwash 1961 (5), Glass Papers. The two documents that confirm Brennan's reporting suggest the presence of many more, from Brennan and others. See Donald Brennan, "Notes on COWSA VIII: Stowe, Vermont," 30 November 1961, box 382, 14G 43d, Pugwash Conference, A1 3008A, RG 59, NARA II; and Herbert Scoville to Brennan, 20 May 1963, CIA-RDP66R00546R00010001 00041-9, CREST (a friendly thank-you note from the CIA, addressed to "Dear Don").

45. Glass, "Scientists in Politics."

46. Linus Pauling to Hans Bethe, 15 June 1962, folder 58.8, and Pauling to Homer A. Jack, 16 June 1962, folder 58.4, both in box 3.058, Pauling Papers.

47. Linus Pauling to Eugene Rabinowich, 10 May 1962, and Pauling to Bentley Glass, 3 July 1962, both in box 35, Correspondence 1962–1965, Glass Papers; Bertrand Russell to Glass, 8 April 1963, Glass to Russell, 16 April 1963, and Russell to Glass, 22 April 1963, all in box 56, Pauling Papers; Rabinowich to Glass, 26 August 1963, box 201, Pugwash, 1962–1963; Pauling to Joseph Rotblat, 10 May 1962, box 201, Pugwash Continuing Committee; and Glass to Rotblat, 5 March 1963, box 202, Pugwash Continuing Committee 1963, all in Glass Papers; Pauling to Rabinowich, 14 September 1962, box 3.058, folder 58.8, and Russell to Rabinowich, 8 August 1963, "Linus Pauling's Safe Drawer 2," folder 2.002, both in Pauling Papers.

48. Joseph Rotblat, "Memorandum on Future Activities and Organization," July 1962, box 201, Pugwash Continuing Committee, Glass Papers.

49. The only historian to have written about SADS describes it as a separate entity from Pugwash. From an administrative standpoint, this is true. The AAAS, which managed the finances of both organizations, maintained separate accounts and reports for Pugwash and the Committee on International Studies of Arms Control (which oversaw SADS). Evangelista, *Unarmed Forces*, 35–39; "Reports at Annual Meeting 8 May 1963," *Records of the Academy (American Academy of Arts and Sciences)* 1962/63 (1963): 103–62. The process of receiving permission to travel abroad was complicated and centralized for Soviet scientists during this time, even for Party members. See Maria A. Rogacheva, *The Private World of Soviet Scientists from Stalin to Gorbachev* (New York: Cambridge University Press, 2017), 161–65.

50. Donald F. Chamberlain to Deputy Director for Science and Technology [SECRET/declassified]," 28 May 1964, Document Number 0001303416, CIA Electronic Reading Room, https://www.cia.gov/library/readingroom/docs/DOC_0001303416.pdf. For the institutional obstacles to holding meetings, see "Reports at Annual Meeting 8 May 1963"; "Reports at Annual Meeting 13 May 1964," *Records of the Academy (American Academy of Arts and Sciences)* 1963/64 (1964): 113–55; Bernard Feld, "Report of the Committee on Pugwash Conferences on Science and World Affairs," *Records of the Academy (American Academy of Arts and Sciences)* 1964/65 (1965): 15; Paul Doty, "Report of the Committee on International Studies of Arms Control," *Records of the Academy (American Academy of Arts and Sciences)* 1965/66 (1966): 15–17.

51. "Debriefing Paul Doty and Marshall Shulman on Belgrade Pugwash," 15 October 1963, box 2, Pugwash, A1 3008E, RG 59, NARA II; Llewellyn Thompson to Bernard Feld, 1 December 1964, box 199, COWSA 1964, Glass Papers.

52. "Treaty Banning Nuclear Weapon Tests in the Atmosphere, in Outer Space and Under Water," US–UK–USSR, 5 August 1963, http://www.state.gov/t/isn/4797.htm.

53. Maddock, *Nuclear Apartheid*, 285.

54. Hager, *Force of Nature*, 545–59.

55. Evangelista, *Unarmed Forces*, 39–40, 82–84; David Holloway, "Physics, the State, and Civil Society in the Soviet Union," *Historical Studies in the Physical and Biological Sciences* 30, no. 1 (January 1999): 173–92, https://doi.org/10.2307/27757823. I return to this issue in chapter 9.

7 : Developing Scientific Minds

This chapter is adapted from Audra J. Wolfe, "What's in a Zone? Biological Order versus National Identity in the Biological Sciences Curriculum Study," in *Global Transformations in the Life Sciences, 1945–1980*, edited by Patrick Manning and Mat Savelli. © 2018. Reprinted by permission of the University of Pittsburgh Press.

1. Lyman Hoover to the President [Asia Foundation], "Project Proposal: Biological Sciences Curriculum Study Yellow Version Biology Adaptation Project," 17 March 1966, box P-246, Science BSCS (Grobman) VII, TAF.

2. John L. Rudolph, *Scientists in the Classroom: The Cold War Reconstruction of American Science Education* (New York: Palgrave, 2002); William V. Mayer, "Biology Education in the United States during the Twentieth Century," *Quarterly Review of Biology* 61, no. 4 (1986): 481–507, https://doi.org/10.1086/415145.

3. FOIA Doc_0001088621 (email), 6 February 1990, http://www.foia.cia.gov/sites/default/files/document_conversions/89801/DOC_0001088621.pdf. Ongoing information on the Asia Foundation's overall budget is hard to come by, but its 1967 budget was $8.5 million. "Statement by Haydn Williams, President, the Asia Foundation, before the Rusk Commission, June 23, 1967," 23 June 1967, COVA, Packet #1, box 192, White House Central Files, Lyndon Baines Johnson Presidential Library (hereafter LBJWHCF).

4. Hal Goodwin especially pushed the idea that science textbooks should be given equal weight as "impressive technology" in US propaganda efforts. See, for example, Harold Goodwin to Mr. Harkness, " 'Project Proposals' Included in the Paper on Development of a Science Program," 28 February 1958, box 3, IOP/A Basic Paper 1958, P 243, RG 306, NARA II.

5. James S. Lay, "Technological Superiority [CONFIDENTIAL/declassified]," 21 June 1956, box 12, Strengthening the Free World Position in Science and Technology, A1 1568F, RG 59, NARA II.

6. Jurgen Herbst, *The Once and Future School: Three Hundred and Fifty Years of American Secondary Education* (New York: Routledge, 1996); Andrew Hartman, *Education and the Cold War: The Battle for the American School* (New York: Palgrave Macmillan, 2008), 8.

7. *General Education in a Free Society: Report of the Harvard Committee* (Cambridge, MA: Harvard University Press, 1945); US Department of Education, *Life Adjustment Education for Every Youth*, Bulletin no. 22 (Washington, DC: Government Printing Office, 1951); Hartman, *Education and the Cold War*, 56–71.

8. James B. Conant, *On Understanding Science* (New Haven, CT: Yale University Press, 1947); *Education in a Divided World: The Function of Public Schools in Our Unique Society* (Cambridge: Harvard University Press, 1948); *Science and Common Sense* (New Haven, CT: Yale University Press, 1951); *Modern Science and Modern Man* (New York: Columbia University Press, 1952); and *Education and Liberty: The Role of the Schools in a Modern Democracy* (Cambridge, MA: Harvard University Press, 1953). Quotation from James B. Conant, *The American High School Today: A First Report to Interested Citizens* (New York: McGraw Hill, 1959), 15.

9. Rudolph, *Scientists in the Classroom*; Christopher J. Phillips, *The New Math: A Political History* (Chicago: University of Chicago Press, 2015); Jerome S. Bruner, *The Process of Educa-*

tion (Cambridge, MA: Harvard University Press, 1960), 14. On Conant's looming influence over these studies, see George Reisch, "When *Structure* Met Sputnik: On the Cold War Origins of *The Structure of Scientific Revolutions*," in Oreskes and Krige, *Science and Technology in the Global Cold War*, 371–92. For a longer history of approaches to teaching science in US schools, see John L. Rudolph, *How We Teach Science: What's Changed and Why It Matters* (Cambridge, MA: Harvard University Press, 2019).

10. Divine, *Sputnik Challenge*, 89–93; Barbara Barksdale Clowse, *Brainpower for the Cold War: The Sputnik Crisis and the National Defense Education Act of 1958* (Westport, CT: Greenwood Press, 1981).

11. Rudolph, *Scientists in the Classroom*; Jack S. Goldstein, *A Different Sort of Time: The Life of Jerrold R. Zacharias* (Cambridge, MA: MIT Press, 1992), 155–90.

12. Walter Auffenberg, "BSCS Newsletter," no. 1 (1959): 1; Mayer, "Biology Education in the United States during the Twentieth Century"; Biological Sciences Curriculum Study, *BSCS Green Version High School Biology* (Chicago: Rand McNally, 1963); Biological Sciences Curriculum Study, *Biological Science: Molecules to Man [BSCS Blue Version]* (Boston: Houghton Mifflin, 1963); Biological Sciences Curriculum Study, *Biological Science: An Inquiry into Life [BSCS Yellow Version]* (New York: Harcourt, Brace, and World, 1963).

13. Arnold B. Grobman, "The Biological Sciences Curriculum Study," *AIBS Bulletin*, no. 2 (1959): 21–23.

14. BSCS, "Impact Test; Also a Study of Student Opinion," BSCS, 1962, box 174, Evaluation Committee (2), Glass Papers.

15. "Objectives to Be Tested with the BSCS Impact Test," n.d. 1962, box 174, Evaluation Committee (1), and "Objectives of the BSCS Impact Test," 25 May 1962, box 174, Evaluation Committee (6), both in Glass Papers.

16. Harry Kelly, "Press Release," 14 August 1961, box 149, BSCS July–Dec 1961, and John Moore, "Different Objectives of Elementary, High School, and College Education," 16 July 1959, box 137, BSCS, 1959, both in Glass Papers.

17. Emphasis in original. Bentley Glass, "The Pervading Biological Themes," 21 June 1961, box 147, Guidelines and Themes, Glass Papers.

18. BSCS, "High School Biology: Blue Version, Parts 1–3," 1961, box 146, BSCS Textbook Drafts, Glass Papers.

19. Ibid., 28–2. (Note that page numbers are nonconsecutive in this experimental edition of the Blue Version.)

20. Ibid., 28–3, 19–2, 20–22, 20–25, 20–10, 20–24.

21. BSCS, "High School Biology: Blue Version, The Laboratory (Teachers' Guide), Parts 1–3," 1961, n.p. [ix], box 146, BSCS Textbook Drafts, Glass Papers.

22. BSCS, "High School Biology: Blue Version, The Laboratory, Parts 1–3," 1961, xv, 201, 205, box 146, BSCS Textbook Drafts, Glass Papers.

23. Harold Goodwin to Elbert P. Little, 4 March 1958, box 5, MIT, P 243, RG 306, NARA II; Goodwin to Harkness, "Project Proposals"; Arnold B. Grobman, "Notes on the BSCS and a Trip to the Orient," 10 October 1962, box 135, BSCS, Glass papers; BSCS International Cooperation Committee, "Condensed Proceedings of the Meeting of November 11 and 12, 1965, at the Mayflower Hotel," 13 January 1966, box P-246, Science BSCS (Grobman) VII, TAF.

24. "Support of Scientific Activities of Several Countries by U.S. Private Foundations," 1961, box 5, National Science Foundation, A1 3008E, RG 59, NARA II; PSAC, "The Ford Foundation Programs," 27 November 1961, box 51, PSAC Panel on International Science, NSFW, RG 307, NARA II; Ford Foundation, "Latin American and Caribbean Program, List of Grants," 15 April 1965, series 200, RG 2.1965, box 16, folder 1, RF; Rockefeller Foundation, "Grant Files RF 59191, 60214-MNS, 62121-MNS," 25 January 1960, series 200, RG 1.2, box 5, folder 36, RF. I am not suggesting that the Ford and Rockefeller foundations were operating with covert government funds. Rather, the point is that the boards and policies of these two foundations were so deeply intertwined with US foreign policy that the exact funding streams are not of great importance for the purposes of cultural diplomacy. See, especially, Inderjeet Parmar, *Foundations of the American Century: The Ford, Carnegie, and Rockefeller Foundations in the Rise of American Power* (New York: Columbia University Press, 2012).

25. "Status Report: Committee for a Free Asia," 9 October 1951; John Devine to Edward W. Barrett [SECRET/declassified], 5 November 1951; and James L. Moader to Mr. Edward W. Barrett [SECRET/declassified]," 14 December 1951, all in box 66, Committee for a Free Asia, A1 1587M, RG 59, NARA II.

26. "The Asia Foundation: A Statement of the Foundation's Purposes and Activities," November 1954, box 80, folder 7, IACF; Jesse M. MacKnight to Mr. Vaughn DeLong, "List of TAF Field Representatives [SECRET/declassified]," 8 June 1955, box 66, Committee for a Free Asia, 1955–1957, A1 1587M, RG 59, NARA II.

27. Robert Blum, "Committee for a Free Asia Executive Committee Report: August 1954 [SECRET/declassified]," 18 August 1954, box 66, Committee for a Free Asia, 1954, A1 1587M, RG 59, NARA II; "Summary of BSCS Participation in Asian Biology," 1968, box P-478, BSCS Materials X, TAF. For the broader phenomenon of books as US propaganda, see Barnhisel, "Cold Warriors of the Book."

28. Marty Fleischmann to Robert S. Schwantes and William L. Eilers, 25 September 1963, and Fleischmann to the Representative, Japan, the Philippines, and the Republic of China, 4 October 1963, both in box P-246, folder: Science BSCS (Grobman) IV, TAF.

29. This assessment is based on in-depth browsing of TAF's records on science programming at the Hoover Institution Library and Archives.

30. Arnold B. Grobman, "A Proposal Submitted by the American Institute of Biological Sciences on Behalf of the Biological Sciences Curriculum Study for Support of a Program to Adapt Recent Revisions in Biological Curricula for Potential Use in Several Foreign Countries," 15 March 1961, box 35, BSCS Foreign Utilization Program, 1961–1963" (35:11), Archie F. Carr Papers, George A. Smathers Library, University of Florida (hereafter Carr Papers); James G. Dickson to Hiden T. Cox, 15 May 1961, and Dickson to Philip W. Hemily, 19 May 1961, both in box 171, BSCS Committee on Foreign Utilization, Glass Papers.

31. Arnold B. Grobman, "International Utilization of Curriculum, BSCS," February 1961, box 169, BSCS Steering Committee; Grobman [Untitled document summarizing BSCS funding status], 1963, box 154, BSCS 1963 (2); Grobman to Dr. John M. Weir, 22 February 1963, box 168, Glass-3; Glen E. Peterson, "Current Status of BSCS Program of International Cooperation," 30 June 1966, box 177, BSCS 1966; and BSCS International Cooperation Committee, "BSCS International Cooperation," 17 September 1968, box 135, International Coopera-

tion Committee, all in Glass Papers; BSCS International Cooperation Committee, "Agenda: International Cooperation Committee Meeting, May 13–14, 1971," 3 May 1971, box P-478, BSCS Materials XI, TAF; William Mayer, "Request for a General Services Grant for the International Cooperation Program of the Biological Sciences Curriculum Study," n.d. 1965, box 177, BSCS, 1964–1966, Glass Papers.

32. Arnold B. Grobman, "Summary of the Tenth BSCS Steering Committee Meeting, January 10–11, 1964," 24 February 1964, box 155, BSCS 1963–1964, Glass Papers.

33. Arnold B. Grobman, "Status of Adaptations of BSCS Versions in Overseas Nations as of November, 1963," 26 November 1963, box 154, BSCS 1963, Glass Papers; Grobman, "Extent of BSCS International Activities," 16 April 1964, box 35, BSCS International Cooperation Committee, 1962–1964 (35:14), Carr Papers.

34. Arnold B. Grobman, "BSCS International News Notes No. 2," 17 June 1963, box 169, Glass-1, Glass Papers.

35. Arnold B. Grobman to Sherman A. Hoslett, 26 March 1963, box P-246, Science BSCS (Grobman) III, TAF.

36. Feng Kuo-Ao to Arnold B. Grobman, 22 April 1963, box 157, BSCS, April–June 1963, Glass Papers; Robert E. Levine, "Memo to the File Re: Record of Visit with Miss Sheng-Hing Lee and Dr. Ting-Pong Koh [sic]," 26 May 1965; Koh Ting-pong to John F. Sullivan, 24 June 1965; and Koh to Clare Humphrey, 12 October 1965, all in box P-246, Science BSCS (Grobman) VI, TAF.

37. William Mayer to Edith S. Coliver, 17 December 1965, box P-246, Science BSCS (Grobman) VI, TAF.

38. Hoover to the President [Asia Foundation], "Project Proposal"; Taipei to San Francisco, telegram, 21 April 1966, and "Approved Project No. 1030: Publication of BSCS Yellow Version Adaptation," 22 April 1966, both in box P-246, Science BSCS (Grobman) VII, TAF; The Representative, Republic of China to the President [Asia Foundation], 9 November 1970, box P-478, BSCS Materials XI, TAF.

39. Arnold Grobman to Keith Kelson, 27 January 1964, box 154, BSCS, 1964 (2), Glass Papers.

40. Grobman, "Summary of the Tenth BSCS Steering Committee Meeting"; The Asia Foundation, "Request for Payment to the American Institute of Biological Sciences," 9 December 1960, and "Bio Data for Ralph Waldo Gerard," 1961, both in box P-253, US Individuals, 1962–1967, Gerard, Dr. Ralph, TAF; William Mayer to Archie Carr, "Minutes of the International Cooperation Committee Meeting of February 17, 1964, in Long Beach, California," 10 March 1964, box 35, BSCS International Cooperation Committee, 1962–1964 (35:14), Carr Papers.

41. Arnold B. Grobman, "Memo on Relaxed BSCS Adaptation Policies," 16 March 1964, box 155, BSCS 1963–1964, Glass Papers; Glen E. Peterson, "Summary of International Cooperation Committee Meeting, June 4 and 5, 1965," 23 June 1965, box 170, BSCS Executive Committee 1965, Glass Papers.

42. Scholarly essays on the purpose of biology education abound; an excellent introduction is Philip J. Pauly, *Biologists and the Promise of American Life: From Meriwether Lewis to Alfred Kinsey* (Princeton, NJ: Princeton University Press, 2000), 171–93. For thinking through the mismatch between US and Third World goals for development, I have found the following

works helpful: Kapil Raj, "Beyond Postcolonialism . . . and Postpositivism: Circulation and the Global History of Science," *Isis* 104, no. 2 (2013): 337–47, https://doi.org/10.1086/670951; Suman Seth, "Putting Knowledge in Its Place: Science, Colonialism, and the Postcolonial," *Postcolonial Studies* 12, no. 4 (December 2009): 373–88, https://doi.org/10.1080/1368879090 3350633; essays in Robert J. McMahon, ed., *The Cold War in the Third World* (New York: Oxford University Press, 2013); and essays in Carol E. Harrison and Ann Johnson, eds., "National Identity: The Role of Science and Technology," *Osiris* 24 (2009).

43. Richard R. Tolman, "Minutes of the Meeting of the BSCS International Cooperation Committee, November 6–7, 1969," 19 November 1969, box P-478, BSCS Materials 7/67, TAF; V. Basnayake, "Report on Study Tours by Dr. V. Basnayake: Parts II and III: BSCS Adaptations," 1964, box P-246, Science BSCS (Grobman) IV, TAF; Evelyn Wares to the Representative, Washington, 26 September 1968, box P-478, BSCS Materials X, TAF; Abena Dove Osseo-Asare, " Scientific Equity: Experiments in Laboratory Education in Ghana," *Isis* 104 no. 4 (2013): 713–41, https://doi.org/10.1086/674941.

44. Tolman, "Minutes of the Meeting of the BSCS International Cooperation Committee."

45. Philip K. Page to Lin Sloan, 31 March 1966, box P-246, Science BSCS (Grobman) VII, TAF. For more on US views of Asians, see Christina Klein, *Cold War Orientalism: Asia in the Middlebrow Imagination, 1945–1961* (Berkeley: University of California Press, 2003).

46. The Asia Foundation, "TAF/Hornig Report Study," July 1967, box P-478, General Information Papers Occasional Papers, 7/67, TAF.

47. This book cannot address the question of "wittingness," that is, the extent to which the Asia Foundation's employees knew of its relationship to the CIA. Correspondence in the foundation's archives show that few if any of the country representatives resigned their positions after the ties to the CIA were revealed in the *New York Times* (Wallace Turner, "Asia Foundation Got CIA Funds," *NYT*, 22 March 1967), which suggests that the news wasn't exactly surprising. The country representatives did, however, seem to suffer a crisis of confidence in the immediate aftermath of the revelations when the home office indicated that foundation employees would have to work more closely with USAID and the State Department to ensure the foundation's continued existence. See, for example, the discussion in Louis Lazaroff, "Staff Memorandum No. 69/69: Research Proposal to AID," 28 May 1969, 69, box P-479, Staff Memorandum 51–100, 1969, TAF.

48. For a study of this phenomenon that focuses specifically on education, see Osseo-Asare, "Scientific Equity." India's Nehru made particularly good use of scientific nationalism, although his most high-profile investment—the Indian atomic bomb—was achieved largely without foreign assistance. See Robert S. Anderson, *Nucleus and Nation: Scientists, International Networks, and Power in India* (Chicago: University of Chicago Press, 2010); and Itty Abraham, *The Making of the Indian Atomic Bomb: Science, Secrecy, and the Postcolonial State* (New York: Zed Books, 1998). For the situation in Taiwan, see J. Megan Greene, *The Origins of the Developmental State in Taiwan: Science Policy and the Quest for Modernization* (Cambridge, MA: Harvard University Press, 2008). China and the Soviet Union engaged in their own competition for Third World loyalties via modernization programs; see Jeremy Friedman, *Shadow Cold War: The Sino-Soviet Competition for the Third World* (Chapel Hill: University of North Carolina Press, 2015).

49. "Summary of BSCS Participation in Asian Biology"; BSCS, "Status of BSCS Adaptation Program," 3 May 1971, box P-478, BSCS Materials XI, TAF; Turner, "Asia Foundation Got CIA Funds"; Edith S. Coliver to Director, Development, via Mr. Schwantes, 19 January 1971, box P-478, BSCS Materials XI, TAF; Robert Goeckermann to William V. Mayer, 31 October 1973, box 6, Mayer, William (2), Arnold B. Grobman Papers, Smithsonian Institution Archives (hereafter Grobman Papers).

50. BSCS, "Status of BSCS Adaptation Program."

51. For the USIA's changing goals, contrast US Information Agency, "Basic Guidance and Planning Paper No. 4: Science and Technology," 18 November 1962, box 12, USIA 1962, with Simon Bourgin, "Agency Policy Paper on Science and Technology," 7 March 1968, box 15, Science and Technology Paper, both in P 243, RG 306, NARA II.

8 : An Unscientific Reckoning

1. For background on the National Student Association and *Ramparts*, respectively, see Karen M. Paget, *Patriotic Betrayal: The Inside Story of the CIA's Secret Campaign to Enroll American Students in the Crusade against Communism* (New Haven, CT: Yale University Press, 2015); and Peter Richardson, *A Bomb in Every Issue: How the Short, Unruly Life of Ramparts Magazine Changed America* (New York: New Press, 2009).

2. E. W. Kenworthy, "Triple Pass: How the CIA Shifts Funds," *NYT*, 19 February 1967; "House Panel to Reopen Tax-Exempt Fund Study," *NYT*, 22 February 1967; Braden, "I'm Glad the CIA Is 'Immoral,'" 14.

3. Saunders, *Cultural Cold War*, 135.

4. United Press International, "Patman Attacks 'Secret' CIA Link," *NYT*, 1 September 1964; US House Select Committee on Small Businesses, *Tax-Exempt Foundations: Their Impact on Small Business. Hearings before Subcommittee No. 1 on Foundations*, 21 September 1964, https://www.cia.gov/library/readingroom/docs/CIA-RDP67B00446R000300020094-4.pdf. Though rumored for years, it took the shock of the Watergate scandal to bring the CIA's relationship with the media fully to light. Two essential accounts of that relationship are Carl Bernstein, "The CIA and the Media," *Rolling Stone*, 20 October 1977; and *The CIA and the Media: Hearings before the Subcommittee on Oversight of the Permanent Select Committee on Intelligence, House of Representatives, Ninety-Fifth Congress, First and Second Sessions . . .* (Washington, DC: Government Printing Office, 1978). For a readable account of the personal relationships on which these arrangements were built, see Gregg Herken, *The Georgetown Set: Friends and Rivals in Cold War Washington* (New York: Knopf, 2014).

5. For a concise account of the NSA's relationship with the CIA, see Wilford, *Mighty Wurlitzer*, 128–48, 238–39. For a more richly detailed study, including a list of witting and unwitting NSA officers, see Paget, *Patriotic Betrayal*.

6. Richardson, *Bomb in Every Issue*, 4, 49–50, 74–81.

7. Douglas Martin, "Nicholas Katzenbach, 1960s Political Shaper, Dies at 90," *NYT*, 9 May 2012; Nicholas Katzenbach to Mr. President, 13 February 1967, box 44, Ramparts—NSA—CIA, National Security Files, Lyndon Baines Johnson Presidential Library (hereafter LBJNSF).

8. Department of State to All American Diplomatic Posts, "State 137161 Circular," 14 February 1967; "State 138199 Circular," 15 February 1967; "State 140124 Circular," 17 February 1967; and "State 142988 Circular," 23 February 1967, all in box 44, Ramparts—NSA—CIA, LBJNSF.

9. Department of State to All Diplomatic Posts, "State 142988 Circular." See also Appendix C, "Coordination and Policy Approval of Covert Operations [SECRET/declassified]," attached to Nicholas Katzenbach to Mr. President, "Report of Your Committee on CIA Relations with Private Voluntary Organizations [SECRET/declassified]," 24 March 1967, box 193, COVA, Packet #4, LBJWHCF.

10. Department of State to All American Diplomatic and Consular Posts, "Soviet Propaganda Treatment of CIA Funding Disclosures," 14 April 1967, box 44, Ramparts—NSA—CIA, LBJNSF. Books that credit/blame the CIA for driving US adventurism continue to find reputable publishers.

11. Hoover Institution Archives Staff, "File List to the Asia Foundation Records, US and International Series, 1962–," 2011, Hoover Institution Library and Archives, Stanford, CA.

12. Katzenbach to Mr. President, "Report of Your Committee"; The White House, "Statement by the President," 29 March 1967, box 44, Ramparts—NSA—CIA, LBJNSF.

13. "Presentation to the Rusk Committee: The Problem and Alternatives," 6 May 1967, box 192, COVA, Packet #2, LBJWHCF; Peter Jessup to Mr. President, "The Future of Radio Free Europe and Radio Liberty, a Summary [SECRET/declassified]," 25 September 1967, box 44, Radio Free Europe, LBJNSF.

14. Turner, "Asia Foundation Got CIA Funds"; "Extract: Minutes of the 303 Committee Meeting, 8 July 1966 [SECRET/declassified]," 8 July 1966, and Peter Jessup to Bill Moyers and W. W. Rostow [SECRET/declassified], 26 July 1966, both in Intelligence Files, box 2, 303 Committee, LBJNSF.

15. Trustees of the Asia Foundation, "Trustees' Statement," 21 March 1967, series 200, RG 2.1967, box 12, folder 699, RF.

16. Nicholas Katzenbach to Dean Rusk, telegram, 21 March 1967, box 44, Ramparts—NSA—CIA, LBJNSF; "Press Clipping: Press Institute to Return Asia Foundation Grant," 26 April 1967, series 200, RG 2.1967, box 12, folder 699, RF; "Document 176: Memorandum from the Central Intelligence Agency to the 303 Committee (April 12, 1967) [SECRET; EYES ONLY/declassified]," in *FRUS, 1964–1968*, vol. 10; "Statement by Haydn Williams, President, The Asia Foundation, before the Rusk Commission, June 23, 1967," box 192, COVA, Packet #1, LBJWHCF.

17. "Agenda—Meeting of the Executive Members of the Rusk Committee," 3 May 1967; Dean Rusk to Mr. President, "Committee on Overseas Voluntary Activities," 10 July 1967; "Interviews Conducted by the Members and Staff of the Committee on Overseas Voluntary Activities," n.d. 1967; and "Minutes of the Committee on Overseas Voluntary Organizations, July 22, 1967," 22 July 1967, all in box 192, COVA, LBJWHCF.

18. "Alternatives for Supporting U.S. Private Voluntary Activities Overseas [CONFIDENTIAL/declassified]," 3 May 1967, box 192, COVA, Packet #2, LBJWHCF.

19. "Report to the President of the Committee on Overseas Voluntary Associations," 4 June 1968, box 192, COVA, Packet #1, LBJWHCF.

20. "Minutes of the Committee on Overseas Voluntary Organizations, July 22, 1967."

21. For the most extreme category, political assassinations, see Frank Church, Chairman, US Senate, *Alleged Assassination Plots Involving Foreign Leaders: An Interim Report of the Select Committee to Study Governmental Relations with Respect to Intelligence Agencies* (Washington, DC: Government Printing Office, 1975). For a masterly overview of the Cold War beyond the US-Soviet contest, see Westad, *The Cold War*. Good starting points for the literature on US development policies include Michael E. Latham, *The Right Kind of Revolution: Modernization, Development, and U.S. Foreign Policy from the Cold War to the Present* (Ithaca, NY: Cornell University Press, 2011); Nils Gilman, *Mandarins of the Future: Modernization Theory in Cold War America* (Baltimore: Johns Hopkins University Press, 2003); David Ekbladh, *The Great American Mission: Modernization and the Construction of an American World Order* (Princeton, NJ: Princeton University Press, 2010); and Wolfe, *Competing with the Soviets*, 55–73. On Rostow's role in all of this, see David Milne, *America's Rasputin: Walt Rostow and the Vietnam War* (New York: Hill and Wang, 2008).

22. For a history of the Vietnam War that looks out from North Vietnam, see Lien-Hang Nguyen, *Hanoi's War: An International History of the War for Peace in Vietnam* (Chapel Hill: University of North Carolina Press, 2012). On the 1968 riots and their aftermath, see Clay Risen, *A Nation on Fire: America in the Wake of the King Assassination* (Hoboken, NJ: John Wiley and Sons, 2009); for white Americans' response to the later civil rights movement, see Carol Anderson, *White Rage: The Unspoken Truth of Our Racial Divide* (New York: Bloomsbury, 2016), 99–118.

23. Peter Jessup, "Minutes of the Meeting of the 303 Committee, 15 December 1967," 15 December 1967, Intelligence File, box 2, 303 Committee, LBJNSF; "Document 209: Memorandum from the Deputy Director of Coordination, Bureau of Intelligence and Research, to the Assistant Secretary of State for East Asian and Pacific Affairs (June 27, 1968) [SECRET; EYES ONLY/declassified]," in *FRUS 1964–1968*, vol. 10; James N. Hyde to Datus C. Smith, 26 June 1968, series 3.1, box 114, folder 775, Rockefeller Brothers Fund Archives, Rockefeller Archive Center, Sleepy Hollow, NY; Haydn Williams to U. Alexis Johnson, 4 April 1969, box 94, CIA Orphans—Rusk Committee '69 (Asia Foundation), A1 1550, RG 59, NARA II; "A Report to the President on the Future of the Asia Foundation," 1 August 1969, box 45, Asia Foundation, [1969], A1 1550, RG 59, NARA II; Williams to Hyde, 29 January 1970, and Hyde to John D. Rockefeller III, 1 April 1970, both in series 3.1, box 114, folder 776, Rockefeller Brothers Fund Archives; FOIA Doc_0001088621 (email), 6 February 1990; Asia Foundation, "Annual Report 2016," http://asiafoundation.org/about/annual-report-2016.

24. Seymour M. Hersh, "Huge C.I.A. Operation Reported in U.S. against Anti-War Forces, Other Dissidents in Nixon Years," *NYT*, 22 December 1974; Frank Church, Chairman, US Senate, *Intelligence Committees and the Rights of Americans: Book II, Final Report of the Select Committee to Study Governmental Operations with Respect to Intelligence Activities* (Washington, DC: Government Printing Office, 1976), 6; Weiner, *Legacy of Ashes*, 354–67.

25. Tom Wicker et al., "C.I.A.: Maker of Policy, or Tool?" *NYT*, 25 April 1966; Tom Wicker et al., "How C.I.A. Put 'Instant Air Force' into Congo," *NYT*, 26 April 1966; Tom Wicker et al., "C.I.A. Spies from 100 Miles Up; Satellites Probe Secrets of Soviet," *NYT*, 27 April 1966; Tom Wicker et al., "C.I.A. Operation: A Plot Scuttled," *NYT*, 28 April 1966; Tom

Wicker et al., "The C.I.A.: Qualities of Director Viewed as Chief Rein on Agency," *NYT*, 29 April 1966.

26. Wicker et al., "C.I.A. Spies from 100 Miles Up"; John Kenneth Galbraith et al., "Letter to the Editor," *NYT*, 9 May 1966.

27. Michael Polanyi to Nicolas Nabokov, 9 May 1966, box 318, folder 1, IACF.

28. Francisco Siobil José to Ivan Kats, 22 June 1966, and Denis de Rougemont and Nicolas Nabokov, "Press Release to the Arab Press," 23 June 1966, both in box 317, folder 12, IACF; Elizabeth M. Holt, "'Bread or Freedom': The Congress for Cultural Freedom, the CIA, and the Arabic Literary Journal Hiwar (1962–67)," *Journal of Arabic Literature* 44, no. 1 (2013): 83–102, https://doi.org/10.1163/1570064x-12341257; Berghahn, *America and the Intellectual Cold Wars*, 235–40.

29. Braden, "I'm Glad the CIA Is 'Immoral'"; Michael Polanyi to Raymond Aron, 9 May 1967; Michael Goodwin to M. Polanyi, 31 May 1967; and Shepard Stone to M. Polanyi, 11 December 1967, all in box 6, folder 10, Michael Polanyi Papers, Joseph Regenstein Library, University of Chicago, Chicago, IL.

30. Eugene Rabinowitch to Bernard Feld, 27 February 1967, box 1, F, Rabinowitch Papers (UI).

31. "Edith Coliver, International and S.F. Activist, Dies at 79," *Jewish News of Northern California*, 4 January 2002, https://www.jweekly.com/2002/01/04/edith-coliver-international -and-s-f-activist-dies-at-79; Edith Coliver to Director, Development, via Mr. Schwantes, 19 January 1971, box P-478, BSCS Materials XI, TAF; Arnold B. Grobman to Coliver, 16 August 1972, box 2, Coliver, Edith, Grobman Papers; Grobman to Robert S. Schwantes, 17 May 1971, box 8, S (3), Grobman Papers.

32. Jean Fowler to James Green, 30 August 1965, and Glen Bowersox to Edith S. Coliver, "Foundation Programs in Science," 16 November 1966, both in box P-246, Science, General, TAF; Robert B. Sheeks to James J. Dalton, 24 June 1963, box P-246, Science Reference Materials Program, TAF; Robert B. Sheeks, "Summary Report on Findings during Science Development Field Trip, April and May, 1958," 12 March 1958, box P-128, Social and Economic: Science Trip Reports (Sedwick and Sheeks), TAF; "Notice of Personnel Changes in the Office of the Foreign Secretary," 1 April 1968, box 124, NAS Correspondence, Glass Papers; William L. Eilers to Robert Blum, 14 October 1959, box P-128, Social and Economic: Science Consultant, Robert Stollberg I, TAF. As was typical during the period in which the CIA's covers were in existence, the NAS acknowledged the Asia Foundation's support. See, for example, Office of the Foreign Secretary, National Academy of Sciences, *International Activities of the National Academy of Sciences National Research Council 1964/1965* (Washington, DC: National Academy of Sciences, 1966), 70. On the Pacific Science Board, see Kroll, "The Pacific Science Board."

33. Harrison Brown, "Report of the Foreign Secretary to the Annual Meeting of the National Academy of Sciences, April 1967," 31 March 1967, box 38, NAS, 104th Annual Meeting; Merle Tuve, "Minutes of the Business Session of the National Academy of Sciences Annual Meeting, April 25, 1967," 25 April 1967, box 124, NAS Meeting Minutes; and "Annual Meeting of the Pacific Science Board, Office of the Foreign Secretary, 16 January 1970," 16 January

1970, box 43, Pacific Science Board, all in Glass Papers; Walter Sullivan, "Harold Coolidge, Expert on Exotic Mammals," *NYT*, 16 February 1985.

34. On scientific activism within professional societies in general, see Moore, *Disrupting Science*, 130–57; Sarah Bridger, *Scientists at War: The Ethics of Cold War Weapons Research* (Cambridge, MA: Harvard University Press, 2015), 194–221; and Rohde, *Armed with Expertise*, 90–115. For the special case of anthropology, see Price, *Cold War Anthropology*, 276–300.

35. Wolfe, *Competing with the Soviets*, 23–39, 111, 114.

36. Todd Gitlin, *The Sixties: Years of Hope, Days of Rage* (New York: Bantam Books, 1987); John McMillian and Paul Buhle, eds., *The New Left Revisited* (Philadelphia: Temple University Press, 2003); Matthew Wisnioski, *Engineers for Change: Competing Visions of Technology in 1960s America* (Cambridge, MA: MIT Press, 2012).

37. The essay, originally titled "The Cultural Cold War: A Short History of the Congress for Cultural Freedom," is reprinted in Christopher Lasch, *The Agony of the American Left* (New York: Knopf, 1969), 98, 112.

38. Wolfe, *Competing with the Soviets*, 105–20; Bridger, *Scientists at War*, 155–221; Rohde, *Armed with Expertise*, 90–115; Moore, *Disrupting Science*, 130–57; Matt Wisnioski, "Inside 'the System': Engineers, Scientists, and the Boundaries of Social Protest in the Long 1960s," *History and Technology* 19, no. 4 (December 2003): 313–33, https://doi.org/10.1080/0734151 032000181077.

39. Moore, *Disrupting Science*, 158–89; Sigrid Schmalzer, Daniel S. Chard, and Alyssa Botelho, eds., *Science for the People: Documents from America's Movement of Radical Scientists* (Amherst: University of Massachusetts Press, 2018); Alice Bell, "Science for the People!," *Mosaic* (blog), 27 January 2015, https://mosaicscience.com/story/science-people. On the Black Panther Party's scientific and medical work, see Alondra Nelson, *Body and Soul: The Black Panther Party and the Fight against Medical Discrimination* (Minneapolis: University of Minnesota Press, 2011).

40. Jon Agar, "What Happened in the Sixties?," *British Journal for the History of Science* 41, no. 4 (2008): 570, https://doi.org/10.1017/S0007087408001179.

41. William H. Tucker, *The Funding of Scientific Racism: Wickliffe Draper and the Pioneer Fund* (Urbana: University of Illinois Press, 2002), 140–56. Shockley published his views in many reputable and semi-reputable journals; I choose not to cite them here. For the broader history of racist thought, including scientific racism, in the United States, see Ibram X. Kendi, *Stamped from the Beginning: The Definitive History of Racist Ideas in America* (New York: Nation Books, 2016).

42. "Racial Studies: Academy States Position on Call for New Research," *Science* 158, no. 3803 (967): 892, https://doi.org/10.1126/science.158.3803.892. See also "Documents Related to William Shockley and the NAS," 1969, box 66, Shockley, Glass Papers.

43. Vermont Royster, "The Lysenko Syndrome," *Wall Street Journal*, 22 May 1968; "Documents Related to William Shockley and the NAS."

44. Allen V. Astin, "Minutes of the Business Session of the National Academy of Sciences, Autumn Meeting, 1971," 25 October 1971, box 129, NAS, 1968–1972, Glass Papers; "Biologist, Scoring Secrecy, Quits Academy of Science," *NYT,* 30 April 1971.

45. Bentley Glass to Robert L. Sinsheimer, 26 January 1973, box 56, Correspondence, 1973, and Glass to John J. Valter, 14 June 1974, box 39, Heredity, Race, IQ, both in Glass Papers.

9 : Scientists' Rights Are Human Rights

1. Edward Clinton Ezell and Linda Neuman Ezell, *The Partnership: A History of the Apollo-Soyuz Test Project*, NASA History Series, NASA SP-4209 (Washington, DC: NASA, 1978), 279, 337; NASA, *Apollo-Soyuz Docking: July 17, 1975*, NASA.gov video, 2013, https://www.youtube.com/watch?v=es7Br9kJBbo.

2. Melvyn P. Leffler, *For the Soul of Mankind: The United States, the Soviet Union, and the Cold War* (New York: Hill and Wang, 2007), 248–52; Organization for Security and Cooperation in Europe, "Helsinki Final Act," 1 August 1975, http://www.osce.org/helsinki-final-act.

3. Sarah B. Snyder, *Human Rights Activism and the End of the Cold War: A Transnational History of the Helsinki Network* (New York: Cambridge University Press, 2011).

4. Holloway, "Physics, the State, and Civil Society"; Charles Rhéaume, "Western Scientists' Reactions to Andrei Sakharov's Human Rights Struggle in the Soviet Union, 1968–1989," *Human Rights Quarterly* 30, no. 1 (2008): 1–20, https://doi.org/10.1353/hrq.2008.0004.

5. Jay Bergman, *Meeting the Demands of Reason: The Life and Thought of Andrei Sakharov* (Ithaca, NY: Cornell University Press, 2009), 249; Andrei Sakharov, *Memoirs*, trans. Richard Lourie (New York: Knopf, 1990), 432–36.

6. Benjamin Nathans, "The Dictatorship of Reason: Aleksandr Vol'pin and the Idea of Rights under 'Developed Socialism,'" *Slavic Review* 66, no. 4 (2007): 630–63, https://doi.org/10.2307/20060376; Snyder, *Human Rights Activism*, 56–57.

7. Nathans, "Dictatorship of Reason"; Bergman, *Meeting the Demands of Reason*, 163–66; Barbara J. Keys, *Reclaiming American Virtue: The Human Rights Revolution of the 1970s* (Cambridge, MA: Harvard University Press, 2014), 106–9.

8. Mordechai Altschuler, *Religion and Jewish Identity in the Soviet Union, 1941–1964*, trans. Saadya Sternberg (Waltham, MA: Brandeis University Press, 2012), 1–6, 78–80.

9. Martin Gilbert, *Shcharansky: Hero of Our Time* (New York: Viking, 1986), 13–17, 429; Natan Sharansky, *Fear No Evil*, trans. Stefani Hoffman (New York: Random House, 1988), xii. After his emigration to Israel in 1986, Shcharansky changed his name to Natan Sharansky. I use the spelling "Shcharansky" throughout this chapter for clarity.

10. Soviet dissidents left behind an unusually rich stash of autobiographical writing, almost all of which features repeated scenes with foreign journalists. I have found the following memoirs particularly useful: Sakharov, *Memoirs*; Elena Bonner, *Alone Together*, trans. Alexander Cook (New York: Knopf, 1986); Sharansky, *Fear No Evil*; and Yuri Orlov, *Dangerous Thoughts: Memoirs of a Russian Life*, trans. Thomas P. Whitney (New York: William Morrow, 1991). Dissident memoirs in English published *before* 1989 should be regarded as a variant of *tamizdat* and therefore treated with more than the usual caution. See Benjamin Nathans, "Talking Fish: On Soviet Dissident Memoirs," *Journal of Modern History* 87, no. 3 (2015): 579–614, https://doi.org/10.1086/682413.

11. "Outspoken Soviet Scientist; Andrei Dmitriyevich Sakharov," *NYT*, 22 July 1968; Andrei Sakharov, "Text of Essay by Russian Nuclear Physicist Urging Soviet-American Coopera-

tion; Joint Action by Two Nations Viewed as Essential to Avert Perils Facing Mankind; Basis for Hope Seen in Rapprochement between Socialist and Capitalist Systems [Reflections on Progress, Peaceful Coexistence, and Intellectual Freedom]," trans. *New York Times, NYT,* 22 July 1968 (hereafter, "Reflections"); Bergman, *Meeting the Demands of Reason,* 135–38; Peter Finn and Petra Couvée, *The Zhivago Affair: The Kremlin, the CIA, and the Battle over a Forbidden Book* (New York: Pantheon, 2014), 17.

12. Bergman, *Meeting the Demands of Reason,* 188–210; Sakharov, *Memoirs,* 380–81.

13. Snyder, *Human Rights Activism,* 57–59; Sakharov, *Memoirs,* 456–57; Orlov, *Dangerous Thoughts,* 188–96.

14. Snyder, *Human Rights Activism,* 60–70.

15. Loren Graham, "Summer Review of the Interacademy Exchange between the US and the USSR: A Feasibility Study of a Review of US–USSR Scientific Exchanges and Relations," 30 August 1975, and NAS, "Minutes: BISE Evaluation Panel on US/Soviet Exchange Programs, June 17, 1975," 17 June 1975, both in ADM: IR: Board on International Scientific Exchange Panel: Review of US-USSR Exchanges and Relations: 1975, NAS Archives. For an overview of how the exchange programs evolved over time, see Glenn E. Schweitzer, *Scientists, Engineers, and Track-Two Diplomacy: A Half-Century of U.S.-Russian Interacademy Cooperation* (Washington, DC: National Academies Press, 2004). In 1965, the academy began establishing bilateral exchange programs with other Communist countries. Glenn E. Schweitzer, *Interacademy Programs between the United States and Eastern Europe, 1967–2009: The Changing Landscape* (Washington, DC: National Academies Press, 2009).

16. Harrison Brown to Jerome Wiesner, 20 May 1964, ADM: IR: Exchange Programs: USSR: Interacademy Exchange Agreement: 1964, NAS Archives; Graham, "Summer Review."

17. Commission on International Relations, National Research Council, *Review of the US/ USSR Agreement on Cooperation in the Fields of Science and Technology* (Washington, DC: National Academies Press, 1977); the text of the science and technology agreement is reproduced on pp. 99–102. For a primer on détente, see Gaddis, *Strategies of Containment,* 272–341.

18. Lawrence C. Mitchell to Members, Advisory Committee on USSR & Eastern Europe, "Recent Problems of American Scientists in the USSR," 3 June 1971, ADM: IR: Exchange Programs: USSR: Difficulties of American Scientists in the USSR: 1971, NAS Archives.

19. Ibid.; D. R. Viglierchio to S. G. Korneev, 18 May 1971; and A. Grachev and Yu. Bobrov, "Mr. Russell and Others [As Translated for the NAS]," trans. J. Blanshei, *Literaturnaya Gazeta,* 5 May 1971, all in ADM: IR: Exchange Programs: USSR: Difficulties of American Scientists in the USSR: 1971, NAS Archives.

20. Harrison Brown to Academician M. D. Millionshchikov, 10 June 1971, ADM: IR: Exchange Programs: USSR: Difficulties of American Scientists in the USSR: 1971, NAS Archives.

21. D. R. Viglierchio to Lawrence C. Mitchell, 18 May 1971, ADM: IR: Exchange Programs: USSR: Difficulties of American Scientists in the USSR: 1971, NAS Archives.

22. Brown to Millionshchikov, 10 June 1971.

23. "Status Report of Scientific Exchanges under the 1972–73 Agreement, July 1 to September 30, 1973" (National Academy of Sciences, 1 October 1973), ADM: IR: Exchange Programs: USSR: Status Report on Exchanges: 1973, NAS Archives; W. Murray Todd to Philip Handler, George S. Hammond, and Members of the Council Committee on National Science

Policy, "The National Academy of Sciences: Human Rights and Allied Issues," 1 April 1976, ADM: IR: Human Rights: NAS Position: Background Information: 1976, NAS Archives.

24. "[DRAFT] Minutes of the BISE Review Panel on US-USSR Scientific Exchanges and Relations, Second Meeting, September 16, 1975," 3 October 1975, ADM: IR: Board on International Scientific Exchange Panel: Review of US-USSR Exchanges and Relations: 1975, NAS Archives; "Minutes: BISE Evaluation Panel on US/Soviet Exchange Programs, June 17, 1975."

25. Carl Kaysen, "Results of the Panel's Summer Study and Recommendations for Completion of the Panel's Task in Winter 1975–1976," 29 December 1975, ADM: IR: Board on International Scientific Exchange Panel: Review of US-USSR Exchanges and Relations: Interim Report: 1975, NAS Archives.

26. Ibid.

27. Keys, *Reclaiming American Virtue*; Todd to Handler et al., "Human Rights and Allied Issues."

28. Todd to Handler et al., "Human Rights and Allied Issues."

29. "Council Minutes: Report of the Foreign Secretary," 25 April 1976; US National Academy of Sciences, "An Affirmation of Freedom of Inquiry and Expression," 27 April 1976; and W. Murray Todd to Council of the National Academy of Sciences, 23 June 1976, all in ADM: IR: Human Rights: NAS Position: Affirmation of Freedom of Inquiry and Expression: Responses: General, NAS Archives; Todd to Philip Handler, George Hammond, and Members of the NAS Advisory Committee on Human Rights, 6 December 1976, ADM: IR: Committee on Human Rights Advisory to the NAS Foreign Secretary: 1976, NAS Archives.

30. Philip Handler to Christian Anfinsen, 3 April 1976, ADM: IR: Human Rights: NAS Position: Affirmation of Freedom of Inquiry and Expression: Responses: General, NAS Archives.

31. George S. Hammond to Philip Handler, 3 May 1976, ADM: IR: Committee on Human Rights Advisory to the NAS Foreign Secretary: 1976, NAS Archives; Robert Kates to The NAS Advisory Committee on Human Rights, "Subject of the NAS Council January 23," 7 February 1977; Kates to Members of the National Academy of Sciences, 29 March 1977; and National Academy of Sciences, "Press Release: Academy Announces Actions to Aid Eight Dissident Scientists," 27 April 1977, all in ADM: IR: Committee on Human Rights: General: 1977, NAS Archives.

32. Philip Handler to Robert Kates, 28 March 1977, ADM: IR: Committee on Human Rights: Meetings: 1977, NAS Archives; Christian Philip Peterson, "The Carter Administration and the Promotion of Human Rights in the Soviet Union, 1977–1981," *Diplomatic History* 38, no. 3 (2014): 628–56, https://doi.org/10.1093/dh/dht102; "Document 5: Letter from President Carter to Andrei Sakharov (February 5, 1977)," in *FRUS 1977–1980*, vol. 6; Jimmy Carter, "Inaugural Address," The American Presidency Project, 20 January 1977, http://www.pres idency.ucsb.edu/ws/?pid=6575.

33. Orlov, *Dangerous Thoughts*, 204–24.

34. Gilbert, *Shcharansky*, 185–86.

35. NAS Committee on Human Rights, "Yuriy Fedorovich Orlov," December 1977, Committee on Human Rights: Cases: USSR: Shcharansky A (1977), NAS Archives; "Academy to Campaign Publicly for Oppressed Scientists," *Science* 196, no. 4291 (1977): 741–743, https://

doi.org/10.1126/science.196.4291.741; Herbert R. J. Grosch to Roman A. Rudenko, 4 July 1977, Committee on Human Rights: Cases: USSR: Shcharansky A (1977), NAS Archives; Gilbert, *Shcharansky*, 169–80, 191.

36. Peter S. Pershan to Philip Handler, 21 June 1977; Handler to Leonid Brezhnev, telegram, 15 December 1977; and Robert C. Brewster to Handler, 27 December 1977, all in Committee on Human Rights: Cases: USSR: Shcharansky A (1977), NAS Archives; "U.S. Academy Acted to Aid Shcharansky," *NYT*, 25 December 1977.

37. "Prominent American Physicists Threaten to Boycott Soviets over the Orlov Case," Information Bulletin (New York: Khronika Press, 20 March 1978); Philip Handler to Valentin Turchin, 22 March 1978; and Turchin to Handler, 15 March 1978, all in Committee on Human Rights: Cases: USSR: Orlov (1978), NAS Archives; Valentin Turchin, "Why You Should Boycott the Russians," *Nature* 273 (25 May 1978): 256–57.

38. Kevin Klose, "Orlov Trial Opens; Westerners, Defense Witnesses Barred," *WaPo*, 16 May 1978; Orlov, *Dangerous Thoughts*, 224–31; "U.S. Scientists Cancel Soviet Trip in Protest over the Trial of Orlov," *NYT*, 21 May 1978; "Statement by Philip Handler, President, National Academy of Sciences," 19 May 1978, Committee on Human Rights: Cases: USSR: Orlov (1978), NAS Archives.

39. Gilbert, *Shcharansky*, 275; Charles Petit, "U.S. Science 'Boycott' over Soviet Dissidents," *San Francisco Chronicle*, 25 July 1978; Morris Pripstein to Supporters of Scientists for Shcharansky, 28 July 1978, box 6, Memoranda, 1978–1979, Scientists for Sakharov, Orlov, and Shcharansky Papers, Hoover Institution Library and Archives (hereafter SOS Papers); Scientists for Orlov and Shcharansky to Dear Colleague, 4 October 1978, box 13, General Chairman's Correspondence, 1979–1980, SOS Papers; Gerson Sher to W. Murray Todd, "Aftermath of the Orlov Trial," 22 June 1978, Committee on Human Rights: Cases: USSR: Orlov: Reactions to Orlov Trial (1978), NAS Archives.

40. George S. Hammond to Harvey Averch, 28 February 1978, ADM: IR: Exchange Programs: USSR: Interacademy Exchange Agreement: 1978, NAS Archives.

41. Jack Minker, "Opinion Paper: Science, Shcharansky, and the Soviets," *Journal of the American Society for Information Science*, September 1978, 219–24. For a broader discussion of scientists' attitudes toward boycotting the Soviet Union, see Paul Rubinson, "'For Our Soviet Colleagues': Scientific Internationalism, Human Rights, and the Cold War," in *The Human Rights Revolution: An International History*, ed. Akira Iriye, Petra Goedde, and William I. Hitchcock (New York: Oxford University Press, 2012), 245–64.

42. Scientists for Orlov and Shcharansky, "Background Information," 1 March 1979, box 6, Policy Files, SOS Papers; US National Academy of Sciences, "Press Release: NAS Calls upon World Scientists to Urge Release of Three Soviets But 'Does Not Endorse' Boycott Actions," 21 August 1978, box 26, NAS Press Releases, SOS Papers.

43. "Agreement on Scientific Exchange and Cooperation between the National Academy of Sciences of the USA and the Academy of Sciences of the USSR in 1979 and 1980," 7 February 1979, ADM: IR: Exchange Programs: USSR: Interacademy Exchange Agreement: 1979, NAS Archives; US National Academy of Sciences, "Press Release: NAS Notifies Soviet Academy: All Joint Symposia Suspended for Six Months," 25 February 1980, box 26, NAS Press Releases, SOS Papers; Scientists for Sakharov, Orlov, and Shcharansky, "Press Release: 7900

Scientists from 44 Countries Halt Exchange with Soviets," 16 October 1980, box 6, Memoranda, 1980–1981, SOS Papers. On Sakharov's arrest and exile, see Bergman, *Meeting the Demands of Reason*, 279–301; and Sakharov, *Memoirs*, 507–23.

44. Schweitzer, *Scientists, Engineers, and Track-Two Diplomacy*, 10.

45. Bonner, *Alone Together*, 58, 126–28.

46. Snyder, *Human Rights Activism*, 167–70; Sharansky, *Fear No Evil*, 399–416; Orlov, *Dangerous Thoughts*, 288–305.

47. Orlov, *Dangerous Thoughts*, 301; Bergman, *Meeting the Demands of Reason*, 333–34, 347–60, 383–91.

48. For a Westerner's experience of these seminars, see Glenn E. Schweitzer, *Techno-Diplomacy: U.S. Soviet Confrontations in Science and Technology* (New York: Plenum Press, 1989), 238–39.

49. Holloway, "Physics, the State, and Civil Society"; Nathans, "Dictatorship of Reason"; Rogacheva, *Private World of Soviet Scientists*, 101–106; Paul R. Josephson, *New Atlantis Revisited: Akademgorodok, the Siberian City of Science* (Princeton, NJ: Princeton University Press, 1997), 299.

50. Rogacheva, *Private World of Soviet Scientists*, 108–151. For a revealing picture of the trade-offs in one of the Soviet Union's atomic cities, see Kate Brown, *Plutopia: Nuclear Families, Atomic Cities, and the Great Soviet and American Plutonium Disasters* (New York: Oxford University Press, 2013).

51. Bergman, *Meeting the Demands of Reason*, 55, 60, 66–68; Sakharov, *Memoirs*.

52. Bergman, *Meeting the Demands of Reason*, 95–97; Sakharov, *Memoirs*, 233–35; Medvedev, *Rise and Fall*, 221–27.

53. Bergman, *Meeting the Demands of Reason*, 121–25, 148–49; Sakharov, "Reflections."

54. Walter C. Clemens, Jr., "Sakharov: A Man for Our Times," *BAS*, December 1971. On Sakharov's familiarity with *BAS*, see Evangelista, *Unarmed Forces*, 207. On the Russian Research Center's ties to the CIA, see Sigmund Diamond, *Compromised Campus: The Collaboration of Universities with the Intelligence Community, 1945–1955* (Oxford: Oxford University Press, 1992); and Engerman, *Know Your Enemy*, 43–70.

55. Minker, "Opinion Paper: Science, Shcharansky, and the Soviets," 219.

Epilogue

1. See, for example, David E. Sanger, "No. 2 Negotiators in Iran Talks Argue Physics Is behind Politics," *NYT*, 28 March 2015.

2. *New Frontiers in Science Diplomacy* (London and Washington, DC: The Royal Society and the American Association for the Advancement of Science, 2010). *Science & Diplomacy*, an online journal published by the American Association for the Advancement of Science, is an excellent source for insights into the operating assumptions of practitioners of contemporary science diplomacy. http://www.sciencediplomacy.org.

3. Audra Wolfe, "When Scientists Do What Diplomats Can't," *Atlantic* (online), 26 September 2015, https://www.theatlantic.com/science/archive/2015/09/science-diplomacy/407455. A good introduction to SESAME that simultaneously reproduces the rhetoric of scientific in-

ternationalism is Dennis Overbye, "A Light for Science, and Cooperation, in the Middle East," *NYT*, 8 May 2017. For the role of CERN in US foreign policy, see Krige, *American Hegemony*, 57–73.

4. Jason Zengerle, "Rex Tillerson and the Unraveling of the State Department," *NYT Magazine*, 17 October 2017; Partnership for Public Service, "Political Appointee Tracker," https://ourpublicservice.org/issues/presidential-transition/political-appointee-tracker.php (accessed 19 December 2017); Bill Faries and Mira Rojanasakul, "At Trump's State Department, Eight of Ten Top Jobs Are Empty," *Bloomberg Politics*, 2 February 2018 (updated 13 March 2018), https://www.bloomberg.com/graphics/2018-state-department-vacancies/.

5. Lev Facher, "With Trump in the White House, Obama Science Experts Operate Shadow Network to Press Their Positions," *STAT* (online), 7 August 2017, https://www.statnews.com/2017/08/07/white-house-ostp-obama-trump; Amy B. Wang, "Trump's Science Envoy Quits in Scathing Letter with an Embedded Message: I-M-P-E-A-C-H," *WaPo*, 23 August 2017.

6. US House Committee on Science, Space, and Technology, "@Breitbart News: Global Temperatures Plunge. Icy Silence from Climate Alarmists," 1 December 2016, https://twitter.com/HouseScience/status/804402881982066688. Bannon resigned his position at the White House on 18 August 2017.

7. Brady Dennis, "Scientists Are Frantically Copying U.S. Climate Data, Fearing It Might Vanish under Trump," *WaPo*, 13 December 2016; Steven Mufson and Juliet Eilperon, "Trump Transition Team for Energy Department Seeks Names of Employees Involved in Climate Meetings," *WaPo*, 9 December 2016; Joe Davidson, "Energy Dept. Rejects Trump's Request to Name Climate-Change Workers, Who Remain Worried," *WaPo*, 13 December 2016.

8. Robin McKie, "Death Threats, Intimidation and Abuse: Climate Change Scientist Michael E. Mann Counts the Cost of Honesty," *The Guardian*, 3 March 2012.

9. J. Ama Mantey, "#MarginSci: The March for Science as a Microcosm of Liberal Racism," *The Root* (online), 20 April 2017, http://www.theroot.com/marginsci-the-march-for-science-as-a-microcosm-of-lib-1794463442; Francie Diep, "When Did Science Become Apolitical?," *Pacific Standard* (online), 13 March 2017, https://psmag.com/news/when-did-science-become-apolitical; March for Science, "Our Mission," https://www.marchforscience.com/our-mission (accessed 7 March 2017).

10. For an analysis of how this process played out in practice, see Amy E. Slaton, *Race, Rigor, and Selectivity in U.S. Engineering: The History of an Occupational Color Line* (Cambridge, MA: Harvard University Press, 2010). More broadly speaking, the perspective here that racism is a "normal" part of American life, baked into US legal and social structures, derives from critical race theory. For entry points, see Kimberlé Crenshaw et al., eds., *Critical Race Theory: The Key Writings That Formed the Movement* (New York: New Press, 1995).

11. W. E. B. Du Bois, "The Negro Scientist," *The American Scholar* 8, no. 3 (1939): 314, http://www.jstor.org/stable/41204425. I am grateful to Evelynn Hammonds, whose Distinguished Lecture at the History of Science Society Meeting in November 2016 helped me articulate the problem with postwar scientific "objectivity." Evelynn Hammonds, "'The Negro Scientist': W. E. B. Du Bois and the Diversity Problem in Science and the History of Science" (History of Science Society Meeting, Atlanta, GA, 2016).

12. Du Bois, "Negro Scientist," 320.

13. Wolfe, *Competing with the Soviets*, 24–25.

14. A thoughtful take on the relationship of science studies to social movements is Agar, "What Happened in the Sixties?" Foundational works for the sociology of scientific knowledge are David Bloor, *Knowledge and Social Imagery* (Boston: Routledge and Kegan Paul, 1976); and Bruno Latour and Steve Woolgar, *Laboratory Life: The Construction of Scientific Facts* (1979; Princeton, NJ: Princeton University Press, 1986). Two of the most influential feminist histories of science are Anne Fausto-Sterling, *Myths of Gender: Biological Theories about Women and Men* (New York: Basic Books, 1985); and Londa Schiebinger, *Nature's Body: Gender in the Making of Modern Science* (Boston: Beacon, 1993). For feminist philosophy of science, start with Helen Longino, *Science as Social Knowledge: Values and Objectivity in Scientific Inquiry* (Princeton, NJ: Princeton University Press, 1990); for cultural studies, Donna Haraway, *Primate Visions: Gender, Race, and Nature in the World of Modern Science* (New York: Routledge, 1989).

15. Richard Severo, "William Proxmire, Maverick Democratic Senator from Wisconsin, Is Dead at 90," *NYT*, 16 December 2005; Naomi Oreskes and Erik M. Conway, *Merchants of Doubt: How a Handful of Scientists Obscured the Truth on Issues from Tobacco Smoke to Global Warming* (New York: Bloomsbury Press, 2010).

16. Alan D. Sokal, "Transgressing the Boundaries: Toward a Transformative Hermeneutics of Quantum Gravity," *Social Text*, no. 46/47 (1996): 217, https://doi.org/10.2307/466856; Alan Sokal, "A Physicist Experiments with Cultural Studies," *Lingua Franca*, May/June 1996, 62–64.

17. Sokal, "Transgressing the Boundaries," 229; Dorothy Nelkin, "The Science Wars: Responses to a Marriage Failed," *Social Text*, no. 46/47 (1996): 98, https://doi.org/10.2307/46 6846.

18. Oreskes and Conway, *Merchants of Doubt*; Bruno Latour, "Why Has Critique Run out of Steam? From Matters of Fact to Matters of Concern," *Critical Inquiry* 30, no. 2 (January 2004): 227, https://doi.org/10.1086/421123. I confess to having taken part in this new ritual of self-flagellation.

Selected Bibliography

Archival Sources

American Philosophical Society, Philadelphia, PA
    Milislav Demerec Papers
    Leslie C. Dunn Papers
    Genetics Society of America Records
    H. Bentley Glass Papers
California Institute of Technology Archives, Pasadena, CA
    Oral history interviews
George A. Smathers Library, University of Florida, Gainesville, FL
    Archie F. Carr, Jr., Papers
Harry Ransom Center at the University of Texas–Austin
    Michael Josselson Papers
Harry S. Truman Library, Independence, MO
    Oral history interviews
Hoover Institution Library and Archives, Stanford, CA
    Asia Foundation Records
    Sidney Hook Papers
    Scientists for Sakharov, Orlov, and Shcharansky Papers
Joseph Regenstein Library, University of Chicago, Chicago, IL
    Committee on Science and Freedom Records
    International Association for Cultural Freedom Records
    Michael Polanyi Papers
    Eugene Rabinowitch Papers
Library of Congress, Washington, DC
    Lloyd V. Berkner Papers
    Wallace R. Brode Papers
Lilly Library, Indiana University, Bloomington, IN
    Ralph Cleland Papers
    Hermann J. Muller Papers
Lyndon Baines Johnson Presidential Library, Austin, TX
    Donald F. Hornig Papers
    National Security Files
    White House Central Files

National Academy of Sciences Archives, Washington, DC
Niels Bohr Library and Archives, American Institute of Physics, College Park, MD
    Oral history interviews
Oregon State University Special Collections and Archives, Corvallis, OR
    Ava Helen and Linus Pauling Papers
Rockefeller Archive Center, Sleepy Hollow, NY
    Detlev Bronk Papers
    Ford Foundation Archives
    Rockefeller Brothers Fund Archives
    Rockefeller Foundation Archives
    Nelson A. Rockefeller Papers
Smithsonian Institution Archives, Washington, DC
    Arnold B. Grobman Papers
University of Illinois Archives, Urbana, IL
    Ludwig F. Audrieth Papers
    Eugene Rabinowitch Papers
US National Archives II, College Park, MD
    RG 59, Records of the Department of State
        A1 1549, Lot 58D606 S/SA International Conferences and IGY Records
        A1 1550, Records of U. Alexis Johnson
        A1 1568A, Executive Secretariat of the OCB Records, Chronological File
        A1 1568C, Executive Secretariat of the OCB Records, Subject and Special Files
        A1 1568F, Executive Secretariat of the OCB Records, Subject Files
        A1 1587M, Bureau of Public Affairs, Misc. Records
        A1 1950, Office of the Assistant Secretary for Public Affairs, Office of Public Affairs,
           Office of the Director, Subject Files, 1944–52
        A1 3008A, General Records Relating to Atomic Energy Matters
        A1 3008E, Bureau of International Scientific and Technical Affairs
        P 83, Bureau of International Scientific and Technical Affairs, Science Attaché Files
        UD 2472, Paris Embassy Records
    RG 306, Records of the US Information Agency
        P 243, Office of the Advisor for Science, Space, and the Environment
    RG 307, Records of the National Science Foundation
        NSF Historian's Files
        Subject Files of Alan T. Waterman

Documentary Sources for US Government Materials

American Presidency Project, University of California, Santa Barbara, www.presidency.ucsb.edu
Central Intelligence Agency

CREST (CIA Records Search Tool)
Freedom of Information Act Electronic Reading Room, www.cia.gov/library
US Department of State
Freedom of Information Act Virtual Reading Room, foia.state.gov
Foreign Relations of the United States (FRUS), history.state.gov
FRUS 1945–1950, Emergence of the Intelligence Establishment
FRUS 1948, Volume 4
FRUS 1950, Volume 1
FRUS 1950, Central and Eastern Europe; The Soviet Union, Volume 4
FRUS 1951, Volume 1
FRUS 1950–1955, The Intelligence Community
FRUS 1952–1954, National Security Affairs, Volume 2
FRUS 1955–1957, Foreign Economic Policy; Foreign Information Program
FRUS 1955–1957, United Nations and General International Matters, Volume 11
FRUS 1955–1957, Soviet Union, Eastern Mediterranean, Volume 24
FRUS 1964–1968, Volume 10
FRUS 1977–1980, Volume 6, Soviet Union

Periodicals

Bulletin of the AIBS
Bulletin of the Atomic Scientists
Chemical and Engineering News
New York Times
Records of the Academy (American Academy of Arts and Sciences)
Saturday Review of Literature (*Saturday Review* after 1952)
Science
Science and Freedom
Time
US News and World Report
Washington Post

Articles, Books, and Theses

Abraham, Itty. *The Making of the Indian Atomic Bomb: Science, Secrecy, and the Postcolonial State*. New York: Zed Books, 1998.

Adams, Mark B. "Introduction: Theodosius Dobzhansky in Russia and America." In *The Evolution of Theodosius Dobzhansky: Essays on His Life and Thought in Russia and America*, edited by Mark B. Adams, 3–11. Princeton, NJ: Princeton University Press, 1994.

———. "Networks in Action: The Khrushchev Era, the Cold War, and the Transformation of Soviet Science." In *Science, History, and Social Activism: A Tribute to Everett Mendelsohn*, edited by Garland E. Allen and Roy M. MacLeod, 255–76. Dordrecht, Netherlands: Springer-Science+Business Media, 2001.

Adler, Les K., and Thomas G. Paterson. "Red Fascism: The Merger of Nazi Germany and Soviet Russia in the American Image of Totalitarianism, 1930's–1950's." *American Historical Review* 75, no. 4 (1970): 1046–64. https://doi.org/10.1086/ahr/75.4.1046.

Agar, Jon. "What Happened in the Sixties?" *British Journal for the History of Science* 41, no. 4 (2008): 567–600. https://doi.org/10.1017/S0007087408001179.

Ajzenberg-Selove, Fay. *A Matter of Choices: Memoirs of a Female Physicist.* New Brunswick, NJ: Rutgers University Press, 1994.

Altshuler, Mordechai. *Religion and Jewish Identity in the Soviet Union, 1941–1964.* Translated by Saadya Sternberg. Waltham, MA: Brandeis University Press, 2012.

Anderson, Carol. *White Rage: The Unspoken Truth of Our Racial Divide.* New York: Bloomsbury, 2016.

Anderson, Robert S. *Nucleus and Nation : Scientists, International Networks, and Power in India.* Chicago: University of Chicago Press, 2010.

Aronova, Elena. "The Congress for Cultural Freedom, *Minerva*, and the Quest for Instituting 'Science Studies' in the Age of Cold War." *Minerva* 50, no. 3 (September 2012): 307–37. https://doi.org/10.1007/s11024-012-9206-6.

Badash, Lawrence. "Science and McCarthyism." *Minerva* 38, no. 1 (2000): 53–80. https://doi.org/10.1023/A:1026516820931

Barnes, Trevor. "The Secret Cold War: The CIA and American Foreign Policy in Europe, 1946–1956. Part I." *Historical Journal* 24, no. 2 (1981): 399–415. https://doi.org/10.1017/S0018246X00005537.

Barnhisel, Greg. "Cold Warriors of the Book: American Book Programs in the 1950s." *Book History* 13, no. 1 (2010): 185–217. https://doi.org/10.1353/bh.2010.0010.

Barrett, Edward W. *Truth Is Our Weapon.* New York: Funk and Wagnalls, 1953.

Bartel, Fritz. "Surviving the Years of Grace: The Atomic Bomb and the Specter of World Government, 1945–1950." *Diplomatic History* 39, no. 2 (2015): 275–302. https://doi.org/10.1093/dh/dhu005.

Bell, Alice. "Science for the People!" *Mosaic*, January 27, 2015. https://mosaicscience.com/story/science-people.

Belmonte, Laura A. *Selling the American Way: U.S. Propaganda and the Cold War.* Philadelphia: University of Pennsylvania Press, 2008.

Berghahn, Volker R. *America and the Intellectual Cold Wars in Europe: Shepard Stone between Philanthropy, Academy, and Diplomacy.* Princeton, NJ: Princeton University Press, 2001.

Bergman, Jay. *Meeting the Demands of Reason: The Life and Thought of Andrei Sakharov.* Ithaca, NY: Cornell University Press, 2009.

Berkner, Lloyd V., Douglas Merritt Whitaker, and National Research Council (US). *Science and Foreign Relations: International Flow of Scientific and Technological Information.* General Foreign Policy Series 30. Washington, DC: US Department of State, 1950.

Bernal, J. D. "The Biological Controversy in the Soviet Union and Its Implications." *Modern Quarterly* 4, no. 3 (1949): 203–17.

———. *The Social Function of Science*. Cambridge, MA: MIT Press, 1967. First published in 1939 by George Routledge & Sons, Ltd. (London).

Bernstein, Carl. "The CIA and the Media." *Rolling Stone*, October 20, 1977.

Biological Sciences Curriculum Study. *BSCS Green Version High School Biology*. Chicago: Rand McNally, 1963.

———. *Biological Science: An Inquiry into Life [BSCS Yellow Version]*. New York: Harcourt, Brace, and World, 1963.

———. *Biological Science: Molecules to Man [BSCS Blue Version]*. Boston: Houghton Mifflin, 1963.

Bird, Kai, and Martin J. Sherwin. *American Prometheus: The Triumph and Tragedy of J. Robert Oppenheimer*. New York: Knopf, 2005.

Bloor, David. *Knowledge and Social Imagery*. Boston: Routledge and Kegan Paul, 1976.

Bonner, Elena. *Alone Together*. Translated by Alexander Cook. New York: Knopf, 1986.

Boyer, Paul. *By the Bomb's Early Light: American Thought and Culture at the Dawn of the Atomic Age*. New York: Pantheon, 1984.

Braden, Thomas W. "I'm Glad the CIA Is 'Immoral.'" *Saturday Evening Post*, May 20, 1967.

Brazinsky, Gregg A. *Winning the Third World: Sino-American Rivalry during the Cold War*. Chapel Hill: University of North Carolina Press, 2017.

Bridger, Sarah. *Scientists at War: The Ethics of Cold War Weapons Research*. Cambridge, MA: Harvard University Press, 2015.

Brink, Frank, Jr. *Detlev Wulf Bronk, 1897–1975*. Biographical Memoirs. Washington, DC: National Academy of Sciences, 1978.

Brown, Andrew. *Keeper of the Nuclear Conscience: The Life and Work of Joseph Rotblat*. Oxford: Oxford University Press, 2012.

Brown, Kate. *Plutopia: Nuclear Families, Atomic Cities, and the Great Soviet and American Plutonium Disasters*. New York: Oxford University Press, 2013.

Bruner, Jerome S. *The Process of Education*. Cambridge, MA: Harvard University Press, 1960.

Bukharin, N. I., et al., eds. *Science at the Cross Roads: Papers from the Second International Congress of the History of Science and Technology 1931*. Routledge Revivals. New York: Routledge, 2013.

Bush, Vannevar. *Modern Arms and Free Men: A Discussion of the Role of Science in Preserving Democracy*. New York: Simon and Schuster, 1949.

———. *Science the Endless Frontier: A Report to the President*. Washington, DC: Government Printing Office, 1945. https://www.nsf.gov/od/lpa/nsf50/vbush1945.htm.

Callahan, Angelina Long. "Sustaining Soviet-American Collaboration, 1957–1989." In *NASA in the World: Fifty Years of International Collaboration in Space*, by John Krige, Angelina Long Callahan, and Ashok Maharaj, 127–51. New York: Palgrave Macmillan, 2013.

Carlson, Elof Axel. *Genes, Radiation, and Society: The Life and Work of H. J. Muller*. Ithaca, NY: Cornell University Press, 1981.

Cline, Ray S. *Secrets, Spies, and Scholars: Blueprint of the Essential CIA*. Washington, DC: Acropolis Books, 1976.

Clowse, Barbara Barksdale. *Brainpower for the Cold War: The Sputnik Crisis and the National Defense Education Act of 1958*. Westport, CT: Greenwood Press, 1981.

Cochrane, Rexmond C. *The National Academy of Sciences: The First Hundred Years, 1863–1963*. Washington, DC: National Academy of Sciences, 1978. https://doi.org/10.17226/579.

Coe, Sophia Dobzhansky. "Theodosius Dobzhansky: A Family Story." In *The Evolution of Theodosius Dobzhansky: Essays on His Life and Thought in Russia and America*, edited by Mark B. Adams, 13–28. Princeton, NJ: Princeton University Press, 1994.

Coleman, Peter. *The Liberal Conspiracy: The Congress for Cultural Freedom and the Struggle for the Mind of Postwar Europe*. New York: Free Press, 1989.

Commission on International Relations, National Research Council. *Review of the US/USSR Agreement on Cooperation in the Fields of Science and Technology*. Washington, DC: National Academies Press, 1977. https://doi.org/10.17226/13528.

Conant, James B. *The American High School Today: A First Report to Interested Citizens*. New York: McGraw Hill, 1959.

———. *Education and Liberty: The Role of the Schools in a Modern Democracy*. Cambridge, MA: Harvard University Press, 1953.

———. *Education in a Divided World: The Function of Public Schools in Our Unique Society*. Cambridge, MA: Harvard University Press, 1948.

———. *Modern Science and Modern Man*. New York: Columbia University Press, 1952.

———. *On Understanding Science: An Historical Approach*. New Haven, CT: Yale University Press, 1947.

———. *Science and Common Sense*. New Haven, CT: Yale University Press, 1951.

Conquest, Robert. *The Great Terror: A Reassessment*. New York: Oxford University Press, 1990.

Corke, Sarah-Jane. "History, Historians, and the Naming of Foreign Policy: A Postmodern Reflection on American Strategic Thinking during the Truman Administration." *Intelligence and National Security* 16, no. 3 (September 2001): 146–65. https://doi.org/10.1080/02684520412331306250.

———. *US Covert Operations and Cold War Strategy: Truman, Secret Warfare, and the CIA, 1945–1953*. New York: Routledge, 2008.

Creager, Angela N. H. *Life Atomic: A History of Radioisotopes in Science and Medicine*. Chicago: University of Chicago Press, 2013.

Crenshaw, Kimberlé, Neil Gotanda, Gary Peller, and Kendall Thomas, eds. *Critical Race Theory: The Key Writings That Formed the Movement*. New York: New Press, 1995.

Crossman, Richard, ed. *The God That Failed*. New York: Harper and Brothers, 1949.

Cull, Nicholas J. *The Cold War and the United States Information Agency: American Propaganda and Public Diplomacy, 1945–1989*. New York: Cambridge University Press, 2008.

Curry, Helen Ann. *Evolution Made to Order: Plant Breeding and Technological Innovation in Twentieth-Century America*. Chicago: University of Chicago Press, 2016.

Darling, Arthur B. *The Central Intelligence Agency: An Instrument of Government, to 1950.* University Park: Pennsylvania State University Press, 1990.

David-Fox, Michael. *Showcasing the Great Experiment: Cultural Diplomacy and Western Visitors to the Soviet Union, 1921–1941.* New York: Oxford University Press, 2012.

deJong-Lambert, William. *The Cold War Politics of Genetics Research: An Introduction to the Lysenko Affair.* New York: Springer, 2012.

deJong-Lambert, William, and Nikolai Krementsov. "On Labels and Issues: The Lysenko Controversy and the Cold War." *Journal of the History of Biology* 45, no. 3 (2012): 373–88. https://doi.org/10.1007/s10739-011-9292-6.

Diamond, Sigmund. *Compromised Campus: The Collaboration of Universities with the Intelligence Community, 1945–1955.* New York: Oxford University Press, 1992.

Divine, Robert A. *Blowing on the Wind: The Nuclear Test Ban Debate, 1954–1960.* New York: Oxford University Press, 1978.

———. *The Sputnik Challenge.* New York: Oxford University Press, 1993.

Doel, Ronald E. "Roger Adams: Linking University Science with Policy on the World Stage." In *No Boundaries: University of Illinois Vignettes,* edited by Lillian Hoddeson, 124–44. Urbana: University of Illinois Press, 2004.

———. "Scientists as Policymakers, Advisors, and Intelligence Agents: Linking Contemporary Diplomatic History with the History of Contemporary Science." In *The Historiography of Contemporary Science and Technology,* edited by Thomas Söderqvist, 215–44. Amsterdam: Harwood Academic Press, 1997.

Doel, Ronald E., and Allan A. Needell. "Science, Scientists, and the CIA: Balancing International Ideals, National Needs, and Professional Opportunities." In *Eternal Vigilance? 50 Years of the CIA,* edited by Rhodri Jeffreys-Jones and Christopher Andrew, 59–81. Portland, OR: Frank Cass, 1997.

Du Bois, W. E. B. "The Negro Scientist." *American Scholar* 8, no. 3 (1939): 309–20. http://www.jstor.org/stable/41204425.

Dudziak, Mary L. *Cold War Civil Rights: Race and the Image of American Democracy.* Princeton, NJ: Princeton University Press, 2000.

Dupree, A. Hunter. *Science in the Federal Government.* Baltimore: Johns Hopkins University Press, 1986. First published in 1957 by Harvard University Press.

Egan, Michael. *Barry Commoner and the Science of Survival: The Remaking of American Environmentalism.* Cambridge, MA: MIT Press, 2007.

Ekbladh, David. *The Great American Mission: Modernization and the Construction of an American World Order.* Princeton, NJ: Princeton University Press, 2010.

Ellman, Michael. "Soviet Repression Statistics: Some Comments." *Europe-Asia Studies* 54, no. 7 (2002): 1151–72. https://doi.org/10.1080/0966813022000017177.

Engerman, David C. *Know Your Enemy: The Rise and Fall of America's Soviet Experts.* New York: Oxford University Press, 2009.

Erickson, Paul, Judy L. Klein, Lorraine Daston, Rebecca Lemov, Thomas Sturm, and Michael D. Gordin. *How Reason Almost Lost Its Mind: The Strange Career of Cold War Rationality*. Chicago: University of Chicago Press, 2013.

Evangelista, Matthew. *Unarmed Forces: The Transnational Movement to End the Cold War*. Ithaca, NY: Cornell University Press, 1999.

Ezell, Edward Clinton, and Linda Neuman Ezell. *The Partnership: A History of the Apollo-Soyuz Test Project*. NASA History Series, NASA SP-4209. Washington, DC: NASA, 1978.

Fan, Fa-ti. "The Global Turn in the History of Science." *East Asian Science, Technology, and Society* 6, no. 2 (2012): 249–58. https://doi.org/10.1215/18752160-1626191.

Fausto-Sterling, Anne. *Myths of Gender: Biological Theories about Women and Men*. New York: Basic Books, 1985.

Finn, Peter, and Petra Couvée. *The Zhivago Affair: The Kremlin, the CIA, and the Battle over a Forbidden Book*. New York: Pantheon, 2014.

Fretter, William B. *Robert Bigham Brode*. Biographical Memoirs. Washington, DC: National Academy of Sciences, 1992. http://www.nasonline.org/publications/biographical-memoirs/memoir-pdfs/brode-robert-b.pdf.

Friedman, Jeremy. *Shadow Cold War: The Sino-Soviet Competition for the Third World*. Chapel Hill: University of North Carolina Press, 2015.

Gaddis, John Lewis. *Strategies of Containment: A Critical Appraisal of American National Security Policy during the Cold War*. Revised and expanded edition. New York: Oxford University Press, 2005.

Galison, Peter. "Removing Knowledge." *Critical Inquiry* 31, no. 1 (2004): 229–43. https://doi.org/10.1086/427309.

Galison, Peter, and Barton Bernstein. "In Any Light: Scientists and the Decision to Build the Superbomb, 1952–1954." *Historical Studies in the Physical and Biological Sciences* 19, no. 2 (1989): 267–347. https://doi.org/10.2307/27757627.

General Education in a Free Society: Report of the Harvard Committee. Cambridge, MA: Harvard University Press, 1945.

Gilbert, Martin. *Shcharansky: Hero of Our Time*. New York: Viking, 1986.

Gilman, Nils. *Mandarins of the Future: Modernization Theory in Cold War America*. Baltimore: Johns Hopkins University Press, 2003.

Gimbel, John. *Science, Technology, and Reparations: Exploitation and Plunder in Postwar Germany*. Stanford, CA: Stanford University Press, 1990.

Giroud, Vincent. *Nicolas Nabokov: A Life in Freedom and Music*. New York: Oxford University Press, 2015.

Gitlin, Todd. *The Sixties: Years of Hope, Days of Rage*. New York: Bantam Books, 1987.

Glad, John. "Hermann J. Muller's 1936 Letter to Stalin." *Mankind Quarterly* 43, no. 3 (2003): 305–19.

Goldstein, Jack S. *A Different Sort of Time: The Life of Jerrold R. Zacharias*. Cambridge, MA: MIT Press, 1992.

Gordin, Michael D. *Red Cloud at Dawn: Truman, Stalin, and the End of the Atomic Monopoly.* New York: Farrar, Straus, and Giroux, 2009.

Gormley, Melinda. "Geneticist L. C. Dunn: Politics, Activism, and Community." PhD dissertation, Oregon State University, Corvallis, OR, 2007.

———. "Scientific Discrimination and the Activist Scientist: L. C. Dunn and the Professionalization of Genetics and Human Genetics in the United States." *Journal of the History of Biology* 42, no. 1 (December 2009): 33–72. https://doi.org/10.1007/s10739-008-9170-z.

Goudsmit, Samuel A. *ALSOS.* New York: Henry Schuman, 1947.

Graham, Loren R. *Science in Russia and the Soviet Union: A Short History.* New York: Cambridge University Press, 1993.

Greene, Benjamin P. *Eisenhower, Science Advice, and the Nuclear Test Ban Debate, 1945–1963.* Stanford, CA: Stanford University Press, 2006.

Greene, J. Megan. *The Origins of the Developmental State in Taiwan: Science Policy and the Quest for Modernization.* Cambridge, MA: Harvard University Press, 2008.

Hager, Thomas. *Force of Nature: The Life of Linus Pauling.* New York: Simon and Schuster, 1995.

Haldane, J. B. S. "Biology and Marxism." *Modern Quarterly* 3, no.4 (1948): 2–11.

Hamblin, Jacob Darwin. " 'A Dispassionate and Objective Effort': Negotiating the First Study on the Biological Effects of Atomic Radiation." *Journal of the History of Biology* 40, no. 1 (July 2007): 147–77. https://doi.org/10.1007/s10739-005-6531-8.

Hamlin, Christopher. "The Pedagogical Roots of the History of Science: Revisiting the Vision of James Bryant Conant." *Isis* 107, no. 2 (2016): 282–308. https://doi.org/10.1086/687217.

Hammonds, Evelynn. " 'The Negro Scientist': W. E. B. Du Bois and the Diversity Problem in Science and the History of Science." Paper delivered at History of Science Society Meeting, Atlanta, GA, 2016.

Haraway, Donna. *Primate Visions: Gender, Race, and Nature in the World of Modern Science.* New York: Routledge, 1989.

Harrington, Daniel F. *Berlin on the Brink: The Blockade, the Airlift, and the Early Cold War.* Lexington: University Press of Kentucky, 2012.

Harris, Sarah Miller. *The CIA and the Congress for Cultural Freedom in the Early Cold War: The Limits of Making Common Cause.* New York: Routledge, 2016.

Harrison, Carol E., and Ann Johnson, eds. "National Identity: The Role of Science and Technology." *Osiris* 24 (2009).

Hart, Justin. *Empire of Ideas: The Origins of Public Diplomacy and the Transformation of U.S. Foreign Policy.* New York: Oxford University Press, 2013.

Hartman, Andrew. *Education and the Cold War: The Battle for the American School.* New York: Palgrave Macmillan, 2008.

Hecht, Gabrielle, and Paul N. Edwards. *The Technopolitics of Cold War: Toward a Transregional Perspective.* Essays on Global and Comparative History. Washington, DC: American Historical Association, 2007.

Herbst, Jurgen. *The Once and Future School: Three Hundred and Fifty Years of American Secondary Education*. New York: Routledge, 1996.

Herken, Gregg. *The Georgetown Set: Friends and Rivals in Cold War Washington*. New York: Knopf, 2014.

Hershberg, James. *James B. Conant: Harvard to Hiroshima and the Making of the Nuclear Age*. New York: Knopf, 1993.

Higuchi, Toshihiro. "Tipping the Scale of Justice: The Fallout Suit of 1958 and the Environmental Legal Dimension of Nuclear Pacifism." *Peace & Change* 38, no. 1 (2013): 33–55. https://doi.org/10.1111/pech.12002.

Hixson, Walter L. *George F. Kennan: Cold War Iconoclast*. New York: Columbia University Press, 1989.

———. *Parting the Curtain: Propaganda, Culture, and the Cold War, 1945–1961*. New York: St. Martin's, 1997.

Hochgeschwender, Michael. "*Der Monat* and the Congress for Cultural Freedom: The High Tide of the Intellectual Cold War, 1948–1971." In *Campaigning Culture and the Global Cold War: The Journals of the Congress for Cultural Freedom*, edited by Giles Scott-Smith and Charlotte Lerg, 71–89. New York: Palgrave Macmillan, 2017.

Hogan, Michael J. *A Cross of Iron: Harry S. Truman and the Origins of the National Security State, 1945–1954*. New York: Cambridge University Press, 1998.

———. *The Marshall Plan: America, Britain, and the Reconstruction of Western Europe, 1947–1952*. New York: Cambridge University Press, 1987.

Hollinger, David A. "Free Enterprise and Free Inquiry: The Emergence of Laissez-Faire Communitarianism in the Ideology of Science in the United States." *New Literary History* 21, no. 4 (1990): 897–919. https://doi.org/10.2307/469191.

Holloway, David. "Physics, the State, and Civil Society in the Soviet Union." *Historical Studies in the Physical and Biological Sciences* 30, no. 1 (January 1999): 173–92. https://doi.org/10.2307/27757823.

Holt, Elizabeth M. "'Bread or Freedom': The Congress for Cultural Freedom, the CIA, and the Arabic Literary Journal Hiwar (1962–67)." *Journal of Arabic Literature* 44, no. 1 (2013): 83–102. https://doi.org/10.1163/1570064x-12341257.

Homans, Charles. "Track II Diplomacy: A Short History." *Foreign Policy*, June 20, 2011. https://foreignpolicy.com/2011/06/20/track-ii-diplomacy-a-short-history.

Home, R. W., and Morris F. Low. "Postwar Scientific Intelligence Missions to Japan." *Isis* 84, no. 3 (1993): 527–37. https://doi.org/10.1086/356550.

Hook, Sidney. *Letters of Sidney Hook: Democracy, Communism, and the Cold War*, edited by Edward Shapiro. Armonk, NY: M. E. Sharpe, 1995.

———. *Out of Step: An Unquiet Life in the 20th Century*. New York: Harper & Row, 1987.

Iber, Patrick. *Neither Peace Nor Freedom: The Cultural Cold War in Latin America*. Cambridge, MA: Harvard University Press, 2015.

Iida, Kaori. "A Controversial Idea as a Cultural Resource: The Lysenko Controversy and Discussions of Genetics as a 'Democratic' Science in Postwar Japan." *Social Studies of Science* 45, no. 4 (2015): 546–69. https://doi.org/10.1177/0306312715596460.

Ings, Simon. *Stalin and the Scientists: A History of Triumph and Tragedy, 1905–1953.* New York: Atlantic Monthly Press, 2016.

Intondi, Vincent J. *African Americans against the Bomb: Nuclear Weapons, Colonialism, and the Black Freedom Movement.* Stanford, CA: Stanford University Press, 2015.

Jacobsen, Annie. *Operation Paperclip: The Secret Intelligence Program That Brought Nazi Scientists to America.* Boston: Little, Brown, 2014.

Jeffreys-Jones, Rhodri. *The CIA and American Democracy.* New Haven, CT: Yale University Press, 1989.

Johnson, David K. *The Lavender Scare : The Cold War Persecution of Gays and Lesbians in the Federal Government.* Chicago: University of Chicago Press, 2004.

Jones, Vincent C. *Manhattan: The Army and the Atomic Bomb.* Washington, DC: US Army Center of Military History, 1985. http://www.history.army.mil/html/books/011/11-10/CMH_Pub_11-10.pdf.

Joravsky, David. *The Lysenko Affair.* Cambridge, MA: Harvard University Press, 1970.

———. "Soviet Marxism and Biology before Lysenko." *Journal of the History of Ideas* 20, no. 1 (1959): 85–104. https://doi.org/10.2307/2707968.

Josephson, Paul R. *New Atlantis Revisited: Akademgorodok, The Siberian City of Science.* Princeton, NJ: Princeton University Press, 1997.

———. *Physics and Politics in Revolutionary Russia.* Berkeley: University of California Press, 1991.

Kendi, Ibram X. *Stamped from the Beginning: The Definitive History of Racist Ideas in America.* New York: Nation Books, 2016.

Kevles, Daniel J. "Cold War and Hot Physics: Science, Security, and the American State, 1945–56." *Historical Studies in the Physical and Biological Sciences* 20, no. 2 (1990): 239–64. https://doi.org/10.2307/27757644.

———. *The Physicists: The History of a Scientific Community in Modern America.* New York: Knopf, 1977.

Keys, Barbara J. *Reclaiming American Virtue: The Human Rights Revolution of the 1970s.* Cambridge, MA: Harvard University Press, 2014.

Killian, James R., Jr. *Sputnik, Scientists, and Eisenhower: A Memoir of the First Special Assistant to the President for Science and Technology.* Cambridge, MA: MIT Press, 1977.

Klein, Christina. *Cold War Orientalism: Asia in the Middlebrow Imagination, 1945–1961.* Berkeley: University of California Press, 2003.

Koch, Stephen. *Double Lives: Stalin, Willi Münzenberg, and the Seduction of the Intellectuals.* Revised and updated edition. New York: Enigma Books, 2004.

Koestler, Arthur. *Arrow in the Blue: The First Volume of an Autobiography, 1905–1953.* New York: Macmillan, 1952.

———. *Darkness at Noon.* New York: Macmillan, 1941.

———. *The Invisible Writing: The Second Volume of an Autobiography: 1932–1940.* New York: Macmillan, 1954.

———. *The Yogi and the Commissar and Other Essays.* New York: Macmillan, 1945.

Kohler, Robert E. *Lords of the Fly: Drosophila Genetics and the Experimental Life*. Chicago: University of Chicago Press, 1994.

———. *Partners in Science: Foundations and Natural Scientists, 1900–1945*. Chicago: University of Chicago Press, 1991.

Kramer, Paul A. "Power and Connection: Imperial Histories of the United States in the World." *American Historical Review* 116, no. 5 (2011): 1348–91. https://doi.org/10.1086/ahr.116.5.1348.

Krementsov, Nikolai. "Big Revolution, Little Revolution: Science and Politics in Bolshevik Russia." *Social Research* 73, no. 4 (2006): 1173–204.

———. *International Science between the World Wars: The Case of Genetics*. New York: Routledge, 2005.

———. "A 'Second Front' in Soviet Genetics: The International Dimension of the Lysenko Controversy, 1944–1947." *Journal of the History of Biology* 29, no. 2 (1996): 229–50. https://doi.org/10.1007/BF00571083.

———. *Stalinist Science*. Princeton, NJ: Princeton University Press, 1997.

Krige, John. *American Hegemony and the Postwar Reconstruction of Science in Europe*. Cambridge, MA: MIT Press, 2006.

———. "Atoms for Peace, Scientific Internationalism, and Scientific Intelligence." *Osiris* 21, no. 1 (2006): 161–81. https://doi.org/10.1086/507140.

———. *Sharing Knowledge, Shaping Europe. US Technological Collaboration and Nonproliferation*. Cambridge, MA: MIT Press, 2016.

Kroll, Gary. "The Pacific Science Board in Micronesia: Science, Government, and Conservation on the Postwar Pacific Frontier." *Minerva* 41, no. 1 (2003): 25–46. https://doi.org/10.1023/A:1022205821483.

Kuznick, Peter J. *Beyond the Laboratory: Scientists as Political Activists in 1930s America*. Chicago: University of Chicago Press, 1987.

Lambright, W. Henry. *Powering Apollo: James E. Webb of NASA*. Baltimore: Johns Hopkins University Press, 1995.

Lasch, Christopher. *The Agony of the American Left*. New York: Knopf, 1969.

Lassman, Thomas C. "Government Science in Postwar America: Henry A. Wallace, Edward U. Condon, and the Transformation of the National Bureau of Standards." *Isis* 96, no. 1 (2005): 25–51. https://doi.org/10.1086/430676.

Latour, Bruno. "Why Has Critique Run out of Steam? From Matters of Fact to Matters of Concern." *Critical Inquiry* 30, no. 2 (2004): 225–48. https://doi.org/10.1086/421123.

Latour, Bruno, and Steve Woolgar. *Laboratory Life: The Construction of Scientific Facts*. Princeton, NJ: Princeton University Press, 1986. First published 1979 by Sage (Beverly Hills, CA).

Lebovic, Sam. "From War Junk to Educational Exchange: The World War II Origins of the Fulbright Program and the Foundations of American Cultural Globalism, 1945–1950." *Diplomatic History* 37, no. 2 (2013): 280–312. https://doi.org/10.1093/dh/dht002.

Lees, Lorainne. *Keeping Tito Afloat: The United States, Yugoslavia, and the Cold War*. University Park: Pennsylvania State University Press, 1997.

Leffler, Melvyn P. *For the Soul of Mankind: The United States, the Soviet Union, and the Cold War*. New York: Hill and Wang, 2007.

Lexow, Wilton. "The Science Attaché Program." *Studies in Intelligence* 10 (Spring 1966): 21–27. https://www.cia.gov/library/center-for-the-study-of-intelligence/kent-csi/vol10no2/html/v10i2a02p_0001.htm.

Longino, Helen. *Science as Social Knowledge: Values and Objectivity in Scientific Inquiry*. Princeton, NJ: Princeton University Press, 1990.

Lucas, Scott. "Campaigns of Truth: The Psychological Strategy Board and American Ideology, 1951–1953." *International History Review* 18, no. 2 (1996): 279–302. https://doi.org/10.1080/07075332.1996.9640744.

———. *Freedom's War: The American Crusade against the Soviet Union*. New York: New York University, 1999.

Lucas, Scott, and Kaeten Mistry. "Illusions of Coherence: George F. Kennan, US Strategy and Political Warfare in the Early Cold War, 1946–1950." *Diplomatic History* 33, no. 1 (2009): 39–66. https://doi.org/10.1111/j.1467-7709.2008.00746.x.

Lysenko, T. D. *Heredity and Its Variability*. Translated by Theodosius Dobzhanksy. New York: King's Crown Press, 1946.

Macekura, Stephen J. "The Point Four Program and U.S. International Development Policy." *Political Science Quarterly* 128, no. 1 (2013): 127–160. doi.org/10.1002/polq.12000.

Maddock, Shane J. *Nuclear Apartheid: The Quest for American Atomic Supremacy from World War II to the Present*. Chapel Hill: University of North Carolina Press, 2010.

Mandler, Peter. *Return from the Natives: How Margaret Mead Won the Second World War and Lost the Cold War*. New Haven, CT: Yale University Press, 2013.

Manzione, Joseph. "'Amusing and Amazing and Practical and Military': The Legacy of Scientific Internationalism in American Foreign Policy, 1945–1963." *Diplomatic History* 24, no. 1 (2000): 21–55. https://doi.org/10.1111/1467-7709.00197.

Matchett, Karin. "At Odds over Inbreeding: An Abandoned Attempt at Mexico/United States Collaboration to 'Improve' Mexican Corn, 1940–1950." *Journal of the History of Biology* 39, no. 2 (2006): 345–72. https://doi.org/10.1007/s10739-006-0007-3.

Mayer, William V. "Biology Education in the United States during the Twentieth Century." *Quarterly Review of Biology* 61, no. 4 (1986): 481–507. https://doi.org/10.1086/415145.

Mayr, Ernst, and William B. Provine, eds. *The Evolutionary Synthesis: Perspectives on the Unification of Biology*. Cambridge, MA: Harvard University Press, 1980.

McClure, Donald S. *Wallace Reed Brode*. Biographical Memoirs 82. Washington, DC: National Academies Press, 2002. http://www.nasonline.org/publications/biographical-memoirs/memoir-pdfs/brode-wallace-r.pdf.

McCutcheon, Robert A. "The 1936–1937 Purge of Soviet Astronomers." *Slavic Review* 50, no. 1 (1991): 100–117. https://doi.org/10.2307/2500602.

McGucken, William. "On Freedom and Planning in Science: The Society for Freedom in Science, 1940–46." *Minerva* 16, no. 1 (1978): 42–72. https://doi.org/10.1007/BF01102181.

McMahon, Robert J., ed. *The Cold War in the Third World*. New York: Oxford University Press, 2013.

McMeekin, Sean. *The Red Millionaire: A Political Biography of Willi Münzenberg, Moscow's Secret Propaganda Tsar in the West*. New Haven, CT: Yale University Press, 2003.

McMillian, John, and Paul Buhle, eds. *The New Left Revisited*. Philadelphia: Temple University Press, 2003.

Medvedev, Z. A. *The Rise and Fall of T. D. Lysenko*. Translated by I. Michael Lerner. New York: Columbia University Press, 1969.

Milne, David. *America's Rasputin: Walt Rostow and the Vietnam War*. New York: Hill and Wang, 2008.

Mitchell, Mary X. "Test Cases: Reconfiguring American Law, Technoscience, and Democracy in the Nuclear Pacific." PhD dissertation, University of Pennsylvania, Philadelphia, 2016.

Monk, Ray. *Robert Oppenheimer: A Life Inside the Center*. New York: Doubleday, 2012.

Montague, Ludwell Lee. *General Walter Bedell Smith as Director of Central Intelligence, October 1950–February 1953*. University Park: Pennsylvania State University Press, 1992.

Moore, Kelly. *Disrupting Science: Social Movements, American Scientists, and the Politics of the Military, 1945–1975*. Princeton, NJ: Princeton University Press, 2008.

Muir-Harmony, Teasel. "Project Apollo, Cold War Diplomacy and the American Framing of Global Interdependence." PhD dissertation, Massachusetts Institute of Technology, Cambridge, MA, 2014. http://dspace.mit.edu/handle/1721.1/93814.

Muller, H. J. "Lenin's Doctrines in Relation to Genetics." In *Science and Philosophy in the Soviet Union*, ed. Loren R. Graham, 453–69. New York: Knopf, 1972.

———. "Observations of Biological Science in Russia." *Scientific Monthly* 16, no. 5 (1923): 539–552.

———. *Out of the Night: A Biologist's View of the Future*. New York: Vanguard Press, 1935.

Nathans, Benjamin. "The Dictatorship of Reason: Aleksandr Vol'pin and the Idea of Rights under 'Developed Socialism.'" *Slavic Review* 66, no. 4 (2007): 630–63. https://doi.org/10.2307/20060376.

———. "Talking Fish: On Soviet Dissident Memoirs." *Journal of Modern History* 87, no. 3 (2015): 579–614. https://doi.org/10.1086/682413.

Needell, Allan A. *Science, Cold War, and the American State: Lloyd V. Berkner and the Balance of Professional Ideals*. Amsterdam: Harwood Academic Press, 2000.

———. "'Truth Is Our Weapon': Project Troy, Political Warfare, and Government-Academic Relations in the National Security State." *Diplomatic History* 17, no. 3 (1993): 399–420. https://doi.org/10.1111/j.1467-7709.1993.tb00588.x

Nelkin, Dorothy. "The Science Wars: Responses to a Marriage Failed." *Social Text*, no. 46/47 (1996): 93–100. https://doi.org/10.2307/466846.

———. *Selling Science: How the Press Covers Science and Technology*. New York: W. H. Freeman, 1987.

Nelson, Alondra. *Body and Soul: The Black Panther Party and the Fight against Medical Discrimination*. Minneapolis: University of Minnesota Press, 2011.

New Frontiers in Science Diplomacy. London and Washington, DC: The Royal Society and the American Association for the Advancement of Science, 2010. https://www.aaas.org/sites/default/files/New_Frontiers.pdf.

Nguyen, Lien-Hang. *Hanoi's War: An International History of the War for Peace in Vietnam*. Chapel Hill: University of North Carolina Press, 2012.

Ninkovich, Frank. *The Diplomacy of Ideas: U.S. Foreign Policy and Cultural Relations, 1938–1950*. New York: Cambridge University Press, 1981.

Nye, Mary Jo. *Michael Polanyi and His Generation: Origins of the Social Construction of Science*. Chicago: University of Chicago Press, 2011.

Oreskes, Naomi. "Introduction." In *Science and Technology in the Global Cold War*, edited by Naomi Oreskes and John Krige, 1–9. Cambridge, MA: MIT Press, 2014.

Oreskes, Naomi, and Erik M. Conway. *Merchants of Doubt: How a Handful of Scientists Obscured the Truth on Issues from Tobacco Smoke to Global Warming*. New York: Bloomsbury Press, 2010.

Orlov, Yuri. *Dangerous Thoughts: Memoirs of a Russian Life*. Translated by Thomas P. Whitney. New York: William Morrow, 1991.

Osgood, Kenneth. *Total Cold War: Eisenhower's Secret Propaganda Battle at Home and Abroad*. Lawrence: University Press of Kansas, 2006.

Osseo-Asare, Abena Dove. "Scientific Equity: Experiments in Laboratory Education in Ghana." *Isis* 104, no. 4 (2013): 713–41. https://doi.org/10.1086/674941.

Paget, Karen M. *Patriotic Betrayal: The Inside Story of the CIA's Secret Campaign to Enroll American Students in the Crusade against Communism*. New Haven, CT: Yale University Press, 2015.

Parker, Jason C. *Hearts, Minds, Voices: US Cold War Public Diplomacy and the Formation of the Third World*. New York: Oxford University Press, 2016.

Parmar, Inderjeet. *Foundations of the American Century: The Ford, Carnegie, and Rockefeller Foundations in the Rise of American Power*. New York: Columbia University Press, 2012.

Parry-Giles, Shawn J. *The Rhetorical Presidency, Propaganda, and the Cold War, 1945–1955*. Westport, CT: Praeger, 2002.

Passaglia, Elio. *A Unique Institution: The National Bureau of Standards, 1950–1969*. Washington, DC: US Department of Commerce, 1999. https://nvlpubs.nist.gov/nistpubs/Legacy/SP/nistspecialpublication925.pdf.

Paul, Diane B. "Eugenics and the Left." *Journal of the History of Ideas* 45, no. 4 (1984): 567–90. https://doi.org/10.2307/2709374.

———. "H. J. Muller, Communism, and the Cold War." *Genetics* 119, no. 2 (1988): 223–25.

———. "A War on Two Fronts: J. B. S. Haldane and the Response to Lysenkoism in Britain." *Journal of the History of Biology* 16, no. 1 (1983): 1–37. https://doi.org/10.1007/BF00186674.

Paul, Diane B., and Costas B. Krimbas. "Nikolai V. Timoféef-Ressovsky." *Scientific American* 266 (February 1992): 86–92. https://doi.org/10.1038/scientificamerican0292-86.

Paul, Richard. *We Could Not Fail: The First African Americans in the Space Program*. Austin: University of Texas Press, 2015.

Pauly, Philip J. *Biologists and the Promise of American Life: From Meriwether Lewis to Alfred Kinsey*. Princeton, NJ: Princeton University Press, 2000.

Peterson, Christian Philip. "The Carter Administration and the Promotion of Human Rights in the Soviet Union, 1977–1981." *Diplomatic History* 38, no. 3 (2014): 628–56. https://doi.org/10.1093/dh/dht102.

Phillips, Christopher J. *The New Math: A Political History*. Chicago: University of Chicago Press, 2015.

Polanyi, Michael. "The Republic of Science: Its Political and Economic Theory." *Minerva* 1, no. 1 (1962): 54–73. https://doi.org/10.1007/BF01101453.

Pollock, Ethan. "From Partiinost' to Nauchnost' and Not Quite Back Again: Revisiting the Lessons of the Lysenko Affair." *Slavic Review* 68, no. 1 (2009): 95–115. https://doi.org/10.1017/S0037677900000103.

Price, David H. *Anthropological Intelligence: The Deployment and Neglect of American Anthropology in the Second World War*. Durham, NC: Duke University Press, 2008.

———. *Cold War Anthropology: The CIA, the Pentagon, and the Growth of Dual Use Anthropology*. Durham, NC: Duke University Press, 2016.

Raj, Kapil. "Beyond Postcolonialism . . . and Postpositivism: Circulation and the Global History of Science." *Isis* 104, no. 2 (2013): 337–47. https://doi.org/10.1086/670951.

Reisch, George. "When *Structure* Met Sputnik: On the Cold War Origins of *The Structure of Scientific Revolutions*." In *Science and Technology in the Global Cold War*, edited by Naomi Oreskes and John Krige, 371–92. Cambridge, MA: MIT Press, 2014.

Rhéaume, Charles. "Western Scientists' Reactions to Andrei Sakharov's Human Rights Struggle in the Soviet Union, 1968–1989." *Human Rights Quarterly* 30, no. 1 (2008): 1–20. https://doi.org/10.1353/hrq.2008.0004.

Richardson, Peter. *A Bomb in Every Issue: How the Short, Unruly Life of Ramparts Magazine Changed America*. New York: New Press, 2009.

Richelson, Jeffrey T. *The Wizards of Langley: Inside the CIA's Directorate of Science and Technology*. Boulder, CO: Westview Press, 2002.

Richmond, Yale. *Cultural Exchange and the Cold War: Raising the Iron Curtain*. University Park: Pennsylvania State University Press, 2003.

Risen, Clay. *A Nation on Fire: America in the Wake of the King Assassination*. Hoboken, NJ: John Wiley and Sons, 2009.

Rogacheva, Maria A. *The Private World of Soviet Scientists from Stalin to Gorbachev*. New York: Cambridge University Press, 2017.

Rohde, Joy. *Armed with Expertise: The Militarization of American Social Research during the Cold War*. Ithaca, NY: Cornell University Press, 2013.

Roll-Hansen, Nils. "Wishful Science: The Persistence of T. D. Lysenko's Agrobiology in the Politics of Science." *Osiris* 23 (2008): 166–88. https://doi.org/10.1086/591873.

Rosenberg, Emily S. *Financial Missionaries to the World: The Politics and Culture of Dollar Diplomacy, 1900–1930*. Cambridge, MA: Harvard University Press, 1999.

Rossiter, Margaret W. *Women Scientists in America: Before Affirmative Action, 1940–1972*. Baltimore: Johns Hopkins University Press, 1995.

Rotblat, Joseph. *Scientists in the Quest for Peace: A History of the Pugwash Conferences*. Cambridge, MA: MIT Press, 1972.

Rubinson, Paul. "'Crucified on a Cross of Atoms': Scientists, Politics, and the Test Ban Treaty." *Diplomatic History* 35, no. 2 (2011): 283–319. https://doi.org/10.1111/j.1467-7709.2010.00950.x.

———. "'For Our Soviet Colleagues': Scientific Internationalism, Human Rights, and the Cold War." In *The Human Rights Revolution: An International History*, edited by Akira Iriye, Petra Goedde, and William I. Hitchcock, 245–64. New York: Oxford University Press, 2012.

———. *Redefining Science: Scientists, the National Security State, and Nuclear Weapons in Cold War America*. Amherst: University of Massachusetts Press, 2016.

Rudolph, John L. *How We Teach Science: What's Changed and Why It Matters*. Cambridge, MA: Harvard University Press, 2019.

———. *Scientists in the Classroom: The Cold War Reconstruction of American Science Education*. New York: Palgrave, 2002.

"The Russell-Einstein Manifesto." In *Scientists in the Quest for Peace: A History of the Pugwash Conference*, 137–40. Cambridge, MA: MIT Press, 1972.

Sakharov, Andrei. *Memoirs*. Translated by Richard Lourie. New York: Knopf, 1990.

Sapp, Jan. "The Nine Lives of Gregor Mendel." In *Experimental Inquiries: Historical, Philosophical, and Social Studies of Experimentation in Science*, edited by H. E. LeGrand, 137–66. Dordrecht, Netherlands: Kluwer Academic Publishers, 1990.

Saunders, Frances Stonor. *The Cultural Cold War: The CIA and the World of Arts and Letters*. New York: New Press, 1999.

Scammell, Michael. *Koestler: The Literary and Political Odyssey of a Twentieth-Century Skeptic*. New York: Random House, 2009.

Schiebinger, Londa. *Nature's Body: Gender in the Making of Modern Science*. Boston: Beacon, 1993.

Schlesinger, Arthur, Jr. *The Letters of Arthur Schlesinger, Jr.*, edited by Andrew Schlesinger and Stephen Schlesinger. New York: Random House, 2013.

———. *A Life in the Twentieth Century: Innocent Beginnings, 1917–1950*. Boston: Houghton Mifflin, 2000.

Schmalzer, Sigrid, Daniel S. Chard, and Alyssa Botelho, eds. *Science for the People: Documents from America's Movement of Radical Scientists*. Amherst: University of Massachusetts Press, 2018.

Schrecker, Ellen. *No Ivory Tower: McCarthyism and the Universities*. New York: Oxford University Press, 1986.

Schweitzer, Glenn E. *Interacademy Programs between the United States and Eastern Europe, 1967–2009: The Changing Landscape.* Washington, DC: National Academies Press, 2009.

———. *Scientists, Engineers, and Track-Two Diplomacy: A Half-Century of U.S.–Russian Interacademy Cooperation.* Washington, DC: National Academies Press, 2004.

———. *Techno-Diplomacy: U.S. Soviet Confrontations in Science and Technology.* New York: Plenum Press, 1989.

Scott-Smith, Giles. *The Politics of Apolitical Culture: The Congress for Cultural Freedom, the CIA, and Post-War American Hegemony.* Routledge / PSA Political Studies Series. New York: Routledge, 2002.

Scott-Smith, Giles, and Charlotte Lerg, eds. *Campaigning Culture and the Global Cold War: The Journals of the Congress for Cultural Freedom.* New York: Palgrave Macmillan, 2017.

Selya, Rena. "Defending Scientific Freedom and Democracy: The Genetics Society of America's Response to Lysenko." *Journal of the History of Biology* 45, no. 3 (2012): 415–42. https://doi.org/10.1007/s10739-011-9288-2.

Seth, Suman. "Putting Knowledge in Its Place: Science, Colonialism, and the Postcolonial." *Postcolonial Studies* 12, no. 4 (December 2009): 373–88. https://doi.org/10.1080/13688 790903350633.

Sharansky, Natan. *Fear No Evil.* Translated by Stefani Hoffman. New York: Random House, 1988.

Shelley, Becky. *Democratic Development in East Asia.* New York: RoutledgeCurzon, 2005.

Shephard, Ben. *The Long Road Home: The Aftermath of the Second World War.* New York: Knopf, 2011.

Shetterly, Margot Lee. *Hidden Figures : The American Dream and the Untold Story of the Black Women Mathematicians Who Helped Win the Space Race.* New York: William Morrow, 2016.

Siddiqi, Asif A. "Korolev, Sputnik, and the International Geophysical Year." In *Reconsidering Sputnik: Forty Years since the Soviet Satellite*, edited by Roger D. Launius, John M. Logsdon, and Robert W. Smith, 43–72. Amsterdam: Harwood Academic Press, 2000.

———. "The Rockets' Red Glare: Technology, Conflict, and Terror in the Soviet Union." *Technology and Culture* 44, no. 3 (2003): 470–501. https://doi.org/10.1353/tech.2003.0133.

Siff, Sarah Brady. "Atomic Roaches and Test-Tube Babies: Bentley Glass and Science Communication." *Journalism & Communication Monographs* 17, no. 2 (2015): 88–144. https://doi.org/10.1177/1522637915577107.

Slaton, Amy E. *Race, Rigor, and Selectivity in U.S. Engineering: The History of an Occupational Color Line.* Cambridge, MA: Harvard University Press, 2010.

Smith, Alice Kimball. *A Peril and a Hope : The Scientists' Movement in America, 1945–47.* Chicago: University of Chicago Press, 1965.

Smocovitis, Vassiliki Betty. *Unifying Biology: The Evolutionary Synthesis and Evolutionary Biology.* Princeton, NJ: Princeton University Press, 1996.

Snyder, Sarah B. *Human Rights Activism and the End of the Cold War: A Transnational History of the Helsinki Network*. New York: Cambridge University Press, 2011.

Sokal, Alan. "A Physicist Experiments with Cultural Studies." *Lingua Franca*, May/June 1996, 62–64.

———. "Transgressing the Boundaries: Toward a Transformative Hermeneutics of Quantum Gravity." *Social Text*, no. 46/47 (1996): 217–52. https://doi.org/10.2307/466856.

Solomon, Susan Gross, and Nikolai Krementsov. "Giving and Taking across Borders: The Rockefeller Foundation and Russia, 1918–1928." *Minerva* 39, no. 3 (2001): 265–98. https://doi.org/10.1023/A:1017909113410.

Tucker, William H. *The Funding of Scientific Racism: Wickliffe Draper and the Pioneer Fund*. Urbana: University of Illinois Press, 2002.

US Congress. House. Committee on Intelligence. *The CIA and the Media: Hearings before the Subcommittee on Oversight of the Permanent Select Committee on Intelligence*. 95th Cong., 1st and 2nd sess. December 27, 28, 29, 1977; January 4, 5 and April 20, 1978. Washington, DC: Government Printing Office, 1978.

US Congress. House. Committee on Small Businesses. *Tax-Exempt Foundations: Their Impact on Small Business: Hearings before Subcommittee No. 1 on Foundations*. 88th Cong., 2nd sess. July 21, 22, 23; August 10, 31; and September 1, 4, 1964. https://www.cia.gov/library/readingroom/docs/CIA-RDP67B00446R000300020094-4.pdf.

US Congress. Senate. Committee on Intelligence. *Alleged Assassination Plots Involving Foreign Leaders: An Interim Report of the Select Committee to Study Governmental Relations with Respect to Intelligence Agencies*. Washington, DC: Government Printing Office, 1975.

———. *Intelligence Activities and the Rights of Americans: Book II, Final Report of the Select Committee to Study Governmental Operations with Respect to Intelligence Activities*. Washington, DC: Government Printing Office, 1976.

———. *Supplementary Reports on Intelligence Activities: Book VI, Final Report of the Select Committee to Study Governmental Operations with Respect to Intelligence Activities*. Washington, DC: Government Printing Office, 1976.

US Congress. Senate. Committee on the Judiciary. *The Pugwash Conferences: A Staff Analysis Prepared for the Subcommittee to Investigate the Administration of the Internal Security Act and Other Internal Security Laws of the Committee on the Judiciary, United States Senate*. 87th Cong., 1st sess. Washington, DC: Government Printing Office, 1961.

US Department of Education. *Life Adjustment Education for Every Youth*. Bulletin No. 22. Washington, DC: Government Printing Office, 1951.

US Department of State. *Documents on Disarmament, 1945–1959*. Volume 1. Washington, DC: US Department of State, 1960.

Walker, Mark. *Nazi Science: Myth, Truth, and the German Atomic Bomb*. New York: Plenum Press, 1995.

Wall, Wendy. "America's 'Best Propagandists': Italian Americans and the 'Letters to Italy' Campaign." In *Cold War Constructions: The Political Culture of United States Imperialism*,

1945–1966, edited by Christian G. Appy, 89–109. Amherst: University of Massachusetts Press, 2000.

Wang, Jessica. *American Science in an Age of Anxiety: Scientists, Anticommunism, and the Cold War.* Chapel Hill: University of North Carolina Press, 1999.

———. "Scientists and the Problem of the Public in Cold War America, 1945–1960." *Osiris* 17 (2002): 323–47. https://doi.org/10.1086/649368.

Wang, Zuoyue. *In Sputnik's Shadow: The President's Science Advisory Committee and Cold War America.* New Brunswick, NJ: Rutgers University Press, 2008.

———. "The Politics of Big Science in the Cold War: PSAC and the Funding of SLAC." *Historical Studies in the Physical and Biological Sciences* 25, no. 2 (1995): 329–56. https://doi.org/10.2307/27757748.

Warner, Michael. "The CIA's Office of Policy Coordination: From NSC 10/2 to NSC 68." *International Journal of Intelligence and CounterIntelligence* 11, no. 2 (June 1998): 211–20. https://doi.org/10.1080/08850609808435373.

———. "Origins of the Congress for Cultural Freedom." *Studies on Intelligence* 38, no. 5 (1995): 89–98. https://www.cia.gov/library/center-for-the-study-of-intelligence/kent-csi/vol38 no5/pdf/v38i5a10p.pdf.

Weber, Karl H. *The Office of Scientific Intelligence, 1948–1968.* Volume 1. Washington, DC: Central Intelligence Agency, 1972. https://www.cia.gov/library/readingroom/docs/osi _1949_68_volume_1.pdf.

———. *The Office of Scientific Intelligence, 1948–1968.* Volume 2. Washington, DC: Central Intelligence Agency, 1972. https://www.cia.gov/library/readingroom/docs/DOC_0000 629785.pdf.

Weiner, Tim. *Legacy of Ashes: The History of the CIA.* New York: Doubleday, 2007.

Weiss, Sheila Faith. *The Nazi Symbiosis: Human Genetics and Politics in the Third Reich.* Chicago: University of Chicago Press, 2010.

Weissberg, Alex. *The Accused.* Translated by Edward Fitzgerald. New York: Simon and Schuster, 1951.

Weisskopf, Victor. *The Joy of Insight: Passions of a Physicist.* New York: Basic Books, 1991.

Westad, Odd Arne. *The Cold War: A World History.* New York: Basic Books, 2017.

———. *The Global Cold War.* New York: Cambridge University Press, 2005.

Wilford, Hugh. *The CIA, the British Left, and the Cold War: Calling the Tune?* New York: Routledge, 2013. First published 2003 by Frank Cass (Portland, OR).

———. *The Mighty Wurlitzer: How the CIA Played America.* Cambridge, MA: Harvard University Press, 2008.

Wisnioski, Matt. *Engineers for Change: Competing Visions of Technology in 1960s America.* Cambridge, MA: MIT Press, 2012.

———. "Inside 'the System': Engineers, Scientists, and the Boundaries of Social Protest in the Long 1960s." *History and Technology* 19, no. 4 (December 2003): 313–33. https://doi.org /10.1080/0734151032000181077.

Wittner, Lawrence S. *Resisting the Bomb: A History of the World Disarmament Movement, 1954–1970*. Volume 2: *The Struggle against the Bomb*. Stanford, CA: Stanford University Press, 1997.

Wolfe, Audra J. "The Cold War Context of the Golden Jubilee; Or, Why We Think of Mendel as the Father of Genetics." *Journal of the History of Biology* 45, no. 3 (2012): 389–414. https://doi.org/10.1007/s10739-011-9291-7.

———. *Competing with the Soviets: Science, Technology, and the State in Cold War America.* Baltimore: Johns Hopkins University Press, 2013.

———. "The Organization Man and the Archive: A Look at the Bentley Glass Papers." *Journal of the History of Biology* 44, no. 1 (March 2011): 147–51. https://doi.org/10.1007/s10739 -011-9276-6.

———. "*Science and Freedom*: The Forgotten Bulletin." In *Campaigning Culture and the Global Cold War: The Journals of the Congress for Cultural Freedom*, edited by Giles Scott-Smith and Charlotte Lerg, 27–44. New York: Palgrave Macmillan, 2017.

———. "What Does It Mean to Go Public? The American Response to Lysenkoism, Reconsidered." *Historical Studies in the Natural Sciences* 40, no. 1 (February 2010): 48–78. https:// doi.org/10.1525/hsns.2010.40.1.48.

———. "What's in a Zone? Biological Order versus National Identity in the Biological Sciences Curriculum Study." In *Global Transformations in the Life Sciences, 1945–1980*, edited by Patrick Manning and Mat Savelli, 146–59. Pittsburgh: University of Pittsburgh Press, 2018.

Yarrow, Andrew L. "Selling a New Vision of America to the World: Changing Messages in Early U.S. Cold War Print Propaganda." *Journal of Cold War Studies* 11, no. 4 (2009): 3–45. https://doi.org/10.1162/jcws.2009.11.4.3.

Zachary, G. Pascal. *Endless Frontier: Vannevar Bush, Engineer of the American Century*. New York: Free Press, 1997.

Index

Breitbart News, 201
Brennan, Donald G., 129
British Royal Society, 199
British Society for Freedom in Science, 82
British Society for Social Responsibility in Science, 173–74
Brode, Wallace, 39–40, 41–42, 51, 99–100, 106, 108, 110–11
Bronk, Detlev, 10, 14, 50, 106, 110, 111
Brown, Harrison, 129, 184, 185, 187
Brown, Irving, 76
Bruner, Jerome, 66, 138
BSCS. *See* Biological Sciences Curriculum Study
Bukharin, Nikolai, 21
Bulletin of the Atomic Scientists: B. Glass and, 113–14, 129, 130; Rabinowitch as editor of, 14, 80, 117, 121; scientific internationalism and, 14, 115; in Soviet nuclear installations, 133, 196
Bundy, McGeorge, 169
Burnham, James, 65, 76, 78
Bush, George W., 207
Bush, Vannevar, 10, 31–32, 39, 204

Carlson, Elof, 20
Carothers, Neil, 54
Carter, Jimmy, 188, 190
CCF. *See* Congress for Cultural Freedom
Central Intelligence Agency (CIA): activities of, 167–68; Brode and, 41–42; covert operations and, 59; creation of, 39; cultural diplomacy of, 5–6; funding system of, 158, 163; International Organizations Division, 77–78; Italian elections and, 58; media relationship of, 159; *New York Times* articles on, 168; Office of Scientific Intelligence, 36, 42–43; organizations funded by, 157; orphans, 165, 167; Pugwash Conferences and, 122, 125; records of, 218, 219; science attaché contact with, 53; Scientific Branch of, 39–40, 42; Soviet-American Disarmament Study Group and, 131; Soviet Russia Division, 181. *See also* Asia Foundation; Committee on Science and Freedom; Congress for Cultural Freedom; Farfield Foundation
Central Intelligence Group, 38–39

Chadwell, Marshall, 46
Chetverikov, Sergei, 20
Chevalier, Haakon, 35
China, People's Republic of, 5, 102, 107, 108, 123
China, Republic of. *See* Taiwan
Christian, George, 160, 161
CIA. *See* Central Intelligence Agency
Clark, Ramsey, 164
Cleland, Ralph, 29, 70
Clemens, Walter, Jr., 196
Cold War: dates and proxy battles of, 3; references to, 56; scientific freedom and, 1–2; US as winner of, 208
Coliver, Edith, 150, 170
Committee for Free Asia, 145–46. *See also* Asia Foundation
Committee of Concerned Scientists, 189–90, 191
Committee on Science and Freedom, 74, 75, 82–84, 88, 89
Communist Information Bureau (Cominform), 57, 63–65, 75
Communist science, 2
Compton, Arthur, 80
Compton, Karl, 38, 42, 70
Conant, James Bryant, 32–33, 51, 94, 137–38
Condon, Edward U., 40–41
Conference on Science and World Affairs, 120. *See also* Pugwash Conferences
Congress for Cultural Freedom (CCF): AAAS and, 124; activities of, 5, 6; CIA and, 168, 235n25; funding for, 78, 158; Josselson and, 88–89; Manifesto for Freedom, 77, 79–80; *Minerva* journal, 14, 88, 89; public face and location of, 77; responses by targets of propaganda of, 87–88; *Science and Freedom* magazine, 74, 83–85, 86–88, 89; "Science and Freedom" meeting, 80–81; science in, 89; scientific freedom and, 74–75; speakers at, 76–77, 78–79. *See also* Committee on Science and Freedom
contact paranoia, 161–62
Cook, Robert, 70, 80
Coolidge, Harold, 13, 171
covert operations, US, 43, 58–60, 93, 167. *See also* Office of Policy Coordination; 10/2 Committee; 303 Committee